AI Techniques for Reliability Prediction for Electronic Components

Cherry Bhargava
Lovely Professional University, India

A volume in the Advances in Computational Intelligence and Robotics (ACIR) Book Series

Published in the United States of America by
IGI Global
Engineering Science Reference (an imprint of IGI Global)
701 E. Chocolate Avenue
Hershey PA, USA 17033
Tel: 717-533-8845
Fax: 717-533-8661
E-mail: cust@igi-global.com
Web site: http://www.igi-global.com

Library of Congress Cataloging-in-Publication Data

Names: Bhargava, Cherry, 1982- editor.
Title: AI techniques for reliability prediction for electronic components /
  Cherry Bhargava, editor.
Description: Hershey, PA : Engineering Science Reference, an imprint of IGI
  Global, 2020. | Includes bibliographical references and index. |
  Summary: "This book explores the theoretical and practical aspects of
  prediction methods using artificial intelligence and machine learning in
  the manufacturing field"-- Provided by publisher.
Identifiers: LCCN 2019030347 (print) | LCCN 2019030348 (ebook) | ISBN
  9781799814641 (hardcover) | ISBN 9781799814658 (paperback) | ISBN
  9781799814665 (ebook)
Subjects: LCSH: Electronic apparatus and appliances--Reliability. |
  Electronic apparatus and appliances--Testing--Data processing. |
  Electronic apparatus and appliances--Service life. | Artificial
  intelligence--Industrial applications.
Classification: LCC TK7870.23 .A44 2020 (print) | LCC TK7870.23 (ebook) |
  DDC 621.38150285/63--dc23
LC record available at https://lccn.loc.gov/2019030347
LC ebook record available at https://lccn.loc.gov/2019030348

This book is published in the IGI Global book series Advances in Computational Intelligence and Robotics (ACIR) (ISSN:
2327-0411; eISSN: 2327-042X)

British Cataloguing in Publication Data
A Cataloguing in Publication record for this book is available from the British Library.

For electronic access to this publication, please contact: eresources@igi-global.com.

Advances in Computational Intelligence and Robotics (ACIR) Book Series

Ivan Giannoccaro
University of Salento, Italy

ISSN:2327-0411
EISSN:2327-042X

MISSION

While intelligence is traditionally a term applied to humans and human cognition, technology has progressed in such a way to allow for the development of intelligent systems able to simulate many human traits. With this new era of simulated and artificial intelligence, much research is needed in order to continue to advance the field and also to evaluate the ethical and societal concerns of the existence of artificial life and machine learning.

The **Advances in Computational Intelligence and Robotics (ACIR) Book Series** encourages scholarly discourse on all topics pertaining to evolutionary computing, artificial life, computational intelligence, machine learning, and robotics. ACIR presents the latest research being conducted on diverse topics in intelligence technologies with the goal of advancing knowledge and applications in this rapidly evolving field.

COVERAGE

- Computer Vision
- Adaptive and Complex Systems
- Agent technologies
- Automated Reasoning
- Computational Intelligence
- Robotics
- Cyborgs
- Synthetic Emotions
- Cognitive Informatics
- Intelligent control

IGI Global is currently accepting manuscripts for publication within this series. To submit a proposal for a volume in this series, please contact our Acquisition Editors at Acquisitions@igi-global.com or visit: http://www.igi-global.com/publish/.

Titles in this Series

For a list of additional titles in this series, please visit:
https://www.igi-global.com/book-series/advances-computational-intelligence-robotics/73674

AI and Big Data's Potential for Disruptive Innovation
Moses Strydom (Emeritus, France) and Sheryl Buckley (University of South Africa, South Africa)
Engineering Science Reference • © 2020 • 405pp • H/C (ISBN: 9781522596875) • US $225.00 (our price)

Handbook of Research on the Internet of Things Applications in Robotics and Automation
Rajesh Singh (Lovely Professional University, India) Anita Gehlot (Lovely Professional University, India) Vishal Jain (Bharati Vidyapeeth's Institute of Computer Applications and Management (BVICAM), New Delhi, India) and Praveen Kumar Malik (Lovely Professional University, India)
Engineering Science Reference • © 2020 • 433pp • H/C (ISBN: 9781522595748) • US $295.00 (our price)

Handbook of Research on Applications and Implementations of Machine Learning Techniques
Sathiyamoorthi Velayutham (Sona College of Technology, India)
Engineering Science Reference • © 2020 • 461pp • H/C (ISBN: 9781522599029) • US $295.00 (our price)

Handbook of Research on Advanced Mechatronic Systems and Intelligent Robotics
Maki K. Habib (The American University in Cairo, Egypt)
Engineering Science Reference • © 2020 • 466pp • H/C (ISBN: 9781799801375) • US $295.00 (our price)

Edge Computing and Computational Intelligence Paradigms for the IoT
G. Nagarajan (Sathyabama Institute of Science and Technology, India) and R.I. Minu (SRM Institute of Science and Technology, India)
Engineering Science Reference • © 2019 • 347pp • H/C (ISBN: 9781522585558) • US $285.00 (our price)

Semiotic Perspectives in Evolutionary Psychology, Artificial Intelligence, and the Study of Mind Emerging Research and Opportunities
Marcel Danesi (University of Toronto, Canada)
Information Science Reference • © 2019 • 205pp • H/C (ISBN: 9781522589242) • US $175.00 (our price)

Handbook of Research on Human-Computer Interfaces and New Modes of Interactivity
Katherine Blashki (Victorian Institute of Technology, Australia) and Pedro Isaías (The University of Queensland, Australia)
Engineering Science Reference • © 2019 • 488pp • H/C (ISBN: 9781522590699) • US $275.00 (our price)

701 East Chocolate Avenue, Hershey, PA 17033, USA
Tel: 717-533-8845 x100 • Fax: 717-533-8661
E-Mail: cust@igi-global.com • www.igi-global.com

Table of Contents

Detailed Table of Contents

Chapter 1

Cherry Bhargava, Lovely Professional University, India

As the integration of components are increasing from VLSI to ULSI level. This may lead to damage of electronic system because each component has its own operating characteristics and conditions. So, health prognostic techniques are used that comprise a deep insight into failure cause and effects of all the components individually as well as an integrated technique. It will raise alarm, in case health condition, of the components drift from the desired outcomes. From toy to satellite and sand to silicon, the major key constraint of designing and manufacturing industry are towards enhanced operating performance at less operating time. As the technology advances towards high-speed and low-cost gadgets, reliability becomes a challenging issue.

Chapter 2

Amit Sachdeva, Lovely Professional University, India
Pramod K. Singh, Sharda University, India

The chapter deals with brief introduction to polymers, composites, and nanocomposites along with their reliability. When we talk about polymeric composites, the terms crystallinity and amorphicity play a very important role, and both of these properties are highly affected by variation in temperature condition. On increasing temperature, the crystalline domains of polymers tend to become amorphous, and as we reduce the temperature, crystalline domains tend to increase. So the reliability of a particular polymer is widely dependent on temperature conditions.

Chapter 3

Sanjeet Kumar Sinha, Lovely Professional Univeersity, India
Sweta Chander, Lovely Professional University, India

The scaling of devices is a fundamental step for advancing technology in the semiconductor industry. The device scaling allows extra components as well as devices on a single chip, which provides large functionality and application for each integrated circuit (IC). The ultimate goal of device scaling is to make each IC smaller, faster, cheaper, and consumes low power. In today's nanoscale technology, the scaling has been continued and follows Moore's law in the initial phase of fabrication and also shows an exponential growth in ICs. The silicon-based semiconductor industry has reached its scaling limits due to tunneling and quantum-mechanical effects in deep nanometer level. The physics of such devices is not going to continue and hold true. This makes nanoelectronics the leading future of the semiconductor industry. The carbon nanotubes and nanowires are the most promising candidates to make illustrated devices.

Chapter 4

Ravinder Kumar, lovely Professional University, India
Hanumant P. Jagtap, Zeal College of Engineering and Research, India
Dipen Kumar Rajak, Sandip Institute of Technology and Research Centre, India
Anand K. Bewoor, Cummins College of Engineering for Women, India

At present, optimization techniques are popular to solve typical engineering problems. It is the action of making the best or most effective use of a situation or resources. In order to survive in the competitive market, each organization has to follow some optimization technique depending on their requirement. In each optimization problem, there is an objective function to minimize or maximize under the given restrictions or constraints. All techniques have their own advantages and disadvantages. Traditional method starts with the initial solution and with each successive iteration converges to the optimal solution. This convergence depends on the selection of initial approximation. These methods are not suited for discontinuous objective function. So, the need of non-traditional method was felt. Some non-traditional methods are called nature-inspired methods. In this chapter, the authors give the description of the optimization techniques along with the comparison of the traditional and non-traditional techniques.

Chapter 5

Pardeep Kumar Sharma, Lovely Professional University, India
Cherry Bhargava, Lovely Professional University, India

Electronic systems have become an integral part of our daily lives. From toy to radar, system is dependent on electronics. The health conditions of humidity sensor need to be monitored regularly. Temperature can be taken as a quality parameter for electronics systems, which work under variable conditions. Using various environmental testing techniques, the performance of DHT11 has been analysed. The failure of humidity sensor has been detected using accelerated life testing, and an expert system is modelled using various artificial intelligence techniques (i.e., Artificial Neural Network, Fuzzy Inference System, and Adaptive Neuro-Fuzzy Inference System). A comparison has been made between the response of actual and prediction techniques, which enable us to choose the best technique on the basis of minimum error and maximum accuracy. ANFIS is proven to be the best technique with minimum error for developing intelligent models.

Chapter 6

Pardeep Kumar Sharma, Lovely Professional University, India
Cherry Bhargava, Lovely Professional University, India

A humidity sensor detects, measures, and reports the content of moisture in the air. Using low cost composite materials, a humidity sensor has been fabricated. The characterization has been done using various techniques to prove its surface morphology and working. The fabricated sensor detects relative humidity in the range of 15% to 65%. The life of the sensor has been calculated using different experimental and statistical methods. An expert system has been modeled using different artificial intelligence techniques that predict failure of the sensor. The failure prediction of fabricated sensor using Fuzzy Logic, ANN, and ANFIS are 81.4%, 97.4%, and 98.2% accurate, respectively. ANFIS technique proves to be the most accurate technique for prediction of reliability.

Walking is very important exercise. Walking is characterized by gait. Gait defines the bipedal and forward propulsion of center of gravity of the human body. This chapter describes the role of artificial neural network (ANN) for prediction of gait parameters and patterns for human locomotion. The artificial neural network is a mathematical model. It is computational system inspired by the structure, processing method, and learning ability of a biological brain. According to bio-mechanics perspective, the neural system is utilized to check the non-direct connections between datasets. Also, ANN model in gait application is more desired than bio-mechanics strategies or statistical methods. It produces models of gait patterns, predicts horizontal ground reactions forces (GRF), vertical GRF, recognizes examples of stand, and predicts incline speed and distance of walking.

The size of the power system is growing exponentially due to heavy demand of power in all the sectors (e.g., agricultural, industrial, and commercial). Due to this, the chance of failure of individual units leading to practical or complete collapse of power supply is common to be encountered. The reliability of power system is therefore the most important feature to be maintained above some acceptable threshold value. Furthermore, the maintenance of individual units can also be planned and implemented once the level of reliability for given instant of time is known. The proposed research therefore aims at determining the threshold reliability of generation system. The generation system consists of boiler, water, blade angle in turbine, shaft coupling, excitation system, generator winding, circuit breaker, and relay. This chapter presents the mathematical model of reliability of individual components and equivalent reliability of the entire generation system. It suggests the approach to determine the critical reliability of both individual and equivalent reliability of the generation system.

Reverse engineering (RE) has become a serious threat to the silicon industry. To overcome this threat, the ICs need to be made secure and non-obvious in order to find their functionality with their architecture. Real-time signal processing algorithms need to be faster and more reliable. Adding up additional circuits for increasing the security of the IC is not permittable due to increase in overhead of the IC. In this chapter, the authors introduce a few high-level transformations (HLT) that are used to make the circuit more reliable and secure against the reverse engineering without having overhead on the IC.

Chapter 10

An efficient design for testability (DFT) has been a major thrust of area for today's VLSI engineers. A poorly designed DFT would result in losses for manufacturers with a considerable rework for the designers. BIST (built-in self-test), one of the promising DFT techniques, is rapidly modifying with the advances in technology as the device shrinks. The increasing complexities of the hardware have shifted the trend to include BISTs in high performance circuitry for offline as well as online testing. Work done here involves testing a circuit under test (CUT) with built in response analyser and vector generator with a monitor to control all the activities.

Chapter 11

This chapter proposes a classical controller to control the industrial processes with time delay. A new population-based metaheuristic technique, called moth flame optimization (MFO) algorithm, is employed to tune the parameters of the classical proportional-integral-derivative (PID) controller for achieving the desired set point and load disturbance response. MFO-based PID controller may deal with wide ranges of processes, which includes integrating and inverse response as well as it may control the processes of any order with time delay. The transient step response profile obtained from the proposed MFO-based PID controller is juxtaposed with those obtained from other methods for optimizing the gains of the PID controller to control the processes with time delay. The proposed controller is analyzed by implementing step disturbance in the process at a specific simulation time. For few time delay processes, reference models are employed for better transient response as well as to analyze the controller for controlling the overall system with reference model.

Chapter 12

Wireless communication/networks are developing into very complex systems because of different requirements and applications of consumers. Today, mobile terminals are equipped with multi-channel and multiple access interfaces for different kinds of applications (or services). The combination of these access technologies needs an intelligent control to interface the best channel, interface/access or link for best services. In interface management, an arrangement is used to assign channels to interfaces in the multi-channel multi-interface environment. Artificial intelligence is one of the upcoming areas with different techniques which is used now a days to meet user's requirements. Quality of service (QoS) and quality of experience (QoE) are the performance parameters on which the success of any technique depends upon. Reliability of any system plays an important role in user satisfaction. This chapter shows some of the artificial techniques that can be used to make a reliable system.

Chapter 13

The UCM (universal compressor-based multiplier) architecture promises to provide faster multiplication operation in supply voltage as low as 0.6 V. The basic component of UCM architecture is a universal compressor architecture that replaces the conventional Wallace tree algorithm. To extend the work further, in this chapter, a detailed PVT (process-voltage-temperature) analysis is performed using Cadence Virtuoso 90nm technology. The analysis shows that the delay of the UCM has reduced more significantly than the Wallace tree algorithm at extreme process, voltage, and temperature.

Arbiter PUF and RO-PUF are two well-known architectures. Arbiter PUF is a simple architecture and easy to implement while RO-PUF require exponentially large hardware. As shown in this chapter, the digital design of RO-PUF response is 42.85% uniform and 46.25% unique.

This chapter presents linear quadratic regulator (LQR) for tuning the parameters of four-term proportional-integral-derivative plus second order derivative controller for controlling terminal voltage of alternator equipped with automatic voltage regulator (AVR) system. Different optimization techniques are considered for juxtaposition with the proposed controller on the basis of terminal voltage response profiles of the AVR system, and Bode plot analysis is carried out for comparing the frequency responses, and through root locus, the stability of the proposed controller is investigated. On-line responses are obtained by implementing a fast performing Sugeno fuzzy logic technique in the controller for working in off-nominal and on-line situations. The controller has undergone an investigation, while having changed system parameters, for the analysis of the robustness of the proposed controller. It is revealed that the performance of the proposed LQR-based controller exhibits a highly improved robust control system for controlling the AVR in power systems.

Chapter 16

In a modern communication system, voltage-controlled oscillator (VCO) acts as a basic building block for frequency generation. VCO with LC tank is preferred with passive inductor and varactor in radio frequency. Practical tuning range of VCO is low and unsuitable for wideband application. Switched capacitor and inductor can widen but at cost of chip area and complex system architecture. To overcome it, an equivalent circuit of the inductor is created. In this work, inductor-less VCO is implemented with CMOS 90nm technology that has center frequency 40GHz and frequency tuning range 37.7GHz to 41.9GHz.

Preface

In modern age, the artificial intelligence has become the backbone of industry. AI systems have the ability to execute tasks naturally associated with human intelligence, like decision making, life estimation and future forecasting etc.

From daily life applications to military applications and from toys to satellites, the use of electronic components is in extensive. Due to rapid evolution of electronics device technology towards low cost and high performance, the electronics products become more complex, higher in density and speed, and lighter for easy portability.

Reliability prediction of the electronic components used in industrial safety systems requires high accuracy and compatibility with the working environment. Reliability has become the major issue for successful operation of an electronic device. If defect or fault comes in an individual component, it can deteriorate the complete system or even it can be proven very dangerous for human lives, as the defective component is deployed in critical applications like military or aviation applications. So, prediction before failure, can save the entire system and data to be lost. The user can replace faulty component with the accurate one, and system will be saved from complete shutdown.

In general, components thrown out from industries are seen to have much life potential remaining in them unused. The failure prediction on the other hand intimates the user about residual life of component. Considering these aspects, here, an attempt is being made to assess the reuse potential of used materials. Predicting remaining useful life is a step to identify the reuse potential.

INSIDE THIS ISSUE

This issue is dedicated to application of artificial intelligence techniques in reliability analysis of various electronics and electrical products and parameters. In this regard, the first chapter is devoted to the need and challenges of reliability and its techniques. The various empirical, experimental and physics of failure techniques involved in reliability assessment are discussed in this chapter. The second chapter is about the fatigue and reliability of polymer-based nanocomposites and importance of reliability in field of nanoelectronics. The reliability of carbon nanotubes FET and nanowire FET devices are discussed in the Chapter 3. The author elaborates the fabrication process along with characterisation techniques of carbon nanotubes FETs. In the fourth chapter, the role of traditional and non-traditional optimization techniques is stated, which are useful to enhance reliability in process industries.

The fifth chapter introduces the artificial intelligence techniques to assess the reliability of various electronic products. The reliability and residual life of humidity sensor DHT11 is predicted using artificial neural networks. In the succeeding Chapter 6, the reliability issue of nanocomposite-based humidity sensor is discussed along with its fabrication and characterization techniques. The role of artificial neural network for prediction of gait parameters and patterns is discussed in Chapter 7. The importance of reliability in thermal power system is illustrated using modelling analysis and simulation, in Chapter 8. In Chapter 9, the author has discussed the reliability and security issues in DSP architectures along with examples.

Chapter 10 is dedicated to VLSI design and its testability issue. The reliability of high-speed advance multipliers is deduced through this chapter. A novel moth flame algorithm is described in Chapter 11, which is useful for PID controlled processes with time delay. In Chapter 12, the role of artificial intelligence in wireless heterogenous network is discussed. The architecture and PVT analysis of multipliers are elaborated in Chapter 13, which also signifies the notable applications of UCM based on its process-voltage-temperature results. Chapter 14 highlights the random number generator schema based on ring oscillator physical unclonable function and its related applications. The importance of linear quadratic regulator in PID plus second order derivative controller is discussed in Chapter 15. The last chapter is about novel architecture for high frequency generation without use of inductor. It provides sustained frequency which low jitter for reliable and secure applications.

CONCLUSION

Now days, as the level of integration is accelerating on fast pace, the reliability has become a challenging issue. Such discussions on the reliability issue and artificial intelligence techniques have transformed the complex system in reliable and intelligent system. This issue has covered various topics such as nano-electronics, thermal power system, VLSI, electronic components, wireless networks, DSP architecture etc. to assess their reliability level and residual life. It would be helpful for emerging components and devices with the possible quality of reliability and security for future applications.

Chapter 1
Reliability Analysis:
Need and Techniques

Cherry Bhargava
ⓘ https://orcid.org/0000-0003-0847-4780
Lovely Professional University, India

ABSTRACT

As the integration of components are increasing from VLSI to ULSI level. This may lead to damage of electronic system because each component has its own operating characteristics and conditions. So, health prognostic techniques are used that comprise a deep insight into failure cause and effects of all the components individually as well as an integrated technique. It will raise alarm, in case health condition, of the components drift from the desired outcomes. From toy to satellite and sand to silicon, the major key constraint of designing and manufacturing industry are towards enhanced operating performance at less operating time. As the technology advances towards high-speed and low-cost gadgets, reliability becomes a challenging issue.

INTRODUCTION

The electronic packaging engineers are integrating more and more chips and parts into assembly. As electronics industry has accelerated towards miniaturization and increasing functionality, the use of green materials, and the trends toward heterogeneous systems based on system in chip (SiC) system in package (SiP) or system on package (SoP) has also increased(Bailey et al., 2008). In modern era, the main goal of manufacturers is to produce more efficient and cheaper product, in order to attract more and more customers. The complex electronic products need vigorous skill to alleviate and control the deficiencies created. For packaging and testing, the reliability becomes a challenging issue, which proves to be a preventive factor for operating performance as well as cost of that product. The integration of components on a single chip is limited by the concern of reliability. The reliability of the interconnection of electronics packaging has become a critical issue(Kharchenko, 2015).

DOI: 10.4018/978-1-7998-1464-1.ch001

The condition monitoring system suggests an appropriate action to be taken before complete shutdown of the system(Rao, 1996). As complexity and cost of electronics products are increasing, it is very important to predict the reliability of design before physical prototyping, qualification testing and before introducing to the market. Once the product is delivered, if the components fail, it imposes significant cost in terms of additional rework, replacement of faulty products, and mainly costumer perception of the product. So, to prevent the system from faults and failures, a fault prediction technique is needed. Electronics packaging is subjected to mechanical vibration and thermal cyclic loads which lead to fatigue crack initiation, propagation and the ultimate fracture of the packaging(Lee, Kim, & Park, 2008).

USE OF ELECTRONICS IN VARIOUS INDUSTRIES

The electronics industry is complex, consisting of several diverse components, technologies, process, materials and devices, ranging from higher order to Nano-order with multiple faces(Moore, 2006). The use of electronics improves industrial performance, due to their size, price, speed and ability to store information. Use of electronics is increasing on faster pace in every segment of manufacturing and design industry. In fact, we come into contact with them every day in such areas as transportation, communications, entertainment, instrumentation and control, aviation, IT, banking, medical appliances, home appliances, manufacturing etc. In world of technical competition, the effective relationship between electronics lifecycle and reliability is becoming complex and crucial(Martin, 1999). In recent years, electronics have been used for various applications in numerous industries, including the following:

1. Aviation:

Avionic systems include various communications, navigation, control, display and other factors which consist of thousands of electronic components that can fit into an aircraft. This industry demands higher safety, control, reliability prediction and maintenance to continue operations without experiencing failures.

2. Automobile:

With the expansion of the automobile industry, there is a growing need or environmental protection, need of safety and security along with fuel efficiency. As the technology advances, the manufacturers of automobile industry offer variety of electronic systems to their customers. For example, the safety systems depend on electronic circuits. The ECU (engine control unit) provides dashboard information on fuel and oil levels, speed, gearing and engine revolutions (via the tachometer). Other electronic systems include FPGA (field programmable gate array), ASIC (application specific integrated circuits) and SOC(system on chip) (Kulshreshtha, 2009).

3. Entertainment and Communications:

A few decades ago, the main application of electronics in entertainment and communications was in telephony and telegraphy. With the advent of radio waves, however, any message could travel from one place to another without the use of wires, and the communications industry was revolutionised. For memory purpose, wireless communication uses the digital integrated circuits. Even the satellite communication becomes possible with the advent of microelectronics. Due to evolvement of integrated circuits, the connection between communication and computation increases which evolves the concept of digital signal processing.

4. Defence Applications:

Defence applications are totally controlled by electronic circuits. Radar (radio detection and ranging) was first used during the Second World War and has led to many significant developments in electronics. With radar, it is possible to identify and find the exact location of enemy aircraft. Today, radar and anti-aircraft guns can be linked to an automatic control system to make a complete unit. To locate the enemy submarines or fighter planes, the infrared systems as well as Radar have been employed. It controls the gun firing as well as guided missiles. The electronic system has been proven a secret communication channel between military headquarters and individual units.

With the advancement in electronics technology, it has been an integrated part of other manufacturing and design industries. From toy industry to satellite and entertainment to safety industries, the use of electronics is increased, for example, to control moisture of material, to catch the lie of prisoner, automatic fire extinguisher, automatic temperature controller etc. A tiny electronic system can control and produce power of thousand megawatts(Bridges).

5. Medical Services:

The use of electronics in medical science has grown quickly. The emerging field of electronics in bio-medical area, the diagnosis and treatment of crucial diseases are possible with the use of ECG, EEG, X-rays and ultrasound scanners etc. The evolution of digital technology in medical field has made the use of medical apparatus comparatively easier and effective, for example, digital blood pressure monitor, digital thermometer or digital blood sugar instrument etc. Now days, a common man can handle such complicated devices easily and can monitor his health status regularly.

6. Instrumentation:

No research laboratory is complete without electronic instruments. For real time, reliable and effective operations, controlling and measurement of operations are necessary. With the advancement in electronic industry, the industry automation is possible. A variety of electronic instruments e.g. CRO, function generators, digital multimeters etc., the precise measurement of quantities are possible. Now days, computers have become integral part of control and measurement systems. Sensors, controllers and actuators have made the life of a common man much easier.

Major concerns about the use of electronics in industry include safety, security, reliability, maintainability, cost of failure, risk factors etc. When a component breaks down, the consequences could be severe. There is a need to predict reliability to determine a suitable replacement policy for electronic component.

FAILURE, ERROR, FAULT AND DEFECT

Failure is the inability of a component or device to perform its required function. It is due to mishandling of device or unfavourable environmental conditions. Error is deviation of predicted value from actual value. It is due to miscalculation of value or mis-interpretation of input-output parameters. Fault is condition which stops the component or device to perform the required task in stipulated period. The defect is an error that causes the device or component to produce incorrect/unexpected outcomes. when defected component is implemented or used in the device, then it results into failure of the device. In any investigation to determine the cause of a failure from which it is expected to initiate a product liability action, a defect must be detected(Le May, 1998). So, detect is the nonconformity that should not be present in the product if it had been designed, produced, and handled in a correct manner.

FAILURES AND ITS CLASSIFICATION

Reliability engineering is concerned mainly with the evaluation of the chances of occurrences of failures of equipment and with their prevention. Failure can be caused by:

1. Use of component beyond the prescribed limits of stresses
2. Even if the stress is within limit, the inherent weakness of component itself.
3. Deterioration or derating of component with age and continuous use.
4. In a device, if one component fails, it would effect on other's component performance or it may cause failure.

Failure can be classified as sudden failure or slow failure, that means prior experiments may or may not predict such failures. Further, failure can be classified as partial, intermittent or complete, based on its effect(Barnes, 1971). When the component is functional but its characteristics deviates from its prescribed limit, it is called partial failure. An intermitted failure is recurrent process. When due to failure, the complete device or component becomes scrap or shutdown, it is known as complete or total failure. In any application, combination of such failures can also occur(O'connor, O'Connor, & Kleyner, 2012).

A rapid and disastrous failure is labelled as 'catastrophic' failure whereas, slight or partial failure is known as degradation. The classification of the failure is solely depending on the application. For example, a case of an amplifier, in which one of the bipolar junction transistors in push-pull configuration suffers from catastrophic failure, then the performance of the amplifier may be degraded but, yet it is acceptable. Failures are also categorised based on its post effect on system or process, which is observed or regulated by that faulty component. For example, in case of nuclear reactors, if the power supply is lost which controls protection of nuclear reactor, is an expensive failure but it is considered as safe. In other case, if power is lost to artificial heart beat system, it could imperil the life of the patient, which is highly unsafe.

Similarly, burning out a faulty resistor can cause failure of valve, it is safe failure but burning out capacitor used in aviation, can prove dangerous to thousands of human lives. So, severity of failures is categorised based on the applications in which those faulty components are deployed(Fenton & Ohlsson, 2000).

Figure 1. Interrelation of faults and failures

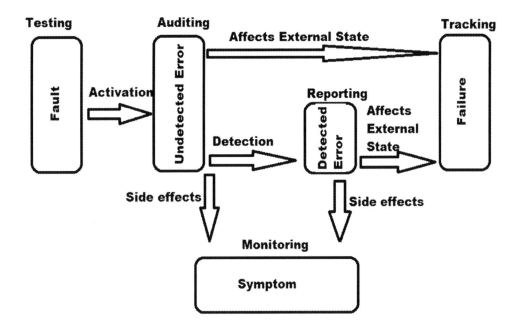

FAILURE RATE

Failure rate is the frequency with which a component or device fails. It is expressed as failure per unit time and denoted by a Greek letter lambda (λ). It is time varying element of the system. Failure rate may be chosen as one of the four possible alternatives:

1. Observed
2. Assessed
3. Predicted
4. Extrapolated

Failure rate depends on the time of use or aging factor of the component or device. For example, the failure rate of a transistor in sixth year of its service is much higher than the failure rate during first year of operation(Fountain, 1965).

Usually, mean time between failure is often stated, rather than failure rate.

$$\text{Mean time between failure} = \frac{1}{\lambda} \qquad (1)$$

The most common measurement of reliability is *mean time between failure* (MTBF). MTBF is a statistical value that indicates how long, on average, an electrical unit can operate before failing. It should be realized that a specified MTBF depends on the method with which it is determined and the method's inherent assumptions. The assumptions vary widely according to the method and can cause very large differences in the final determination. The methods and assumptions used must be fully understood for the specified MTBF to be meaningful (Kullstam, 1981).

MTBF can be determined either by pure calculation or by a more direct demonstration. Demonstrated MTBF gives the most realistic results, but it is usually impractical due to the length of the testing and the high cost involved. If the testing is done before the manufacturing processes are mature, the variations due to manufacturing anomalies can make consistent results difficult. A calculated MTBF, at a specified temperature, is often used as a verification that the design is relatively sound. This can be done using mathematical procedures for summing up the published failure rates of all the individual components to give a failure rate of the composite. Numbers generated in this manner should be used for comparative purposes only. This type of analysis may be useful for meeting contractual goals or marketing purposes, but it cannot provide any insight to corrective actions and often does not reflect the actual reliability measure of the equipment. The actual failure mechanisms in electronic assemblies and their frequency of occurrence are best identified using failure analysis.

For non-repairable components or device, instead of MTBF, another term is used that is called MTTF (mean time to failure). It can be described as: Under prescribed stress conditions, the total time for which component or device was made under observation, divided by total number of failures that occur during that cumulative time. Mean life is described as: Under stated stress conditions, mean value of observed time divided by number of failures occur during that time period. Both MTTF and mean life are same, both are applicable to non-repairable components. Only one difference is there that MTTF is considered for only pre-set time, but mean life is considered for the complete life time of that component. Mean life is calculated when all the samples under observation have failed, but MTTF is calculated when the pre-set time period is over. In such a way, MTTF is different from MTBF, as MTBF is for failure in repairable components whereas MTTF is for failure in repairable components. Mean time to repair is known as MTTR(Singh & Mourelatos, 2010).

Maintenance outlines the bunch of pursuits performed on a component to recollect or retain in to a prescribed condition. Maintenance is further classified as preventive and corrective maintenance. The preventive maintenance is carried out at pre-defined time interval to lessen the deterioration failures i.e. wear out failures. The aspiration behind preventive maintenance is to identify and overhauling the hidden failures, the failures which are not identified. Corrective maintenance does not only identify the failure but also correct the failure and maintain its original state of function. The steps for corrective maintenance are:

1. Detection and localization
2. Replace and re-assemble
3. Re-checking the health condition

NEED OF FAILURE PREDICTION

From daily life applications to military applications and from toys to satellites, the use of electronic components is in extensive. In critical applications like aviation industry, if a component fails before its actual life, it can cost lives of many human beings. So, prediction before failure, can save the entire system and data to be lost. The user can replace faulty component with the accurate one, and system will be saved from complete shutdown(Agarwal, Paul, Zhang, & Mitra, 2007).

During manufacturing process of electronic components/devices, different tests are conducted on these components/devices to check its performance and capabilities. Then data sheets of those components have been framed, which signifies minimum and maximum range of all electrical parameters as well as environmental stress. Then components/devices are released to real market with a warranty period depends on the qualification testing(Vichare & Pecht, 2006). Before purchasing the specific component, user needs assurance about the long life, reliable and satisfactory performance throughout the operation. Moreover, quality assurance cells are becoming strict towards protection of user rights. So, the product manufacturers reacted to these constraints by offering extended warranties and guarantees. The warranty may be in the shape of lifetime i.e. the specific component will work successfully for specific duration of time, that is called lifetime. This document is generally endorsed with datasheet. The manufacturer has to replace or rectify the component free of cost, if it falls under warranty period. If the component under warranty time, could not perform the specified task, it will incur an extra cost on manufacturer to replace it or compensate the user. Replacement will not only become extra financial burden on manufacturer, but also, it will degrade the reputation of manufacturer in competitive market. The warranty service or replacement cost may vary from 2-10%, depending on sale price of that component (Murthy, 2007). The decision on warranty period is directly connected with reliability of the system or component.

So, product reliability is the most important constraint for efficient and successful operation of gadget or product. Not only manufacturer will bear the repairing and additional cost, but also it will degrade the market reputation of the manufacturer, in case component or product fails before its actual failure time or warranty period. So, failure prediction is the most critical issue in fast growing electronic industry.

The performance of next generation U.S. Army operations such as miniature driver-less ground vehicles and driver-less Arial vehicles are intensely dependent on electronics(Habtour, Drake, & Davies, 2011). Due to vibration and shocking environment, these electronic systems may experience variation in their performance. This will lead major damage to packaging and soldering of joints. Nowadays, failure rate prediction is not just a domain for the military. As electronics is used in almost every area of human life, and as a result all of these areas being "safety critical", the prediction of the lifetime of electronic modules becomes an ever-increasing necessity(Jánó & Pitică, 2011). This is especially true for aeronautics and automotive applications, where temperatures can sometimes well exceed the maximum guaranteed operating temperature for a particular component. Performance parameters as well as failure prediction, reliability and safety need to be built in during design and development and retained during operation and production of item.

ELECTRONICS COMPONENTS

An electronic component is defined as a discrete device used to affect electrons or their associated fields. A component has ability to switch and control the current and voltages without any mechanical or nonelectrical instruction; examples such as capacitor, transistors etc. The components may be active components or passive components. Active components rely on source on energy for example diode, integrated circuits, bipolar junction transistors etc. passive components do not rely on any source of energy. They cannot amplify the power of a signal. For example, resistor, capacitor, inductor etc.(Lebby, Blair, & Witting, 1996).

Electronic components are susceptible to various forms of damage that can be attributed to a limited number of physical or chemical mechanisms. Primary failure mechanisms include:

1. Corrosion
2. Distortion
3. Fatigue and fracture
4. Wear

The majority of electronic component failures analyzed are often corrosion related, but other mechanisms considered mechanical in nature also influence the reliability of electronic components. Due to fatigue or fracture, the electronic components face sudden movement or vibration. The electronic components are widely used from semiconductor industry to aviation industry, from manufacturing to processing industry. If defect or fault comes in an individual component, it can deteriorate the complete system or even it can be proven very dangerous for human lives, in case such defective component is deployed in critical applications like military or aviation applications.

Capacitor Failures

A capacitor is an electrical charge storage device. Capacitors are used extensively in all electronic circuits(Becker, 1957). Capacitors perform many functions and are commonly used to:

*   Create a localized ideal voltage source for adjacent components
*   Provide charge storage for timing functions
*   Tune critical circuits
*   Provide an impedance that varies with frequency for signal selection
*   Store charge for lost power conditions

Capacitors have the same basic construction: two conductive surfaces separated by a dielectric material. The energy stored in a capacitor is stored in the dielectric medium. Transfer of charge into and out of a capacitor is achieved through the conductive plates and lead wires(Bar-Cohen, 2014). An equivalent circuit for a capacitor is shown in Figure 2.

There are numerous failure modes for capacitors including:

*   Catastrophic dielectric breakdown resulting in an unintended release of energy
*   Reduction (sometimes intermittent) in capacitance
*   Raise in the value of ESR (equivalent series resistance)
*   An increase in the leakage current that may lead to a thermal runaway condition
*   Loss of an electrical connection resulting in a loss of capacitance

The results of each one of these failures varies from no apparent effect (as in the case of an open failure of a redundant bypass capacitor) to a catastrophic and spectacular explosion and arcing event when a high-voltage electrolytic bus capacitor fails(Harada, Katsuki, & Fujiwara, 1993). The causes for capacitor failures include:

*   Excessive applied voltage
*   Voltage transients
*   Surge currents

Figure 2. Equivalent circuit of a capacitor

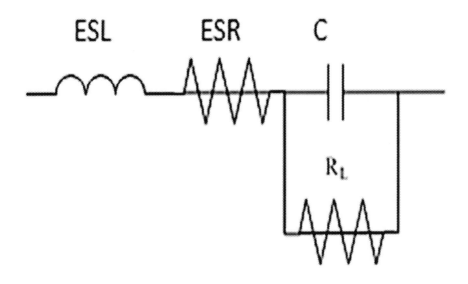

- Excessive power dissipation
- Thermal stress
- Mechanical stress and flexing
- Manufacturing defects
- Contamination

Ceramic Capacitors

Ceramic capacitors come in many different packages, including disk, tubular, and assorted surface mount designs. The dielectric material in a ceramic capacitor is made from earth fired under extreme heat. Ceramic capacitor failure mechanisms include:

1. Mechanical cracking
2. Surge currents
3. Dielectric breakdown

1. Mechanical Cracking: The dielectric material in a ceramic capacitor is extremely inflexible. Combined with the minimal required spacing between capacitor plates to achieve the required capacitance, this inflexibility results in a somewhat fragile structure. Mechanical stresses on ceramic capacitors may result in mechanical cracking. The sources of mechanical stresses include different coefficients of expansion between a ceramic capacitor and PCB materials, mechanical flexing of a PCB, assembly-induced stress, and mechanical shock or vibration.

The effect of a mechanical crack may take time to show up in a ceramic capacitor. For example, if the stress caused by a flexed PCB causes a ceramic capacitor to crack, it may return to its normal position when the flexing force is relaxed. This return to normal can result in no noticeable loss in capacitance or degradation in performance as the severed capacitor plates may physically touch again. However, only a slight misalignment of the parallel interleaved plates would result in a short circuit(Sakata, Kobayashi, Fukaumi, Nishiyama, & Arai, 1995).

2. Surge Currents: Excessive current (called *surge* current) exceeds the instantaneous power dissipation capability of localized regions of the dielectric, resulting in a thermal runaway condition(Qin, Chung, Lin, & Hui, 2008).

3. Dielectric Breakdown: Dielectric breakdown can be caused by overvoltage conditions or manufacturing weaknesses that compromise the dielectric. Manufacturing weaknesses can rarely if ever be documented, as the evidence is inevitably destroyed in the failure event. Dielectric breakdown results in an unintended flow of current between the capacitor terminals, which leads to excessive power dissipation and may result in an explosive failure.

Electrolytic Capacitor

Capacitance (C) is calculated from the formula $C = K$ (a$/$d), where K is the dielectric constant, a is the conductive plate surface area, and d is the distance between the conductive plates. In a never-ending quest for higher capacitance per unit of volume, designers have always tried to reduce the distance between the conductive plates while maximizing the area. The aluminium electrolytic capacitor (most common) is made from a ribbon of aluminium with a thin layer of oxide on its surface. A water-based or other electrolyte, electrically isolated from the insulated aluminium plate, forms the other electrode as it is in contact with a non-oxidized aluminium plate (Gualous, Bouquain, Berthon, & Kauffmann, 2003). The oxide layer on the positive aluminium plate is typically 0.01 µm. A diagram showing typical construction of an electrolytic capacitor is shown in Figure 3.

The oxide insulator behaves as a rectifier that conducts current in one direction while blocking current in the other. Electrolytic capacitors must therefore be used in applications where the voltage across the capacitor will maintain the correct polarity. The application of the incorrect polarity across an aluminium electrolytic capacitor will result in destruction of the capacitor.

Electrolytic capacitors are specified for a maximum dc working voltage. The peak of a ripple voltage riding on a dc level must not be allowed to exceed the maximum dc working voltage during normal continuous operation. However, electrolytic capacitors are specified to withstand an occasional voltage surge beyond their maximum dc working voltage. Exceeding the maximum surge rating will likely result in a catastrophic capacitor failure. Leakage current is relatively high on electrolytic capacitors due to the thin dielectric barrier and the large conductor surface area.

The capacitance of electrolytic capacitors typically is not specified very precisely, as production methods result in large variations in capacitance. Common capacitance values specified would include: −10/+50 percent, −10/+75 percent, and ±20 percent.

The equivalent series resistance (ESR) is a simplified model to account for losses including those caused by lead resistance, foil resistance, and dielectric losses. The equivalent series resistance is the sum of in-phase AC resistance. It includes resistance of the dielectric, plate material, electrolytic solution, and terminal leads at a particular frequency. ESR acts like a resistor in series with a capacitor (thus the name Equivalent Series Resistance). The drastically change of ESR can change the capacitance and hence it may fail the complete electronic product or design.

Figure 3. Typical electrolytic capacitor construction
(Image credit: http://electricalacademia.com/basic-electrical/capacitor-types-construction-and-uses/)

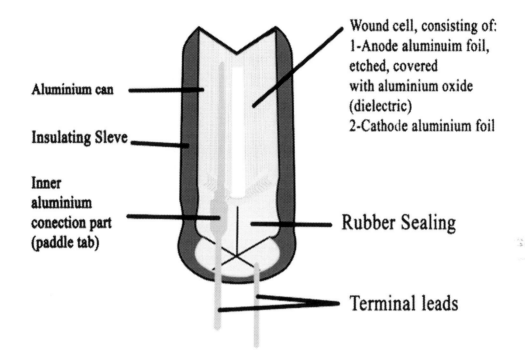

Depending on the application, this simplification may not be appropriate. The ESR model will allow calculation of the expected capacitor power dissipation for a specified ripple condition. Since heating is a major factor in reducing the expected life of an electrolytic capacitor, ripple current limitations must be maintained to avoid overheating of the capacitor. Ripple currents are especially significant (and harder to evaluate) in high-frequency switching applications. Not only do the fundamental frequency ripple currents contribute to losses (heating), but higher-frequency harmonic currents will also contribute.

The key factors in determining premature failure of electrolytic capacitors in service are:

- Elevated temperature (caused by ambient temperature or ripple current)
- Excessive voltage (continuous or momentary)

For example, for a given operating temperature, the life of a typical electrolytic capacitor will be a factor of 5 times longer if it is run at 50 percent of the rated voltage(Pascoe, 2011). Another example is that the life of a typical electrolytic capacitor will be a factor of 17 times longer if the capacitor core temperature is reduced from 65° to 25° C.

Electrolytic capacitors fail in four general ways:

1. Dielectric breakdown resulting in a short between capacitor terminals
2. Loss of capacitance
3. Open circuit
4. Reverse polarity

1. **Dielectric Breakdown**: It can be caused by overvoltage conditions or manufacturing weaknesses that compromise the dielectric. Manufacturing weaknesses can rarely if ever be documented, as the evidence is inevitably destroyed in the failure event. Dielectric breakdown results in an unintended flow of current between the capacitor terminals, which leads to excessive power dissipation and may result in an explosive failure. Diagnosing a dielectric breakdown capacitor is generally readily apparent, as the failures are typically explosive and readily visible. Should the energy available to a failed capacitor not result in a catastrophic failure, testing of a suspect capacitor should be performed using a light bulb in series with the capacitor under test to prevent supply energy to a failed capacitor. Leakage current can more safely be measured using this technique.

2. Loss of capacitance: It occurs when there is a loss of electrolyte. A loss of electrolyte typically occurs when there is a leak in the sealed case of the electrolytic capacitor. A leak in a seal can develop over time due to normal environmental processes. Seal degradation is generally accelerated by some cleaning solvents, increased temperature, vibration, or manufacturing weakness. A typical result of a loss of electrolyte is a loss of capacitance, an increase in ESR, and a corresponding increase in power dissipation(Ma & Wang, 2005).

3. Open circuit: It typically occurs when terminals of connections inside the capacitor degrade and fail. Loss of an electrical connection can occur as a result of corrosion, vibration, or mechanical stress. The result of an open-circuit failure is a loss of capacitance.

4. Reverse polarity: Aluminium electrolytic capacitors are polarized and must be connected in the correct polarity. They can withstand reverse voltages up to 1.5 V. Higher reverse voltage can cause failure by pressure build up and rupture of the capacitor's safety vent structure. Non-polar and semi-polar devices are available that can withstand reverse voltage.

Humidity Sensor Failure

A humidity sensor or hygrometer is a sensor that senses and measures both moisture as well as temperature. Relative humidity is term which is calculated as ration of air moisture to the maximum amount of moisture at a specific temperature. To control the humidity, moisture or temperature, the relative humidity becomes a major constraint (Yamazoe & Shimizu, 1986).

DHT11 digital temperature and *humidity sensor* is a composite Sensor contains a calibrated digital signal output of the temperature and humidity.

The Humidity sensors work by recognizing changes that modify electrical quantities such as current or temperature, noticeable all around in the air. Basically, there are three types of humidity sensors: thermal humidity sensor, capacitive humidity sensor and resistive humidity sensor. To explore humidity in air, these sensors monitors regularly any change in atmosphere.

Types of Humidity Sensor

Based on application and material used, humidity sensor is of various types i.e. thermal humidity sensor, capacitive humidity sensor and resistive humidity sensor. Each one has different principle of operation. The humidity sensor is also known as hygrometer, as it senses humidity as well as moisture.

Figure 4. Humidity sensor
(Image credit: https://electronicsforu.com/resources/electronics-components/humidity-sensor-basic-usage-parameter)

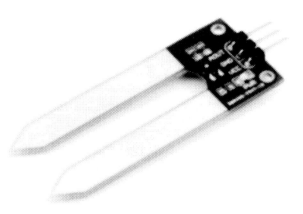

Thermal Humidity Sensor

In thermal humidity sensor, two thermal sensors conduct electricity based upon the humidity of the surrounding air. One sensor is encased in dry nitrogen while the other measures ambient air. The difference between the two measures the humidity.

Capacitive Humidity Sensor

Humidity sensors relying on this principle consists of a hygroscopic dielectric material sandwiched between a pair of electrodes forming a small capacitor. Most capacitive sensors use a plastic or polymer as the dielectric material, with a typical dielectric constant ranging from 2 to 15. In absence of moisture, the dielectric constant of the hygroscopic dielectric material and the sensor geometry determine the value of capacitance. The majority application area of capacitive humidity sensor is weather and commercial and industries.

Resistive Humidity Sensor

Resistive type humidity sensors pick up changes in the resistance value of the sensor element in response to the change in the humidity. As humidity changes, so does the resistance of the electrodes on either side of the salt medium.

When the moisture level increase, the concentration of ions in the dielectric and in turn it decreases the resistance, which effect the voltage and hence these variation gives us relative humidity and temperature reading. Application of humidity sensor can be observed in many practical areas like soil-humidity detection, cold store monitoring, maintaining green-house, lie detector etc.

Failure Mode of Humidity Sensor

Increase in humidity, beyond a certain level can deteriorate the component. Similarly, if the temperature increases beyond a certain point, it will stop detecting. With increase in thermal stress, the melting starts and catastrophic failure will occur(Ramakrishnan, Syrus, & Pecht, 2000).

DHT11 is a sensor that senses and detects humidity as well as moisture. DHT11 is an Arduino compatible sensor, which can be integrated with Arduino board for the experimental reading. The stressor parameter used here is temperature. The normal range if DHT11, is Humidity [20-90] %RH, Temperature [0-50] ° C. That means this is the application range for this particular sensor. For checking the response, various range of temperature have been provided to see the response as well the condition of the sensor until and unless it gets completely dysfunction and deteriorate.

The life estimation methods and modes are depicted in Figure 5, where various techniques of artificial intelligence are shown which are used to predict the residual life of the humidity sensor. These mainly used artificial intelligence techniques are artificial neural networks, fuzzy logic, adaptive neuro fuzzy inference system, particle swarm optimization and genetic programming.

The application era of humidity sensor is wide. It may be used in soil quality prediction, lie detector as well as in industrial applications. The residual life prediction of humidity sensor is an important concern because the failure of malfunctioning of humidity sensor may deteriorate the whole system where it is employed.

Figure 5. Life estimation of a humidity sensor

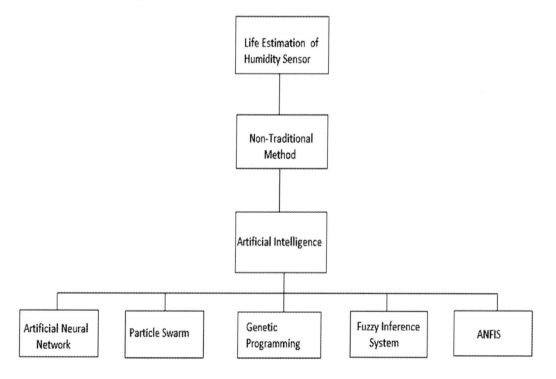

LIFE CYCLE OF A COMPONENT

Reliability is the capability of a component/device or system to perform its intended function consistently, within stipulated time period without failure or degradation. The reliability prediction estimates the mean time between failure (MTBF), mean time to failure (MTTF) and exploring the root cause of the failure. Therefore, it is an important issue that the life cycle model should address both probabilistic and deterministic context of component life cycle(Dylis & Priore, 2001). During life cycle of a component, reliability needs to be accessed at its different stages of life.

1. Reliability Assessment at design level
2. Reliability Assessment at manufacturing level
3. Reliability Assessment at launching level
4. Reliability Assessment at operational level

The 'design reliability phase' is specification of reliability at component level i.e. check of component qualification at design level. But when the component is produced, its reliability differs because of assembly error. The reliability of the produced/manufactured component/product is known as inherent reliability. The product needs to be launched to market for its real-time application. Reliability may distort due to vibration factor during transportation or storage at crude environment (such as excessive temperature or humidity etc.). The extrapolation of reliability at the time of market is known as 'reliability at sale' phase. Once the component/device is sold, either it will be stored off shelf or it will be inserted directly to some device/system for operation/function immediately. Reliability performance may be affected due to operational environment such as (thermal stress, vibration load, excessive electrical parameters etc.), so this phase of reliability is called 'field reliability'.

Figure 6. shows the different phases of reliability and factors effected the consistent performance of the component/device.

Figure 6. Life cycle of a component
(Image credit: https://www.nap.edu/read/18987/chapter/17#223)

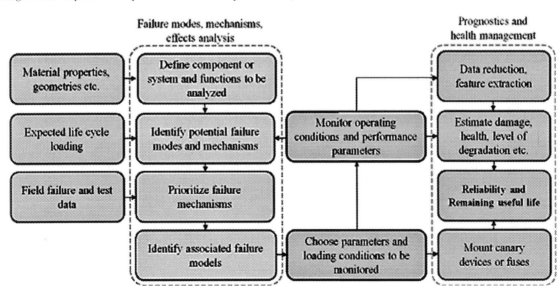

So, the assessing the component performance at different phases are necessary. After analysing all the factors and parameters, warranty period is extracted, which is mentioned along with the data sheet(Solomon, Sandborn, & Pecht, 2000).

The Bath-Tub Curve

Failure can occur in any phase of component life cycle from design phase to product phase or from on-shelf to off-shelf. To analyse the root cause of failure by analysing field/user data and design end results to defecate the failure mechanism is known as the failure analysis(Misra, 2008).

Electronic failures are further classified into three different sections of failure:

1. Infant mortality
2. Random occurring events
3. Wear-out

The defects that usually introduced during manufacturing/assembly process is known as early life failures. A constant failure rate, due to randomly occurring events is having useful life period(Hashemian & Bean, 2011). When the component/ device is exposed to long term effect of environmental stress or overloaded electrical parameters, then derating/deterioration of the component occurs known as the wear out. The life time of a component is graphically represented as "Bath tub curve", as shown in Figure 7.

Initially, an item typically exhibits a high failure rate. This failure rate decreases during the infant mortality period and stabilizes when weak and defective units have been eliminated. Failures occurring in the initial phase of deployment usually result from built-in flaws due to faulty workmanship,

Figure 7. Bath tub curve
(Image credit: https://en.wikipedia.org/wiki/Bathtub_curve)

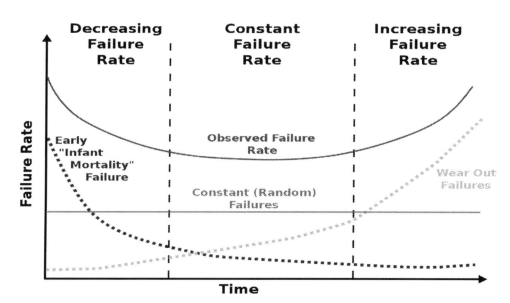

transportation damage or installation errors(Neri, Allen, & Anderson, 1979). Causes of infant-mortality failures include:

1. Inadequate quality control
2. Material deficiencies
3. Improper handling and packages
4. Incorrect procedures
5. Incomplete testing

Infant-mortality failures can be reduced by monitoring and controlling the manufacturing process. Failure analysis is vital during this phase to identify and correct the causes of problems so they can be eliminated prior to deployment(Huston & Clarke, 1992).

The useful life period is normally characterized by relatively infrequent and unpredictable failures. The specifics vary among hardware types and components and will vary based on margins used in the design of an equipment. Failures occurring during the useful life period are referred to as random failures because of this infrequent and unpredictable nature. Causes of useful life failures include:

1. Inadequate component or system design margins
2. Latent component or system defects
3. Misapplication
4. Excessive electrical, thermal or physical stress

The failure rate of a component will increase rapidly in the wear out phase. Wear out failures are due primarily to the deterioration of the design strength of a device resulting from the operation and exposure to environmental fluctuations(Aggarwal, 2012). This deterioration may result from various physical and chemical phenomena that include:

1. Corrosion/oxidation
2. Insulation break down
3. Fatigue or frictional wear
4. Cracking of plastic materials

Wear out failures may be reduced or eliminated through preventive maintenance and by providing adequate design margins to extend the life of components.

ELECTRONIC COMPONENTS FAILURE MODE AND MECHANISM

When designing systems containing electronic components, it is valuable to an engineer to have the knowledge of component failure modes and mechanisms and their probability of occurrence. These factors are also invaluable to an analyst when performing failure analyses and developing recommendations to eliminate the future occurrence of failures. Table 1 presents failure modes and mechanisms for a representative group of electronic components. The data used to prepare this Table was collected by the Reliability Analysis Centre in Rome, NY(Borgovini, Pemberton, & Rossi, 1993).

Dormant Related Failure Mechanism

Dormancy can be important in the design of a product, since a predominant portion of its life cycle may be in a non-operational mode. Moisture is probably the single most important factor affecting long-term nonoperating reliability. All possible steps should be taken to eliminate moisture from electronic devices and products(Kececioglu & Tian, 1998). Hygroscopic materials should be avoided or protected against accumulation of excess moisture. Most failures that occur during nonoperating periods are of the same basic kind as those found in the operating mode, although they are precipitated at a slower rate. Dormancy failures may be related to manufacturing defects, corrosion, and mechanical fracture.

Table 1. Failure Mode Distribution of electronic components

Device type	Failure mode	Device type	Failure mode
Alarm, annunciator	-Misleading indication -Failure to operate on demand -Spurious operation -Degraded alarm	Battery, lead acid	-Degraded output -Short -Intermittent output
Battery, lithium	-Degraded output -Start-up delay -Short -Open	Battery, rechargeable, Ni-Cd	-Degraded output -Zero output
Capacitor, aluminium, electrolytic, foil	-Short -Open -Electrolyte leak -Capacitance reduced	Capacitor, paper	Short Open
Capacitor, ceramic	-Short -Change in value -Open	Capacitor, mica/glass	Short Change in value Open
Capacitor, plastic	-Open -Short -Change in value	Capacitor, tantalum	-Short -Open -Change in value
Capacitor, tantalum, electrolytic	-Short -Open -Change in value	Capacitor, variable, piston	-Change in value -Short -Open
Coil	-Short -Open -Change in value	Connector/connection	-Open -Poor -Intermittent Short
Circuit breaker	-Opens without stimuli -Does not open	Crystal, quartz	-Open -No oscillation
Diode, general	-Short -Open -Parameter change	Diode, rectifier	-Short -Open -Parameter change
Diode, silicon control rectifier (SCR)	-Short -Open	Diode, small signal	-Parameter change -Open -Short
Diode, Triac	-Failed off -Failed on	Diode, thyristor	-Failed off -Short -Open -Failed on

continues on following page

Table 1. Continued

Device type	Failure mode	Device type	Failure mode
Diode, Zener, voltage reference	-Parameter change -Open -Short	Diode, Zener, voltage regulator	-Open -Parameter change -Short
Electric motor, ac	-Winding failure -Bearing failure -Fails to run, after start -Fails to start	Fuse	-Fails to open -Slow to open -Premature open
Hybrid Device	-Open circuit -Degraded output -Short circuit -No output	Keyboard Assembly	-Spring failure -Contact failure -Connection failure -Lock-up
Liquid crystal display	-Dim rows -Blank display -Flickering rows -Missing elements	Meter	-Faulty indication -Unable to adjust -Open -No Indication
Microcircuit, digital, bipolar	-Output stuck high -Output stuck low -Input open -Output open	Microcircuit, digital, MOS	-Input open -Output open -Supply open -Output stuck low -Output stuck high
Microcircuit, interface	-Output stuck low -Output open -Input open -Supply open	Microcircuit, linear	-Improper output -No output
Microcircuit, memory, bipolar	-Slow data transfer -Data bit loss	Microcircuit, memory, MOS	-Data bit loss -Short -Open -Slow data transfer
Microwave amplifier	-No output -Limited voltage gain	Microwave attenuator	-Attenuation increase -Insertion loss
Microwave connector	-High insertion loss -Open	Microwave detector	-Power loss -No output
Microwave, diode	-Open -Parameter change -Short	Microwave filter	-Center frequency drift -No output
Microwave mixer	-Power decrease -Loss of intermediate frequency	Microwave modulator	-Power loss -No output
Microwave oscillator	-No output -Untuned frequency -Reduced power	Microwave voltage-controlled oscillator (VCO)	-No output -Untuned frequency -Reduced power
Microwave phase shifter	-Incorrect output -No output	Microwave polarizer	-Change in polarization
Microwave yttrium from garnet (YIG) resonator	-No output -Untuned frequency -Reduced power	Optoelectronic LED	-Open -Short
Optoelectronic sensor	-Short -Open	Power Supply	-No Output -Incorrect Output

When designing for dormancy, materials sensitive to cold flow and creep, as well as metallized and non-metallic finishes that have flaking characteristics, should be avoided. Also, the use of lubricants should be avoided. If required, use dry lubricants such as graphite. Teflon® gaskets should not be used. Conventional rubber gaskets or silicone-based rubber gaskets are typically preferable.

When selecting components, it is recommended that only components with histories of demonstrated successful aging be used. Semiconductors and microcircuits that contain deposited resistors may exhibit aging effects. Parts that use mono-metallization are recommended to avoid galvanic corrosion. Chlorine or other halogen-containing materials should not be sealed within any circuitry components. Variable resistors, capacitors, inductors, potentiometers, and electromechanical relays are susceptible to failure in dormant environments and are generally not recommended.

It is recommended that components and systems in storage should not be periodically tested. Historical data shows that failures are introduced as a result of the testing process. Causes of many of the failures were test procedures, test equipment, and operator errors.

The important guidelines for the storage purpose are:

1. Disconnect all power
2. Ground all the components
3. Use nitrogen to prevent moisture and corrosion
4. Maintain temperature at $50° F \pm 5° F$ ($10° C \pm 2.7° C$) (drain all equipment of water to prevent freezing or broken pipes).
5. Control relative humidity to $50° F$ ($10° C$) ± 5 percent (reduces corrosion and prevents electrostatic discharge failure).
6. Periodically recharge batteries
7. Protect against rodents: squirrels have chewed cables; mice have nested in electronic cabinets and porcupines have destroyed support structures (wood). Door/window seals, traps/poison, and frequent inspection protect against these rodents.

ENVIRONMENTAL STRESSES CAUSING ELECTRONIC COMPONENTS FAILURE

Failures relating to distortion, fatigue and fracture usually occur as a result of an equipment being operated outside of its design environment or beyond its intended life. Environmental factors that accelerate failures in electronic components and assemblies include vibration, thermal cycling, and thermal shock. Table 3 provides the distribution of failures with respect to environmental factors found in an aircraft environment.

Temperature extremes at or beyond design limits of a component or system, and at high rates of temperature cycling, can occur in the operational environment. In some cases, the equipment might be installed in a temperature-controlled compartment. In others, the equipment will be directly exposed to temperature extremes and cycling. Cold soaking for hours prior to start-up will make equipment subject to thermal shock when it is powered up. The effects of temperature extremes, cycling, and thermal shock include parameter drift, mechanical deformation, chemical reactions, increased contact resistance, dielectric breakdown, and electro-migration. To address the potential problems associated with temperature extremes, cycling and thermal shock, several approaches are commonly used(Ye, Lin, & Basaran, 2002).

Table 2. Dormant part failure mechanism

Type	Mechanism	Percent failure mode	Accelerating factor
Microcircuit	-Surface anomalies -Wire bond -Seal defects	35–70 Degradation 10–20 Open 10–30 Degradation	Moisture, temperature Vibration Shock, vibration
Transistor	-Header defects -Contamination Corrosion	10–30 Drift 10–50 Degradation 15–25 Drift	Shock, vibration Moisture, temperature Moisture, temperature
Diode	-Corrosion -Lead/die contact -Header bond	20–40 Intermittent 15–35 Open 15–35 Drift	Moisture, temperature Shock, vibration Shock, vibration
Resistor	-Corrosion -Film defects -Lead defects	30–50 Drift 15–25 Drift 10–20 Open	Moisture, temperature Moisture, temperature Shock, vibration
Capacitor	-Connection -Corrosion -Mechanical	10–30 Open 25–45 Drift 20–40 Short	Temperature, vibration Moisture, temperature Shock, vibration
RF coil	-Lead stress -Insulation	20–40 Open 40–65 Drift	Shock, vibration Moisture, temperature
Transformer	-Insulation	40–80 Short	Moisture, temperature
Relay	-Contact resistance -Contact corrosion	30–40 Open 40–65 Drift	Moisture, temperature Moisture

Table 3. Environmental stress effect on Aircraft system

Environmental factor	Failure percentage
Sand and dust	6
Moisture	19
Shock	2
Temperature	40
Salt Altitude	4 2
Vibration	27

The basic design can incorporate techniques to promote the even distribution of heat, thereby preventing hot spots and providing adequate cooling. Parts and components capable of operating in high temperatures can be selected. External cooling, insulation, and other design techniques reduce the periodicity and magnitude of temperature cycling and protect against thermal shock(Pecht, 2009). Table 4 summarizes the effects, sources, and corrective actions required to eliminate the effects of temperature on components and systems.

Humidity and moisture can produce undesirable effects in electronics. It is not just those systems at sea or exposed to the weather for which moisture is a problem. Perfectly dry air is an exception. Humidity is nearly an ever-present condition, and it takes little to convert that humidity to moisture. For example, if a vehicle that has been operated in the cold outdoors is parked in a warm garage, condensation quickly occurs. Humidity can cause corrosion, electrical shorts, breakdown in insulation, and changes in resis-

Table 4. Effect of temperature on components

Environment	Principal effects	Corrective action	Sources
High Temperature	• Acceleration of chemical reactions • Deterioration of insulation • Dielectric breakdown • Electro migration • Electrical parameter drift	• Minimize temperature through design and cooling • Eliminate heat source • Provide heat conduction path	Component self- heating, friction of mechanical and electromechanical assemblies, ambient temperature
Low Temperature	• Materials become brittle • Viscosity of lubricants increases • water condensation • Ice formation • Electrical parameter change	• Provide heating • Insulate • Incorporate materials designed for low temperature	Typically experienced in uncontrolled environment
Thermal cycling	• Stress relaxation • Ductile and brittle fractures • material deformation • Dielectric breakdown • Electrical parameter drift	• Minimize coefficient of thermal expansion (CTE) differences between materials • Introducing heating/cooling whatever required	Typically experienced in uncontrolled environment, as a result of power on/off cycling
Thermal shocking	• Material cracking • Seal features • Ruptures • Electrical parameter drift	• Minimize CTE mismatches • Use appropriate mechanical design tolerances	Transition of an equipment between two temperature extremes

tance. Humidity can be controlled through design (hermetically sealed components, desiccants, and so on) and through environmental control systems that remove moisture. Controlling temperature changes can prevent condensation, and drains can be provided to eliminate any moisture that does condense inside equipment. Table 5 summarizes the effects, sources, and corrective actions required to eliminate the effects of humidity and moisture on components and systems.

Equipment located at sea or near coastal areas is subject to salt fog. Moisture and salt can combine to form a serious environmental condition. The effects on equipment reliability caused by moisture are exacerbated by the salt, which can accelerate contamination and corrosion of contacts, relays, and so forth. Equipment can be protected against salt fog by hermetically sealing components or by installing the equipment in environmentally controlled areas or shelters. Table 6 summarizes the effects, sources and corrective actions required to eliminate the effects of salt fog on components and systems.

Sand and dust can interfere with the mechanical components of electronic systems. Switches, relays, and connectors are vulnerable to the effects of sand and dust intrusion.

So, during the presence of moisture, sand and dust may form dangerous acids and clogging of electrochemical components.

Table 5. Effect of humidity and moisture on components

Environment	Principal effects	Corrective action	Sources
• Humidity and moisture	• Corrosion of materials • Breakdown in insulations • Changes in resistance • Fungus growth	• Control environment • Select moisture-resistant or hermetic parts • Use conformal coatings • Provide drains to eliminate condensed moisture	Humidity is present in the natural environment. Rapid cooling of equipment can cause condensation.

Table 6. Effect of salt fog on components

Environment	Principal effects	Corrective action	Sources
Salt fog	• Combines with water to form acid/alkaline solutions • Corrosion of metals • Insulation degradation	• Protective coatings • Eliminate dissimilar metals • Use hermetic seals.	• Salt is found in the atmosphere, the oceans, lakes, and rivers, and on ground surfaces

Table 7. Effect of sand and dust on components

Environment	Principal effects	Corrective action	Sources
Sand and Dust	• Long-term degradation of insulation • Increased contact resistance • Clogging of electromechanical components and connectors • Acids formation when moisture is present • Will collect at high static potential points and can form ionization paths	• Control environment • Incorporate protective coatings and enclosures • Filter air	Sand and dust are prevalent in all uncontrolled environments. Air turbulence and wind can force dust and sand into cracks and crevices in equipment.

Hermetically sealing components or installing equipment in environmentally controlled areas can eliminate or mitigate the effects of sand and dust. Table 7 summarizes the effects, sources, and corrective actions required to eliminate the effects of sand and dust.

Decreases, particularly rapid ones, in atmospheric pressure can cause failure in components if the design does not allow internal pressure to rapidly re-equalize. Decreases in pressure can also lead to electrical arcing, particularly in the presence of high voltage. Proper design and installation in environmentally controlled areas can avoid problems associated with pressure. Table 8 summarizes the effects, sources, and corrective actions required to eliminate the effects of atmospheric pressure on components and systems.

When the equipment is subjected to sudden random or sinusoidal vibrations then it may result into accidental failure of connections and microelectronic components or hardening of mechanical parts. Through proper design, vibration effects can be avoided (avoiding resonant frequencies through stiffening, orientation, and so on) or reduced (use of vibration isolators, and so forth).

The mechanical shock is the application of accidental force on the equipment. Due to which, there may be destroy of interconnections, fracture structures, and break loose particles that can contaminate electronic components that incorporate a cavity in their design. Shock can occur when a mechanic drops the equipment as it is removed for maintenance or when an aircraft makes a hard landing. External or

Table 8: Effects of high altitude on components

Environment	Principal effects	Corrective action	Sources
High altitude	• Induced damage due to pressure differential in a product • Evaporating/drying due to outgassing • Forms corona (arcing) • Ozone generation • Reduction of electrical breakdown voltages • Chemical changes within organic materials	• Pressurize • Increase mechanical strength • Properly insulate high voltage components • Minimize use of organic materials	Aircraft or other applications susceptible to rapid fluctuations in altitude or atmospheric pressure

Table 9. Effect of vibration and mechanical shock on components

Environment	Principal effects	Corrective action	Sources
Vibration	• Intermittent electrical contacts • Touching/shorting of electrical parts • Wire chafing • Loosening of hardware • Component/material fatigue	• Stiffen mechanical structure • Reduce moment of inertia • Control resonant frequencies	Vibration isolation is the controlled mismatch of a product's resonant and natural frequencies. It does not usually provide shock isolation. Shock mounts can increase vibration damage.
Mechanical shock	• Interference between parts • Permanent deformation due to overstress	• Use stronger material (as stiff and as light as possible) • Use shock mounts • Superstructure should be stiffer than supporting structure • Use stiff supporting structure if system natural frequency is >35 Hz Transmit rather than absorb energy	The sudden application of force, measured in Gs of acceleration and milliseconds duration. Can be caused by handling, transportation, gunfire, explosion and/or propulsion.

internal shock mounting, ruggedized packaging, and other techniques can reduce the level of shock actually experienced. Table 9 summarizes the effects, sources, and corrective actions required to eliminate the effects of vibration and mechanical shock on components and systems.

The effects of solar radiation can be a significant thermal load beyond any load from the local ambient environment. This is especially true for automobiles and aircraft on runways. Sheltering and the use of different types of paint can help. Wind loading can be a significant stress, especially on large structures such as antennas.

Associated with electrical and electronic equipment are electrical and magnetic fields and emanations. These fields and emissions may be intentional (radio transmissions, for example) or unintentional (due either to poor design or to unavoidable laws of physics). The fields and emanations of two pieces of equipment can interfere (hence, the term *electromagnetic interference* or EMI) with the proper operation of each. Proper design is, of course, the best and first step in avoiding EMI problems. Shielding, grounding, and other techniques can also reduce EMI.

Other Factors Influencing Electronic Failure

The manufacturing and maintenance environment can also affect the reliability of electronic components. Many of the failures reviewed by a failure analyst are a result of manufacturing errors. Table 10 outlines manufacturing defects typically uncovered during the manufacturing process.

Static electricity is also a serious threat to modern electronic devices and is prevalent in areas of low humidity. Electrostatic discharge (ESD) can degrade the operation of a component, cause component failures, or go undetected and result in latent defects. Certain devices are inherently more vulnerable to ESD damage. Parts should be selected with an eye toward ESD susceptibility.

Precautions should be taken to reduce sources of electrostatic charges. Electrostatic charges can be brought into the assembly, repair, and part storage areas by people. They can also be generated in these areas by normal work movements. Clothing articles of common plastic such as cigarette and candy wrappers, styrofoam cups, part trays and bins, tool handles, packing containers, highly finished surfaces, waxed floors, work surfaces, chairs, processing machinery, and many other prime sources of static charges.

Table 10. Manufacturing defects

Manufacturing defect	Percentage
Open	34
Solder bridges	15
Missing parts	15
Misoriented parts	9
Marginal joints	9
Balls, voids	7
Bad parts	7
Wrong parts	7

These electronic charges can be great enough to damage or cause the malfunction of modern electronic parts, assemblies, and equipment.

Electrostatic discharge can cause intermittent upset failures, as well as hard catastrophic failures, of electronic equipment. Intermittent or upset failures of digital equipment are usually characterized by a loss of information or temporary disruption of a function or functions. In this case, no apparent hardware damage occurs, and proper operation may be resumed after the ESD exposure. Upset transients are frequently the result of an ESD spark in the vicinity of the equipment(Greason & Castle, 1984).

While upset failures occur when the equipment is in operation, catastrophic or hard failures can occur at any time. Most ESD damage to semiconductor devices, assemblies, and electronic components occurs below the human sensitivity level of approximately 4000 V. Other failures may not be catastrophic but may result in a slight degradation of key electrical parameters such as (a) increased leakage current, (b) lower reverse breakdown voltages of P-N junctions or (c) softening of the knee of the V-I curve of P-N junctions in the forward direction. Some ESD part damage is subtler. It can remain latent until additional operating stresses cause further degradation and, ultimately, catastrophic failure. For example, an ESD overstress can produce a dielectric breakdown of a self-healing nature. When this occurs, the part can retest good, but it may contain a weakened area or a hole in the gate oxide. During operation, metal may eventually migrate through the puncture, resulting in a direct short through the oxide layer(Wysocki, Vashchenko, Celaya, Saha, & Goebel, 2009).

The natural environment in which maintenance occurs is often harsh and can expose equipment to conditions not experienced during normal operation. Maintenance itself can involve the disconnection and removal of equipment and subsequent reinstallation and reconnection of equipment. This "remove and reinstall" process is not always done because the equipment has failed. Often, equipment may be removed for test and calibration or to allow access to another failed item. This secondary handling (the primary handling required when the equipment has failed) and the natural environment make up the maintenance environment. The maintenance environment can degrade the reliability of equipment not properly designed for it. Maintenance-induced failures can be a significant factor in the achieved field reliability of an item. Proper design can mitigate both the probability and consequences of maintenance-induced failure.

Rain, ice, and snow can affect the temperature of equipment and can introduce moisture, with all its associated problems. These weather elements are a particular problem when maintenance is performed. During maintenance, access doors and panels are opened, exposing even that equipment located in environmentally controlled areas. Proper packaging, the use of covers or shrouds, and other techniques can be used to protect equipment from the elements. Internal drains, sealants and gaskets, and other design measures can also be used to protect equipment.

Equipment must be capable of surviving the rigors of being shipped via all modes of transport over intercontinental distances. The process of preparing for shipment, the shipment itself, the process of readying the equipment for use after delivery, and the handling associated with all these steps impose stresses sometimes more severe than the operating environment itself.

FAILURE ANALYSIS

Failure analysis is the technique which will analyse the failures and faults in the respective component or device, which prevents it to perform the assigned task. Failure analysis explores the root cause of failure and then suggests the preventive and corrective measures, so that overall performance of the system will not degrade.

Determination of the Failure Cause

The course of action taken to determine the cause(s) of an electronic failure depends on the item to be evaluated (e.g., power supply, populated printed wiring board/module or discrete piece part). The process is outlined in Figure 8. This process can be applied to analyse either a component or printed wiring assembly. The problem analysis path for a discrete component incorporates an external visual examination and electrical test verification. The evaluation of a printed wiring assembly is a more complex procedure, requiring fault analysis to isolate a problem. This isolation often includes the review and evaluation of manufacturing and assembly processes, board materials and construction, component mounting techniques, and connectors. Board-level electrical testing is typically the first procedure used to localize the problem. Also, it is important that the failure analyst interface with the original equipment manufacturer (OEM), since it is important that diagnostic and corrective action interchange, as well as to provide an unbiased review of results.

Higher-level assemblies may require the use of circuit analysis techniques and methods (e.g., SPICE, Monte Carlo simulation) to isolate design problems. These procedures can either provide basic information concerning the sensitivity of a circuit to variability (degradation) in the parameters of its component parts or use actual parameter variability data to simulate real-life situations and determine the probability of whether circuit performance is within tolerance specifications. In addition to providing information on the sensitivity of a circuit to degradation failures, these techniques are capable of giving stress level information of the type needed to identify overstressed parts. Catastrophic failures can often be related to excessive environmental conditions, lack of design margin, or overstress, which can provide valuable insight for corrective actions. These techniques will, therefore, be used to identify problems resulting from circuit designs that are not robust enough to tolerate the specific conditions being encountered.

Finite element analysis (FEA), a graphically oriented, automated technique for analysing the effects of mechanical loads and thermal stress on a component or system, may also be utilized. It is a powerful method for modelling structures and assemblies with various levels of mechanical, vibration, and thermal stress, and producing visual evidence of any resulting deformation, harmonic resonance, or thermal flow and induced stress patterns(Pecht & Gu, 2009).

In contrast to design-related problems, failures can also be due to defective parts. A background investigation is typically conducted to determine if similar problems have been encountered, and whether they are vendor, technology, or batch related. Such investigations on previous findings can often lead to quick solutions.

For exploring the root cause of the problem, rectifying the error and necessary corrective action, the failure analysis of the problem is needed. It indicates the faulty component as well as indicate the main cause of the failure so that it can be avoided in future.

Special emphasis should be given to parts and assemblies that *retest OK* (RTOK). Multiple parts or assemblies are typically removed by a field technician or repair shop during a repair. Some of these part removals may not actually be failures, since it is typically more cost effective not to fault isolate down to a single component or assembly. RTOKs can also result when a component is not tested to an environmental extreme, causing it to fail in a given application. It is often necessary to work with the customer and with the equipment OEMs and device vendor(s) to ascertain the failure cause of devices that are found to be RTOK.

Analysis of Individual Items

The failure analysis process incorporates reliability to identify failure mechanisms for parts and materials. A failure analyst must go beyond the symptomatic defect identified during the failure analysis process and identify and correct the basic underlying part or process problem causing the problem (e.g., the root cause).

Figure 8. Electronic problem analysis design flow [47]

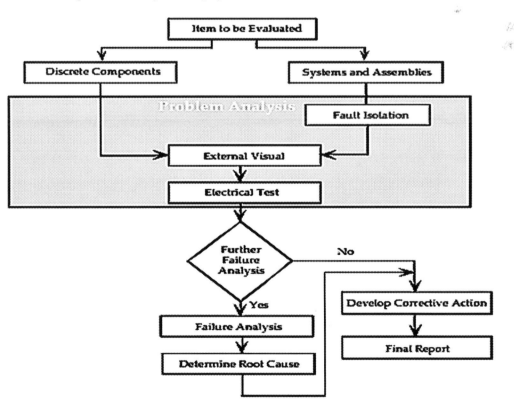

Once a discrete part or process step (product or fabrication) causing a problem has been fault isolated, a number of sophisticated failure analysis techniques may be required to correctly determine solutions to a problem. The specific techniques required to analyse a part or process depend on the problem under study.

When performing or selecting a source for failure analysis, it may be difficult for one laboratory to have the expertise in all failure analysis techniques required to perform an analysis. A given laboratory will often have their own area of expertise. The selection of a failure analysis laboratory needs to be based on the component type and the failure analysis techniques/methods that may be required to perform the analysis. In some instances, the original vendor will perform electrical testing of devices manufactured by them. Vendors and manufacturers are usually willing to analyse their own items at no cost just to determine how well these items function in the field. Additionally, participation by company engineers in the failure analysis process can be beneficial. Experience has shown that knowledge of the circumstances of the problem under study is necessary during the performance of a failure analysis. The elimination of this step has typically led to extraneous testing or the failure to perform key testing, resulting in a requirement for additional follow-up analyses. Also, a review of this step will remove any suspicion that results have been biased by a testing facility.

Failure Analysis Techniques

Failure diagnosis is the part of failure analysis technique. Once a discrete component is tested and it falls under a faulty zone, then various failure analysis techniques have been incorporated to analyse the root cause of the failure. It includes different steps. The first technique is to inspect the component using optical microscopy. This is an effective method for isolating certain faults and failures. Many faults and failures, where vibration is the root cause, can be located using this step. This level of analysis can be subjected to broken leads and wires of electronic board. Using visual inspections, corrosion defect, electrical overstress and packaging defects can be detected.

The next technique is to operate it under electrical testing modes. By using this way of analysis, further failure analysis is channelized effectively and efficiently. Test sets, curve tracers, and point-to-point resistance measurements can be used at this step to determine into which of the following categories the part would fall.

1. A marginal or degradation problem, where parts coming from the tails of a normal product distribution combine to cause malfunctions.
2. A "sport," where a part falls well outside of normal product distribution, or a particular critical parameter was never specified.
3. An environmental problem, where the equipment is stressed electrically, mechanically, thermally or chemically (intentionally or unintentionally) beyond the design limit
4. RTOK, which could result from misapplication of a part, an intermittent problem such as cracked solder or loose particles, improper removal (troubleshooting printed wiring assemblies usually cannot isolate to a single part), poor electrical connections external to the part, or software problems in the printed wiring assembly tests.

Based on experience, most of the parts fall into the last two categories listed above. Typically, over one-third of the problems can be isolated during the first two steps. The test sequence followed for analysing the remaining two-thirds of the parts can take many different paths based on the findings during the first two steps. In any event, benign tests that do not cause irreversible changes in the part should always be used first. Depending on the findings at this point, they could include such tests as X-ray, fine and gross leak, loose particle detection (PIND), shock, vibration, centrifuge, thermal cycling, and dew point.

For hermetic cavity devices, residual gas analysis (RGA) should be run on some samples prior to opening. The analysis of the gaseous materials in the part cavity can often lead to the root causes of a problem or the forming of hypotheses that lead to solutions. This is particularly true if polymeric materials are used(Czanderna, 2012). RGA is also useful for package integrity, surface contamination, and metal plating problems. Immediately after the part is punctured for the RGA, a piece of tape should be placed over the hole for loose particle recovery.

If no cause is found after all non-destructive evaluation has been completed, then the destructive analysis begins. Again, the order of analysis must be carefully planned so that the maximum information can be obtained before each irreversible step is completed. Careful, detailed internal visual examination and photographic documentation is the most effective method for either solving the problem or deciding which analysis technique to use next. Often, the path of the overstress in a printed wiring assembly can be determined, as well as approximate current/voltage levels and whether the source was high or low impedance. Good parts can then be stressed to see if the field failures can be duplicated. Visual inspection can also provide information on contamination, poor electrical interconnections, conductive and nonconductive films, adhesive failures, and manufacturing and handling defects. At this point in the analysis, experience shows that about two-thirds of the part failure causes have been isolated to the point where corrective action can be initiated. For the remaining one-third of the parts, many techniques may be used. These are selected based on information gathered by the previous analyses.

For the successful operation of manufacturing industry, the root cause identification of failure and faults are must. Failure analysis will identify the root cause of problem. This will uncover many causes of failure such as electrostatic discharge, software or maintenance reasons etc. To eliminate the root cause of the failures, a corrective action is needed. Based on the analysis results, feedback from customer, historical data and assistance from manufacturer, the analyst develops the corrective action, so that failures can be removed permanently. If these corrective actions will be implemented on faulty component at early stage, it will automatically reduce the maintenance and replacement cost, furthermore it will enhance the reputation of manufacturer in the competitive market.

A report describing the corrective action recommendation as it pertains to a particular component/printed wiring assembly, system application, manufacturing process, device technology, and so on will be delivered to the customer for each problem studied. This report will include a description of the problem, a summary of failure circumstance information, results of failure testing, references to other test reports and source documents, corrective action recommendations, and a cost/benefit analysis to support the action recommendations, where applicable.

Failure Analysis Using Statistical Distribution

Failure analysis is the technique to identify the failure, determine its root cause, analyse the data and provide the results. Based on results, suitable corrective actions can be taken to eliminate the device failures. Failures may occur at any stage; early stage, mid or final stage.

Statistics are typically used in the failure analysis process to analyse data pertaining to conditions surrounding a specific failure or group of failures. Statistics are of particular importance when modelling the time-to-failure characteristics of a component or system. Several distributions typically used for this purpose include the exponential, Weibull, normal, lognormal, and gamma. The exponential distribution is probably most often used because of its simplicity, and it has been shown in many cases to fit the time-to-failure characteristics of electronic devices. This distribution, however, is not always representative of specific failure causes, and other distributions are required to model the data. A description of distributions typically used to model electronic and electromechanical component time-to-failure characteristics are provided in this section. Discrete distributions such as the binomial and Poisson are not covered but are described in numerous statistics and reliability texts(Kalbfleisch & Prentice, 2011).

1. Exponential Distribution

The exponential distribution is widely used to model electronic reliability failures during the useful life period, which tends to exhibit random failures and a constant failure or hazard rate(Raheja & Gullo, 2012). The probability density function for an exponentially distributed time-to-failure random variable t, is given by:

$$f(t) = \lambda e^{-\lambda t} \tag{2}$$

where λ is the failure rate and $t > 0$. The reciprocal of the failure rate is the mean expected life, and it also referred to as the *mean time between failure* (MTBF).

2. Weibull Distribution

The Weibull distribution is very versatile and can be used to represent many different times to failure characteristics. It continues to gain popularity, particularly in the area of mechanical reliability, due to this versatility. The underlying physical phenomena causing the failure, however, must be clearly understood prior to the selection of this distribution(Xie & Lai, 1996). The two-parameter Weibull probability density function in terms of time, t, is represented by the following equation:

$$f(t) = \frac{\beta}{n^{\beta}} t^{\beta-1} \exp\left[-(\frac{t}{n})^{\beta}\right] (for\, t \geq 0)$$
$$t(t) = 0\, (for\, t < 0) \tag{3}$$

where β is the shape parameter and is the scale parameter, or characteristic life. When $\beta = 1$, the distribution is equivalent to the exponential distribution, and when $\beta = 3.44$, the distribution approximates a normal distribution.

3. Normal Distribution

The normal distribution is often used to model the distribution of chemical, physical, mechanical, or electrical properties of components or systems. Data conforming to a normal distribution is symmetrical about the mean of this distribution(Rausand & Arnljot, 2004). When a value is subject to many sources of variation, irrespective of how the variations are distributed, the resultant composite of these variations has been shown to approach a normal distribution(Høyland & Rausand, 1994). The normal distribution is rarely used to model the time to failure of wearout mechanisms, but it can be used to model component design variations—for example, the tensile strength of materials. The probability density function is given by:

$$f(x) = \frac{1}{\sigma(2\pi)^{1/2}} \exp\left[-\frac{1}{2}(\frac{x-u}{\sigma})^2 \right] \tag{4}$$

where μ is the location parameter (which is equivalent to the mean), and σ is the scale parameter (which is equal to the standard deviation).

4. Lognormal Distribution

The lognormal distribution is a skewed distribution that can be used to model data where a large number of occurrences being measured are concentrated in the tail end of the data range. It is not as versatile as the Weibull distribution but has been shown to be applicable to the modelling of many wear out phenomena(Lu & Meeker, 1993). The probability density function for the lognormal distribution is represented by:

$$f(x) = \frac{1}{\sigma x(2\pi)^{1/2}} \exp\left[-\frac{1}{2}(\frac{\ln x - \mu}{\sigma}^2 \right] (for\ x \geq 0) \tag{5}$$

$$f(x) = 0\ (for\ x < 0)$$

where μ and σ are, respectively, the mean and standard deviation of the data. The lognormal distribution is often used to describe the time-to-failure characteristics of electronic component failure mechanisms such as electro-migration and time-dependent dielectric breakdown.

5. Gamma Distribution

The gamma distribution is used to model situations where a number of failure events must occur before the failure of the item of concern. The probability density function is:

$$f(x) = \frac{\lambda}{T(a)}(\lambda x)^{a-1} \exp(-\lambda x)\ (for\ x \geq 0) \tag{6}$$

$$f(x) = 0\ (for\ x < 0)$$

where λ is the failure rate of a complete failure and *a* is the number of events required to cause a complete failure. Γ(*a*) is the gamma function, which is represented by:

$$T|a| = \int_0^\infty x^{a-1} \exp(-x)dx \tag{7}$$

when $a = 1$, the gamma distribution is equivalent to the exponential distribution.

Selection of a Distribution

The specific statistical distribution best suited for a particular analysis can be determined empirically, theoretically, or by a combination of both approaches. A distribution chosen empirically is done by fitting data to a distribution using probability paper or computer software programs. Goodness-of-fit tests, such as the chi-square and Kolmogorov Smirnov, can also be used to determine when a particular data set fits a distribution. A description of typical goodness-of-fit tests has been provided(Cleves, 2008).

1. **Chi-Square Test**: This test can be used to determine how well theoretical or sample data fits a particular distribution. A measure of the discrepancy between observed and expected values of a distribution is supplied by the chi-square (χ^2) statistic. The test is derived from the number of observations expected to fall into each interval when the correct distribution is chosen(Tabachnick, Fidell, & Osterlind, 2001). The statistic is evaluated by subtracting an expected frequency from an observed frequency and is represented by:

$$x^2 = \sum_{j=1}^{k} \frac{(o_j - e_j)^2}{e_j} \tag{8}$$

where χ = number of events
    o   = observed frequency
    e   = expected frequency

Chi-square distribution Tables are required to compare the value of to a critical value with the appropriate degrees of freedom. When data is limited, e.g., less than five observations from three or fewer groups or classes of data, it is recommended that a different test, such as the Kolmogorov-Smirnov (K-S) goodness-of-fit test be used(Mann, Singpurwalla, & Schafer, 1974).

2. **Kolmogorov-Smirnov (K-S) Test**: This test is a nonparametric test that directly compares an observed cumulative distribution function (cdf) to the theoretical or expected distribution's cdf. This test is easier to use than the chi-square test previously discussed and can provide better results with a small number of data points(Wang, Bannister, Meyer, & Parish, 2017).

The K-S test, instead of quantizing the data first, is a continuous method, since it looks directly at the observed cumulative distribution function and compares it to the theoretical or expected cdf. The cumulative distribution function $F_o(t)$ may be evaluated from data at some point, t^*, by:

$$F_o(t^*) = \frac{Number\ of\ components\ failing\ by\ t^*}{Total\ number\ of\ components\ on\ test + 1} \tag{9}$$

One is added to the denominator to reduce bias. The theoretical cumulative density function $F_E(t)$ may be evaluated by integrating the probability density function $f(t)$. If the observed expected cumulative distribution functions are $F_O(x)$ and $F_E(x)$, then

$$D = \max|F_o(x) - F_E(x)| \tag{10}$$

is the K-S statistic for a one-sample statistic. A two-sample procedure also exists for comparing two samples. The K-S statistic represents the largest deviation of observed values from expected values on the cumulative curve. D is then compared to Tables of critical values and, if it exceeds the critical value at some predetermined level of significance, it is concluded that the observations do not fit the theoretical distribution chosen.

FAILURE PREVENTION AND ELIMINATION

With the increasing emphasis on identifying the source of a problem to ensure that the failure mode is eliminated, the failure analysis definition can be expanded to include failure prevention(Doggett, 2005). Additionally, reliability physics practices can be used to verify the adequacy of a process or component in satisfying its desired application early in the design process. Root cause analysis, properly applied, can identify corrective action for many problems (i.e., a failed semiconductor device, a poor yield machining operation, and a non-robust process).

Determining corrective actions for identified problems should address both catastrophic and chronic failures and defects, as illustrated in Table 11. A catastrophic example would be a change of flux in a standard soldering process. The result could be poor adhesion of component leads to printed wiring board contact pads. Usually, in cases like this, the cause and effect are easily identified, and corrective action determination is straightforward. In the chronic case, yield is usually considered acceptable.

The actual cause in this case is the result of various factors. If each is not addressed, the resulting corrective action will not suffice. Figure 9. illustrates the difference between catastrophic and chronic failures and the result of root cause analysis continuous improvement.

The catastrophic failures are changes in capability resulting in total loss of useful performance. Operating characteristics of a material, product, or system undergo sudden and drastic change. Whereas, the chronic failure is frequent but low-impact failure of a system that takes little time to correct.

Table 11. Catastrophic and Chronic Failures and Defects

Failure	Occurrence	Cause	Yield	Solution
Catastrophic	Sudden	Single	Unacceptable	Straightforward
Chronic	Over time	Compound	Acceptable	Difficult

However, to achieve high yields, low costs, and customer satisfaction, continuous improvement, through minimization of chronic defects, must occur, as illustrated in Figure 9. Improvement starts with Pareto analysis, but chronic defects may not be eliminated, because the defect factor selection is arbitrary and not thorough(Karuppusami & Gandhinathan, 2006).

The traditional "quality control" scenario of an approach to improvement is illustrated in Table 12. In most cases, chronic defects are either not identified or are considered to be in the "noise level" and not addressed(Glaser, 1980).

An approach to the reduction of chronic defects is illustrated in Table 13. The process used is PM Analysis. This approach is described in the Japan Institute of Plant Maintenance PM Analysis Cause training course. "P" stands for phenomenon and physical, and "M" stands for mechanism, machine, man, material, and method. Basically, this is a much more detailed analysis that attempts to detail all defect factors and tries to correct them to minimize chronic defects. It is important that you do not focus on only important items but also include those whose contribution is minor.

A successful continuous improvement plan would be first to apply the conventional quality control approach. If satisfactory results do not occur, then implement the more rigorous PM approach. However,

Figure 9. Difference between Catastrophic and Chronic Failures [59]

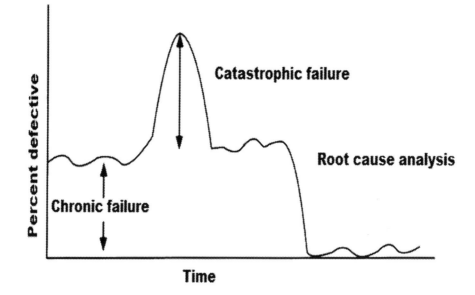

Table 12. Traditional Approach to Improvement

Steps	Description
1. Goal	Defect reduction
2. Philosophy	■ Prioritize ■ Analyse by priority ■ Identify problem factors and take corrective action
3. Approach	Fishbone diagram (brainstorming)
4. Success	Good for catastrophic problems

Table 13. Improvement Approach for Achieving "Zero" Defects

Steps	Description
Goal	Zero defects
Philosophy	■ Minimize prioritization ■ Maximize determination of defect factors ■ Investigate each factor ■ Correct all identified problems ■ Attempt to correct all defect factors ■ If not successful, change factor selection
Approach	PM analysis
Success	Chronic defects

it must be noted that cost and schedule factors can minimize the success of either approach. As improvements are made and yields rise, management support to continue analysis may decrease.

When the root cause of a failure has been determined, a suitable corrective action should be developed which will prevent further recurrence of this failure. Examples of corrective actions include, but are not limited to, changes in design, part derating, test procedures, manufacturing techniques, materials, and packaging.

After a corrective action has been proposed, its effectiveness should be established prior to its universal application. This may be performed on a prototype equipment, on the failed unit, in the engineering laboratory, or by various other appropriate means.

Failure analysis of items is necessary when prior problem analysis/fault isolation does not define the cause of a specific problem. It will also be used to verify problems uncovered by the fault isolation process. The following describes the capability, techniques, and expertise available to perform these failure analyses.

FAILURE PREDICTION TECHNIQUES

Failure prediction is an essential parameter for successful operation of electronics devices and components. Failure prediction warns the user to replace the faulty component or device before it deteriorates the entire system. It explores the residual life of respective component or device in terms of mean time between failure or end of life time. To predict failure of any component or device, an enormous methods and models have been used i.e. empirical method, analytical method, theoretical methods, experimental methods and artificial intelligence techniques etc.

Empirical Methods

These methods are one type of failure prediction method which is based on the historical data collected from users, manufacturers or test data. These are the standard methods which are used to predict failure e.g. military handbook MILHDBK-217F, TELECORDIA, PRISM, BELLCORE etc. These techniques are widely accepted technique in various fields such as military, telecommunication or commercial etc (Cherry Bhargava, Vijay kumar Banga, & Singh, 2014). Proper care must be taken, while comparing field

Figure 10. Methods to predict failure

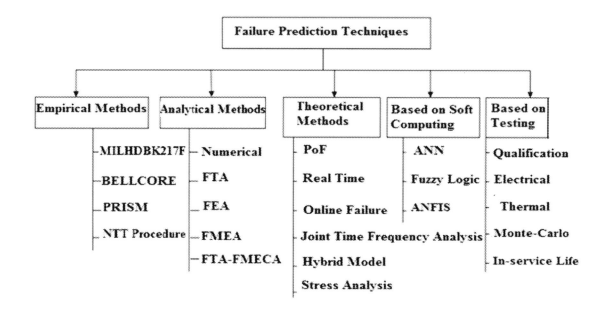

reliability and predicted reliability using empirical methods, as these techniques are based on constant hazard rate. It has been observed that even after component burn-in, the early life failures do occur in the field. These techniques are comparatively easy to use and providing good approximation. But the data is not as per device specification. Moreover, some data is not updated. So perfect approximation cannot be made out(Puente, Pino, Priore, & de la Fuente, 2002).

The uses of reliability models are described below:

- To assess reliability requirements in planning reports, making initial design specifications and requests for proposals, and arranging proposed reliability requirements.
- To compare existing reliability requirements with the latest technology trends and provide guidance in costing and scheduling decisions.
- To provide a basis for item selection among components and vendors.
- To identify and rank dominant problem areas and suggest possible solutions.
- To allocate reliability requirements among the subsystems and lower-level items to attain the reliability target.
- To evaluate the range of proposed parts, technology, materials, and processes.
- To evaluate the design before making a prototype.
- To provide a basis for trade-off studies and to evaluate design alternatives.

The accuracy of reliability modelling and prediction depends on the assumptions, data sources and other relevant influences(Handbook, 1995). The primary value of reliability prediction as a design tool is to assess and compare various possible approaches.

Military handbook MILHDBK217F is standard and widely used model as a reliability prediction for military applications. The failure rate using military handbook can be calculated as:

$$\lambda = \sum_{i=1}^{n} \left(\lambda ref, i * \pi S * \pi T * \pi E * \pi Q * \pi A \right) \tag{11}$$

where:

λref= Reference failure rate
πS=Stress factor,
πT=Temperature factor,
πE=Environmental factor,
πQ=Quality factor and
πA=Adjustment factor

Military handbook data is not updated, so a revised version MILHDBK217F-revised has been launched but due to dissatisfaction, a telecommunication company launched TELECORDIA SR-332 addresses infant mortality rate as well as steady state useful operating life failure rate. The difference between both methods has been summarised in Table 14.

Analytical Methods

To predict the failure, analytical methods provide better solution. Various numerical Models are used to predict failure. Finite Element method (FEM) is accurate technique for fault prediction, as it deals with both normal and faulty characteristics of device. Fault tree analysis (FTA) is numerical approach to detect fault, based on the simulation. Failure mode, mechanisms and effect analysis (FMMEA) is the systematic and anticipatory method that deals with step by step approach to identify the faults during design or in service. A modified approach was discussed, which is hybrid of FTA and FMECA which identifies and estimates the effects of failure(Barlow & Proschan, 1996). Numerical modelling can be used, in order to predict failure of components, which are due to substrate bending. The damage equivalent method can be established using development of damage index, which is caused by solder- joint stress. To estimate the reliability of electronic boards, they are exposed to cyclic temperature changes, which lead to thermal fatigue failures. The explicit FEA method and wavelet decomposition methods have been used(Ortiz, Leroy, & Needleman, 1987).

Table 14. Comparison between MILHDBK and TELECORDIA

MILHDBK-217F	TELECORDIA SR-332
Acts as a reliability prediction model in telecommunication industry	Acts as a reliability prediction model in telecommunication industry
Military handbook uses failure in time (FIT) per million hours	Uses failure in time (FIT) per billion hours
It has 14 environment classifications (3 ground, 8 air, 1 space, 2 sea)	It has 6 environment classifications (4 ground, 1 air, 1 space)
A specific SMT device model	No specific SMT device model
Handles lesser number of gate count IC	Handles larger gate count IC
Provides only steady state useful life failure rate	Provides a measure of infant mortality rate and steady state useful operating life failure rate

Theoretical Methods

Physics of failure understands the failure mechanism and apply different models of physics to the data. For accurate prediction of wear-out failure of components, analysis is costly, complex and detailed manufacturing characteristics of component are needed. Physics of failure can be used both in design stage and production stage(Gu & Pecht, 2008). The communication theory model and hybrid model are used by members of electromagnetic community, in order to investigate difference in disturbed thresholds for single pulse illuminations for complex electronic equipment. To locate and quantify deformation of electronic assemblies, use of topography and deformation measurements (TDM) is new approach.

In critical safety applications, such as aeronautics and electronic controlled vehicles, operating conditions become harsher, which leads to dramatic decrease in component lifetime. In order to perform real time prediction of electronic modules lifetime affected by operating conditions. Qualitative data from PoF approach and quantitative data from the statistical analysis is combined to form a modified physics of failure approach. This methodology overcomes some of the challenges faced by PoF approach as it involves detailed analysis of stress factors, data modelling and prediction. A decision support system is added to this approach to choose the best option from different failure data models, failure mechanisms and failure criteria.

Soft Computing Based Methods

Some authors have also applied a soft computing approach (neuro-fuzzy model) for intelligent online performance monitoring of electronic circuits and devices and classifies the stressors broadly into three main groups: electrical stressors, mechanical stressors and thermal environment of the component. It is seen that the failure of the device occurs either due to rising temperature or high leakage current. Neuro-fuzzy model along with Monte Carlo analysis can be used in the failure prediction of critical electronic systems in power plants.

Apart from monte Carlo analysis, other techniques using artificial intelligence methods can be deployed to predict failure i.e. artificial neural networks, adaptive neuro fuzzy inference systems, fuzzy logic etc.

Experimental Testing Methods

In life data analysis or Weibull analysis, under normal operating conditions reliability is measured using a test which is conducted on large sample of units, time-to-failure are analysed using statistical distribution(Fard & Li, 2009). Due to design or manufacturing, components are not independent, but physics of failure and empirical method assumes that there is no interaction between failures, so for realistic prediction at system level, life testing method is preferred over physics of failure or empirical method. A large number of electronic and electrical components are integrated in consumer products. In order to safe and reliable operation of a module, understanding of component's failure mode is desirable(Miller & Nelson, 1983).

Occasionally, it is hard to put a component or device under real time testing of thermal stress for thousands of hours. For such cases, accelerated life testing is used, in which component is placed in hot chamber for less time but on accelerated pace of temperature, vibration, voltage or current etc. depends on the nature of experiment. It helps to predict life and failure using Arrhenius equation. In this method, the desired component or device is put on a hot chamber or plate for specified units of time and then

Figure 11. Monte Carlo analysis

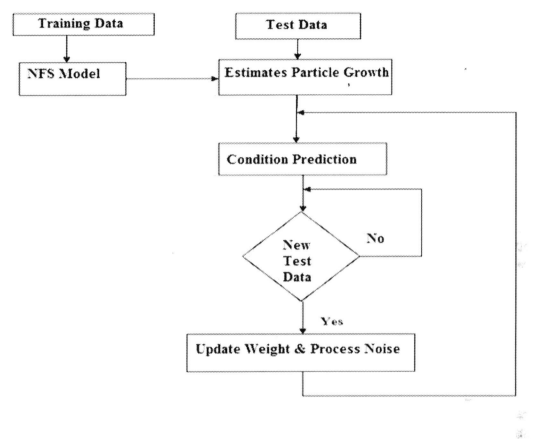

increase temperature. As temperature increases, the heating of component and device accelerates and it will degrade the performance of the component(Yang, 1994).

The results obtained from experimental methods can be compared with outcomes of other methods i.e. theoretical, analytical or empirical etc. In such a way, accurate model can be explored with highest range of accuracy.

FUTURE SCOPE

Successful implementation of reuse strategy depends on the reliability of the methods to assess the useful remaining life of the component or device. Different approaches to predict failure and residual life of the component/materials/devices are available in literature.

Quite often, in electronics industries, the components or devices claimed to have some useful life even after their continuous usage. Sometime, after purchasing the component or device, they are not put in immediate operation and they remain off-shelf. On the other hand, some components put on excessive and continuous use. Due to aging factor or storage conditions, they are prone to sudden failures or faults. Predicting failure or residual life is a step to identify the reuse potential (Gokulachandran & Mohandas, 2015). As such, accurate prediction of failure of the components is of great implication in any electronic or associated industries. This exercise will in turn reduce the overall repairing and replacing cost and help achieve quality productivity. This aspect of the use of the tool has not been discussed sufficiently by researchers. Considering these aspects, in this work an attempt is made to develop a comprehensive methodology and a variety of techniques for failure prediction estimation. Here, electrolytic capacitors and humidity sensors are chosen as an electronic component to highlight the use and methods of failure prediction along with its merits.

The failure prediction has been done using various theoretical and experimental techniques. Artificial intelligence-based methods have been used to estimate the life time or failure. An accurate method has been selected on the basis of error analysis. A GUI based system, using fuzzy logic has been created which describes failure prediction based on above discussed techniques.

Thus, this thesis details the experimental work, techniques to assess failure or residual lifetime, the comparative study of techniques, the major contributions and possible directions for future work.

The major contribution of the work is:

1. Application of Taguchi approach for design of experiments
2. Accelerated life testing for experimental work, thermal stress analysis has been done using Arrhenius equation.
3. Theoretical methods for failure prediction have been explored. A corrective model has been proposed which discusses the impact of humidity on failure of electrolytic capacitors, along with temperature, voltage, frequency and ripple current.
4. Various artificial intelligence methods i.e. ANN, ANFIS have been used for predicting failure.
5. Humidity sensor and capacitor has been fabricated and failure stage has been analysed and then prepared with market available humidity sensor or electrolytic capacitor.
6. Comparative study has been made which will explore the most accurate method.

Such requirements would be highly significant in small scale industry with minimum facilities wherein there is a greater need for reuse of used components. Identification of the reuse potential of the material at the end of its life cycle is a positive step toward achieving green manufacturing goals.

REFERENCES

Agarwal, M., Paul, B. C., Zhang, M., & Mitra, S. (2007). *Circuit failure prediction and its application to transistor aging.* Paper presented at the 25th IEEE VLSI Test Symposium, Berkeley, CA. 10.1109/VTS.2007.22

Bailey, C., Lu, H., Stoyanov, S., Yin, C., Tilford, T., & Ridout, S. (2008). *Predictive reliability and prognostics for electronic components: Current capabilities and future challenges.* Paper presented at the 31st IEEE International Spring Seminar on Electronics Technology, ISSE'08, Budapest, Hungary.

Bar-Cohen, Y. (2014). *High temperature materials and mechanisms.* CRC Press. doi:10.1201/b16545

Barlow, R. E., & Proschan, F. (1996). *Mathematical theory of reliability.* SIAM. doi:10.1137/1.9781611971194

Barnes, F. (1971). *Component Reliability.* Springer.

Becker, H. I. (1957). *Low voltage electrolytic capacitor.* Google Patents.

Bhargava, C., Banga, V. K., & Singh, Y. (2014). *Failure prediction and health prognostics of electronic components: A review.* Paper presented at the IEEE Conference on Recent Advances in Engineering and Computational Sciences (RAECS), Chandigarh, India.

Borgovini, R., Pemberton, S., & Rossi, M. (1993). Failure mode, effects, and criticality analysis (FMECA). *Proceedings of the IEEE.*

Cleves, M. (2008). *An introduction to survival analysis using Stata.* Stata Press.

Czanderna, A. W. (2012). *Methods of surface analysis* (Vol. 1). Elsevier.

Doggett, A. M. (2005). Root cause analysis: A framework for tool selection. *The Quality Management Journal, 12*(4), 34–45. doi:10.1080/10686967.2005.11919269

Dylis, D. D., & Priore, M. G. (2001). *A comprehensive reliability assessment tool for electronic systems.* Paper presented at the IEEE Annual Symposium on Reliability and Maintainability, Philadelphia, PA. 10.1109/RAMS.2001.902485

Fard, N., & Li, C. (2009). Optimal simple step stress accelerated life test design for reliability prediction. *Journal of Statistical Planning and Inference, 139*(5), 1799–1808. doi:10.1016/j.jspi.2008.05.046

Fenton, N. E., & Ohlsson, N. (2000). Quantitative analysis of faults and failures in a complex software system. *IEEE Transactions on Software Engineering, 26*(8), 797–814. doi:10.1109/32.879815

Fountain, J. H. (1965). *A general computer simulation technique for assessments and testing requirements.* Paper presented at the ACM Conference on the SHARE design automation project, New York, NY. 10.1145/800266.810757

Glaser, R. E. (1980). Bathtub and related failure rate characterizations. *Journal of the American Statistical Association, 75*(371), 667–672. doi:10.1080/01621459.1980.10477530

Gokulachandran, J., & Mohandas, K. (2015). Comparative study of two soft computing techniques for the prediction of remaining useful life of cutting tools. *Journal of Intelligent Manufacturing, 26*(2), 255–268. doi:10.100710845-013-0778-2

Greason, W. D., & Castle, G. P. (1984). The Effects of Electrostatic Discharge on Microelectronic Devices A Review. *IEEE Transactions on Industry Applications, IA-20*(2), 247–252. doi:10.1109/TIA.1984.4504404

Gu, J., & Pecht, M. (2008). *Prognostics and health management using physics-of-failure.* Paper presented at the IEEE Annual Symposium on Reliability and Maintainability Symposium (RAMS 2008), Las Vegas, NV 10.1109/RAMS.2008.4925843

Gualous, H., Bouquain, D., Berthon, A., & Kauffmann, J. (2003). Experimental study of supercapacitor serial resistance and capacitance variations with temperature. *Journal of Power Sources, 123*(1), 86–93. doi:10.1016/S0378-7753(03)00527-5

Habtour, E., Drake, G. S., & Davies, C. (2011). *Modeling damage in large and heavy electronic components due to dynamic loading.* Paper presented at the IEEE Annual Symposium on Reliability and Maintainability Symposium (RAMS), Lake Buena Vista.

Handbook, M. S. (1995). MIL-HDBK-217F. Department of Defense, US.

Harada, K., Katsuki, A., & Fujiwara, M. (1993). Use of ESR for deterioration diagnosis of electrolytic capacitor. *IEEE Transactions on Power Electronics, 8*(4), 355–361. doi:10.1109/63.261004

Hashemian, H. M., & Bean, W. C. (2011). State-of-the-art predictive maintenance techniques. *IEEE Transactions on Instrumentation and Measurement, 60*(10), 3480–3492. doi:10.1109/TIM.2009.2036347

Høyland, A., & Rausand, M. (1994). *System reliability theory: models and statistical methods.* Wiley.

Huston, H. H., & Clarke, C. P. (1992). *Reliability defect detection and screening during processing-theory and implementation.* Paper presented at the IEEE 30th Annual Symposium on International Reliability Physics, San Diego, CA.

Jánó, R., & Pitică, D. (2011). *Parameter monitoring of electronic circuits for reliability prediction and failure analysis.* Paper presented at the IEEE 34th International Spring Seminar on Electronics Technology (ISSE), Tratanska Lomnica, Slovakia.

Kalbfleisch, J. D., & Prentice, R. L. (2011). *The statistical analysis of failure time data* (Vol. 360). John Wiley & Sons.

Karuppusami, G., & Gandhinathan, R. (2006). Pareto analysis of critical success factors of total quality management: A literature review and analysis. *The TQM Magazine, 18*(4), 372–385. doi:10.1108/09544780610671048

Kececioglu, D., & Tian, X. (1998). Reliability education: A historical perspective. *IEEE Transactions on Reliability, 47*(3), SP390–SP398. doi:10.1109/24.740556

Kharchenko, V. A. (2015). Problems of reliability of electronic components. *Modern Electronic Materials, 1*(3), 88–92. doi:10.1016/j.moem.2016.03.002

Kullstam, P. A. (1981). Availability, MTBF and MTTR for repairable M out of N system. *IEEE Transactions on Reliability, 30*(4), 393–394. doi:10.1109/TR.1981.5221134

Kulshreshtha, D. C., & Chauhan, D. S. (2009). *Electronics Engineering.* New Age Publications.

Le May, I. (1998). Product liability and failure analysis. *Technology Law and Insurance, 3*(2), 163–171. doi:10.1080/135993798349550

Lebby, M. S., Blair, T. H., & Witting, G. F. (1996). *Electronic book.* Google Patents.

Lee, S.-B., Kim, I., & Park, T.-S. (2008). Fatigue and fracture assessment for reliability in electronics packaging. *International Journal of Fracture, 150*(1-2), 91–104. doi:10.100710704-008-9224-4

Lu, C. J., & Meeker, W. O. (1993). Using degradation measures to estimate a time-to-failure distribution. *Technometrics, 35*(2), 161–174. doi:10.1080/00401706.1993.10485038

Ma, H., & Wang, L. (2005). *Fault diagnosis and failure prediction of aluminum electrolytic capacitors in power electronic converters.* Paper presented at the IEEE 31st Annual Conference on Industrial Electronics Society (IECON 2005), Raleigh, NC.

Mann, N. R., Singpurwalla, N. D., & Schafer, R. E. (1974). *Methods for statistical analysis of reliability and life data.* Wiley.

Martin, P. L. (1999). *Electronic failure analysis handbook: techniques and applications for electronic and electrical packages, components, and assemblies.* McGraw-Hill Professional Publishing.

Miller, R., & Nelson, W. (1983). Optimum simple step-stress plans for accelerated life testing. *IEEE Transactions on Reliability, 32*(1), 59–65. doi:10.1109/TR.1983.5221475

Misra, K. B. (2008). *Handbook of performability engineering.* Springer Science & Business Media. doi:10.1007/978-1-84800-131-2

Moore, G. E. (2006). Cramming more components onto integrated circuits. *IEEE Solid-State Circuits Society Newsletter, 20*(3), 33–35. doi:10.1109/N-SSC.2006.4785860

Murthy, D. (2007). Product reliability and warranty: An overview and future research. *Production, 17*(3), 426–434. doi:10.1590/S0103-65132007000300003

Neri, L., Allen, V., & Anderson, R. (1979). Reliability based quality (RBQ) technique for evaluating the degradation of reliability during manufacturing. *Microelectronics and Reliability, 19*(1-2), 117–126. doi:10.1016/0026-2714(79)90369-X

O'Connor, P. D., O'Connor, P., & Kleyner, A. (2012). *Practical reliability engineering.* John Wiley & Sons.

Ortiz, M., Leroy, Y., & Needleman, A. (1987). A finite element method for localized failure analysis. *Computer Methods in Applied Mechanics and Engineering, 61*(2), 189–214. doi:10.1016/0045-7825(87)90004-1

Pascoe, N. (2011). *Essential Reliability Technology Disciplines in Development.* Northern Telecomm Europe Ltd. doi:10.1002/9780470980101

Pecht, M. (2009). *Product reliability, maintainability, and supportability handbook.* CRC Press. doi:10.1201/9781420009897

Pecht, M., & Gu, J. (2009). Physics-of-failure-based prognostics for electronic products. *Transactions of the Institute of Measurement and Control, 31*(3-4), 309–322. doi:10.1177/0142331208092031

Puente, J., Pino, R., Priore, P., & de la Fuente, D. (2002). A decision support system for applying failure mode and effects analysis. *International Journal of Quality & Reliability Management, 19*(2), 137–150. doi:10.1108/02656710210413480

Qin, Y., Chung, H. S., Lin, D., & Hui, S. (2008). *Current source ballast for high power lighting emitting diodes without electrolytic capacitor.* Paper presented at the IEEE 34th Annual Conference on Industrial Electronics (IECON'08), Orlando, FL.

Raheja, D. G., & Gullo, L. J. (2012). *Design for reliability.* John Wiley & Sons. doi:10.1002/9781118310052

Ramakrishnan, A., Syrus, T., & Pecht, M. (2000). Electronic Hardware Reliability. In The RF and Microwave Handbook (pp. 102-110). Academic Press. doi:10.1201/9781420036879.ch22

Rao, B. (1996). *Handbook of condition monitoring.* Elsevier.

Rausand, M., & Arnljot, H. (2004). *System reliability theory: models, statistical methods, and applications* (Vol. 396). John Wiley & Sons.

Sakata, K., Kobayashi, A., Fukaumi, T., Nishiyama, T., & Arai, S. (1995). *Solid electrolytic capacitor and method for manufacturing the same*. Google Patents.

Singh, A., & Mourelatos, Z. P. (2010). On the Time-Dependent Reliability of Non-Monotonic, Non-Repairable Systems. *SAE International Journal of Materials and Manufacturing, 3*, 425-444.

Solomon, R., Sandborn, P. A., & Pecht, M. G. (2000). Electronic part life cycle concepts and obsolescence forecasting. *IEEE Transactions on Components and Packaging Technologies, 23*(4), 707–717. doi:10.1109/6144.888857

Tabachnick, B. G., Fidell, L. S., & Osterlind, S. J. (2001). *Using multivariate statistics*. Allyn and Bacon.

Vichare, N. M., & Pecht, M. G. (2006). Prognostics and health management of electronics. *IEEE Transactions on Components and Packaging Technologies, 29*(1), 222–229. doi:10.1109/TCAPT.2006.870387

Wang, K., Bannister, M. E., Meyer, F. W., & Parish, C. M. (2017). Effect of starting microstructure on helium plasma-materials interaction in tungsten. *Acta Materialia, 124*, 556–567. doi:10.1016/j.actamat.2016.11.042

Wysocki, P., Vashchenko, V., Celaya, J., Saha, S., & Goebel, K. (2009). *Effect of electrostatic discharge on electrical characteristics of discrete electronic components*. Paper presented at the Prognostics and Health Management Society Annual Conference of the Prognostics and Health, San Diego, CA.

Xie, M., & Lai, C. D. (1996). Reliability analysis using an additive Weibull model with bathtub-shaped failure rate function. *Reliability Engineering & System Safety, 52*(1), 87–93. doi:10.1016/0951-8320(95)00149-2

Yamazoe, N., & Shimizu, Y. (1986). Humidity sensors: Principles and applications. *Sensors and Actuators, 10*(3-4), 379–398. doi:10.1016/0250-6874(86)80055-5

Yang, G.-B. (1994). Optimum constant-stress accelerated life-test plans. *IEEE Transactions on Reliability, 43*(4), 575–581. doi:10.1109/24.370223

Ye, H., Lin, M., & Basaran, C. (2002). Failure modes and FEM analysis of power electronic packaging. *Finite Elements in Analysis and Design, 38*(7), 601–612. doi:10.1016/S0168-874X(01)00094-4

Chapter 2
Reliability Study of Polymers

Amit Sachdeva
Lovely Professional University, India

Pramod K. Singh
Sharda University, India

ABSTRACT

The chapter deals with brief introduction to polymers, composites, and nanocomposites along with their reliability. When we talk about polymeric composites, the terms crystallinity and amorphicity play a very important role, and both of these properties are highly affected by variation in temperature condition. On increasing temperature, the crystalline domains of polymers tend to become amorphous, and as we reduce the temperature, crystalline domains tend to increase. So the reliability of a particular polymer is widely dependent on temperature conditions.

INTRODUCTION

Polymers-An Introduction

Polymers are defined as a combination of small repeated structural unit known as monomers. Two monomers combine to form dimer, three combine to form trimer, four combine to form tetramer and so on to form a polymer. So the term polymer is derived from combination of two terms poly+mer that explains that it is formed by combination of 'n; number of monomers. Polymers can also be a combination of different molecules with variable molecular weight. Variation in molecular weight may be fine or a large difference may exist. Today we have enough evidence to prove that polymers are mixtures of molecules having long chain of atoms. But it was not accepted upto 1930s. Hermann Staudinger was awarded noble prize in chemistry in 1953 on his study on macromolecules/polymers. Before 1953 polymers were regarded as a colloidal aggregate of tiny molecules having an irregular and nonspecific organization (Peacock and Calhoun, 2012; Termonia and Smith,1988).

DOI: 10.4018/978-1-7998-1464-1.ch002

Polymers are classified into various types based on:

1. Source
2. Type of backbone
3. Structure
4. Composition
5. Mode of polymerization
6. Molecular force

Source

1. **Natural Polymer**: Polymers like rubber, cellulose, proteins, starch etc that are mainly found in animals and plants are known as natural polymers.
2. **Semi-Synthetic Polymer**: These polymers includes all the derivatives from cellulose like cellulose nitrate and cellulose acetate (rayon) .
3. **Synthetic Polymer**: They include all the artificial or man made polymers like polyethene, nylon, synthetic rubber

Backbone of the Polymer Chain:

1. **Organic Polymers:** In these type of polymers backbone of polymer chain is made up of carbon atoms. Vacancies on the side of these carbon atoms is made up of low molecular weight atoms like H,O and N.
2. **Inorganic Polymers:** Such polymers like silicon rubber and glass do not have any carbon atom in their backbone.

Structure of Polymers

1. **Linear Polymers**: They are elongated straight chain polymers that are generally soft rubbery materials .On heating, linear polymers either become soft or are melted.Examples of such polymers include PVC, Polyethylene etc.
2. **Branched Polymers**: As compared to linear polymers, branched polymers have certain groups attached on their side chains. Examples of such polymers are low density polyethene.
3. **Cross Linked Polymers**: These polymers are very hard and usually does not get dissolved in any solvent. As compared to linear polymers, cross linked polymers consists of covalently bonded linear polymer chains. Monomers used in polymeric chains can be trifunctional or bifunctional. Examples include urea-formaldehyde resins, vulcanised rubber etc.

Composition of Polymeric Chain

1. **Homopolymers:** Polymers that are synthesised by polymerisation of same type of monomer throughout its chain are known as homopolymers as shown in figure 1.
2. **Heteropolymers or Copolymers:** Polymers that consists of two unlike monomers joined together repeatedly to form a chain like structure are known as copolymers.

Figure 1. Types of polymers

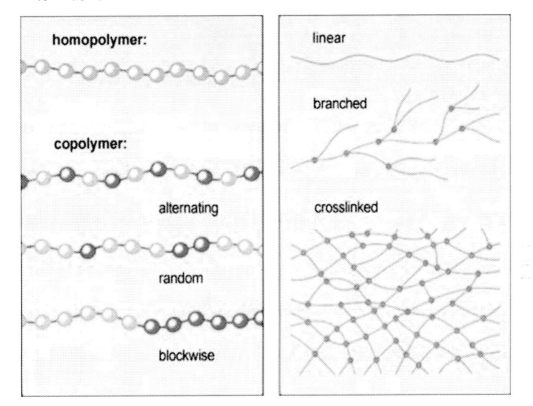

Method to synthesize copolymers is known as copolymerization. Examples of copolymers include ethylene vinyl acetate, styrene isoprene styrene (SIS), ABS plastic, SBR etc. There are various types of copolymers like random copolymers, alternating copolymers, graft polymers and block copolymers.

Alternating copolymers consists of polymers in which two monomers alternate each other along a polymeric chain.

Random copolymers have randomly arranged monomers in a complex polymer structure.

Block copolymers are copolymers in which one type of monomers are assembled together, and the other type are grouped together.

Graft copolymers are branched copolymers with one type of monomers assembled in the main chain and other type of monomer attached at the side chain.(Bauman *et al.*, 2017; Naranjo *et al.*,2006)

Mode of Polymerisation

1. **Addition Polymers:** They are synthesised by continuous addition of monomers without emission of small molecules like water, hydrogen chloride etc.The monomers are generally double or triple bonded molecules. For e.g. polyethylene from ethylene etc. As discussed above complex polymers synthesised by addition of same monomers are known as homopolymers e.g. polyethylene.
2. **Condensation Polymers:** These complex molecules are synthesised by reaction between monomers leading to elimination of small molecules like water, ammonia, hydrogen chloride.Monomer units used can have two or three functional groups. Perfect example of such polymers is nylon 6,6

synthesised by reaction between adipic acid and hexamethylene diamine .Other examples of such type of polymers include terylene or dacron, nylon 6, nylon 6,6 etc.

A variant in condensation polymer is a cross linked polymer that has 2D or 3D structure formed by combination of bi or tri functional group polymers in which atleast one of the monomeric units should have three functional groups.

Classification Based on Molecular Forces

The mechanical strength of polymeric materials is due to intermolecular interaction. These interactions are due to two types of forces i.e vander waal force and hydrogen bonding that exists within the polymeric chain. On the basis of above mentioned interaction polymers can be classified into following types.

1.   Elastomers
2.   Resins
3.   Fibres
4.   Plastics
5.   Thermoplastic
6.   Thermoset

1.   **Elastomers:** They are generally amorphous type of polymers as the polymeric chains are held together by weak intermolecular forces of attraction. This low intermolecular force allows the polymer to be stretched and is the major reason behind its elastic behaviour. On account of its elasticity they are also known as rubber like solids. For perfect elasticity, cross linking atoms are used between polymeric chains. These crosslinks allow these rubbery solids to regain its initial position once the applied force is removed. Examples are Buna S, Buna N, vulcanised rubber, neoprene etc.
2.   **Resins:** Polymeric materials that are utilised as adhesives, sealants, etc. are known as resins. Generally these type of polymers occur in liquid form are expressed as liquid resins. Examples of resins include all the existing polysulphide sealants as well as epoxy adhesives.
3.   **Fibers:** A bulk material is converted into a fibre/long filament, if it's length is 100 times its diameter. Fibres are known to posses very high tensile strength along with elevated values of modulus of elasticity.Fibres are straight chain polymers having a very high crystalline nature on account of strong vanderwall forces holding the straight polymeric chains.Examples are polyesters, nylon 6,6.
4.   **Plastics**: They are semicrystalline polymers having value of vander wall forces more that elastomers and less than fibres.Plastics are known for the hardness and toughness which is achieved by application of pressure and temperature.eg PVC, PMMA, PEMA etc. Plastics are further classified into:
       a.   Thermoplastic polymer
       b.   Thermosetting polymer
              a.   **Thermoplastic Polymers:** These polymers tends to become soft when they are exposed to heat.After that they can be converted into any shape on application of mechanical stress. Important thing is that these polymers maintain their shape even after decreasing the temperature.This process of heating, changing shape and cooling for retaining of the shape can be done any number of times without changing property or chemical composi-

Figure 2. Various types of plastics

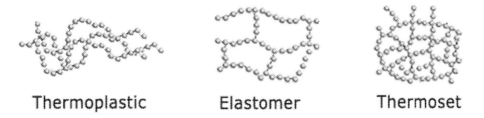

tion of the polymer (Figure 2).All the polymers that softens when heated and becomes stiff on being subjected to cooling are known as thermoplastic polymers.Polymers either consists of straight chains or slightly branched chains. Examples of these polymers are PVC, nylon, polythene, wax etc

b. **Thermosetting Polymers:** These polymers change into infusible and insoluble mass on heating. Perfect analogy is like an egg yolk that sets itself into a solid mass when heated and can't be given any other shape after that.This is because these polymers undergo a chemical change that converts it into infusible mass. Such polymers are known as thermosetting polymers.They consists of highly branched molecules as compared to straight chain molecules in thermoplastics. This branches undergo extensive crosslinking on heating.Common examples are bakelite, urea formaldelyde etc. (Gowariker et al.,2001; Harper, 1996; William, 2004).

Properties of Polymers

The properties of a polymer are highly dependent on its structural arrangement .Some of the important structural characteristic features are:

1. Rigidity of Polymeric chains
2. Intermolecular forces of attraction between polymeric chains.
3. Crystallinity in polymer structure
4. Cross linking among polymer chains

The simplest of above mentioned methods is the cross linking method. Crosslink is defined as a chemical bond that exists between polymeric chains rather than the ends of these chains.

Crosslinks are enormously important in the context of physical properties of polymeric chains as they provide a limit on translational movement of polymeric chains across each other. Also crosslinks as shown in figure 3 increase the molecular weight of the complex molecule. Two crosslinks are required per polymeric to connect the polymeric molecules together in order to produce a large gigantic molecule (Billmeyer, 1984; Prane, 1986; Allcock and Lampe, 1990).

Assume a long linear chained polymer tied in the form of a knot. On application of force or pressure on such a molecule, they will move past each other in the form of a plastic flow if intermolecular forces among polymeric chains is weak. These polymer are usually soluble in solvents which will dissolve small chain molecules that posses chemical structures similar to that of the polymeric chain. On the

Figure 3. Various Crosslinks Joining Polymeric Chains

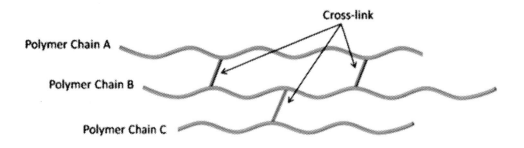

other hand, if intermolecular forces among the polymeric chains are strong enough to prevent sliding of molecules across each other than the existing polymeric structure will be solid at room temperature. But when this polymer is heated it will become weak and will undergo plastic flow. Such type of polymer is known as thermoplastic polymer.

Crosslinks also help in the production of gel type polymers. There are few crosslinks that reduce the solubility of polymer in a solvent but will help in absorption of solvent within the polymeric structure. On absorption of solvent polymer gets swollen and is known as the gel polymer. More is the crosslinking, lesser will be the tendency to form gel polymers as high amount of crosslinking prevents movement of polymeric chain and will not allow the entry of solvent molecules.

Thermosetting polymers are synthesized from comparatively low molecular weight molecules, usually semifluid in nature, which on heating in a mould gets converted into highly cross linked structure. This structure is in the form of insoluble, hard and infusible product having a 3D network of intermolecular bonds connecting the polymeric chains(Reimschuessel,1977; Handerson, and Szwarc, 1994)

Physical Properties of Polymers

Polymers are generally synthesized in large scale for its use as a very important structural materials. Mainly the physical properties of polymers are important in particular its various applications may be as a fuel in rockets, rubber tyres, polymer electrolytes etc. All the polymers without cross links, the properties are dependant on intermolecular forces between polymeric chains.

Let us take an example of polyethene which is made up of number of molecules having 1000 to 1800 CH_2 groups in straight continuous chain. As the complete material consists of different molecules, it will not crystallize by conventional methods. XRD pattern of polyethene shows its crystalline nature. Crystalline portion with ordered arrangement is almost 100 Angstrom in length with ordered CH_2 groups oriented as in perfectly crystalline low molecular weight hydrocarbons. These are known as crystallites. There are certain amorphous regions between the available crystallites that has random arrangement of atoms.

These amorphous regions consists of crystal defects. The attractive forces between polymeric chains in crystallites of a polymer are known as vander Waals forces or dispersion forces. These forces are same that exist among hydrocarbon molecules which may be in the liquid, solid states, and to some extent also exists in vapour state (Sroog, 1976; Noren and Stille, 1971). These relatively weak forces appear through synchronised motions of electrons in when two separate atoms approach each other. The force of attraction is counteracted by repulsive force once the atoms come too close to each other attractive forces between hydrogen pairs in the crystalline region of say polyethene varies between 0.1-0.2 kcal/mole/ pair, but for 1000 CH crystallite units, the amount of these intermolecular interactions can be far more than Carbon Carbon bond strength. So when a crystalline polymer is subjected to large amount of stress such that it fractures than carbon-carbon bonds break and free radicals are generated that can be identified by ESR spectroscopy.

Effect of Temperature

For practical application of polymers, the effect of variation in temperature on physical properties is very important. At low values of temperature, polymers generally become hard and glassy solids because the motility of various segments in a polymeric chain with respect to each other is very slow. Approximate temperature of polymers below which glassy nature is apparent is known as the Glass Transition Temperature as shown in figure 4 also abbreviated as T_g. When a polymer is heated to high temperature such that the crystallites undergo melting, this temperature is known as the Melting Temperature also abbreviated as T_m. Temperature of polymer above the value of melting temperature is known as the molding temperature. There is a gradual decrease in mechanical strength of a polymer as its temperature approach melting temperature. Another temperature having value comparable to Tm is the decomposition temperature, defined as the temperature where thermal breakdown of polymeric chains take place. Value of decomposition temperature is very sensitive to impurities, like presence of oxygen, and is strongly influenced by presence of antioxidants, inhibitors etc. At temperature values 200-400°C more than melting temperature uncatalyzed scission of bonds in a polymeric chain takes place at a large rate. It is impossible to prevent polymeric degradation during such scission process. (Braunsteiner,1974; Webster,1991; McGrath 1991), Physical properties like tensile strength, XRD pattern, elasticity, resistance of polymers to plastic flow and softening of polymers on heating can be explained in a better way by studying polymers in terms of its amorphous regions, crystalline region, the flexibility in the chains, cross links along with the intermolecular forces acting between polymer chains .

Another method to explain structure property correlation is by classifying solid electrolytes based on the arrangement of polymeric chains within the 3D polymer structure.

Structure Property Correlation

The properties of polymers can be clearly explained in terms of their chemical structural arrangement. The linear polymers like polyethene having regular straight chains posses a very low barriers for its

Figure 4. Effect of Glass Transition Temperature on Crytallinity/Structure

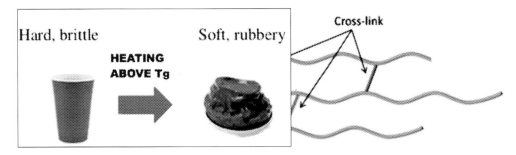

rotation. Such polymers tend to be highly crystalline with extremely high values melting point along with low values of Glass transition temperatures. On the other hand polymers like polyvinyl chloride, polyvinyl fluoride, and polystyrene have less amount of crystallinity and rather posses high values of glass transition temperature that proves the existence of strong attractive force among the polymer chains. The less crystallinity of such polymers is due to irregularity of stereochemical arrangement and irregular configuration of chiral carbons along the chain. Natta's discovery in 1954 that stereochemical arrangement of chiral carbon centres in polymeric is crucial in evaluating the physical properties has produced a enormous impact on theoretical and practical features of polymer chemistry. Based on polymeric structure or arrangement of polymeric chains we classify polymers into following categories.(Fried, 1982)

1. **Amorphous Polymer:** It is the polymer without any crystallites. The attractive forces among the polymer chains is weak and its motion is not restricted by cross linking or any type of rotational barriers. These polymers also have low values of tensile strength and undergo plastic flow when subjected to stress . Chains tend to slip on each other on being stressed.
2. **Unoriented Crystalline Polymer**: It is the polymer that has large amount of crystallized portion the crystallites are irregularly arranged and oriented randomly with respect to each other, as in figure 5. Such polymers on heating melt at one go i.e they show sharp T_m. Above the value of melting temperature, such polymers show amorphous nature. So on heating they show plastic flow, which allows them to get molded into desired shape. All other parameters remaining constant, value of Tm is higher for polymers that posses stiff chains.
3. **Oriented Crystalline Polymer**: In this type of polymers the crystallites are in regular orientation with respect to each other. It is accomplished with the help of cold drawing process. Let us take an example of nylon, that has very strong intermolecular forces between its chains but are in unoriented state. When it is subjected to stress along one direction, it will lead to occurrence of plastic flow in the material. So the polymer will elongate and crystallites will be drawn together leading to perfect orientation along the direction in which stress is applied .This oriented crystalline polymer will have high tensile strength as compared to an unoriented polymer.
4. **Elastomers:** They are basically amorphous polymers. The elastic behaviour of these amorphous polymer is due to high amount of flexibility in polymeric chains due to feeble forces among polymeric chains or due an irregular and unoriented structure.

The ability of crystallization of polymer chains can be reduced by the use of methyl groups. These groups inhibit regularity and proper ordering by steric hindrance (Fried, 1983).

A elastomer must have some cross-linked regions in order to prevent the plastic flow and must be flexible enough to have a low values of T_m. On application of tension in elastomers, the material elongates and all the polymeric chains in amorphous regions gets straighten out become parallel to each other.

At the elastic limit, maximum orientation and crystallization is attained. This crystallization is different from that attained using cold drawing process. Crystallinity of cold drawing process is stable only while the polymer is under tension while in elastomers crystallinity is maintained even after removal of tensile force. The forces among the polymeric chains are too weak to maintain well oriented crystalline state in the absence of tension. So when tensile force is removed, polymer contracts and original amorphous polymer is again produced.

A good quality elastomer should never undergo any plastic flow in any of its states may be stretched or relaxed state. Whenever it is stretched it should have a memory of the conditions while it was in

Figure 5. Various portions of a polymeric structure

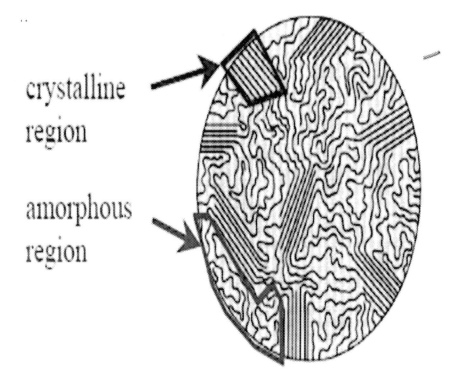

relaxed state. Such conditions are best achieved in case of natural rubber by curing (vulcanising) using sulfur. Natural rubber being tacky generally undergoes plastic flow readily, but on heating with 1-8% (by weight) of sulfur in addition to a accelerator, sulfur cross-links are produced between polymeric chains. These crosslinks helps in reduction of plastic flow and also helps to provide a reference framework like structure for the already stretched polymer to return back to its original position when allowed to relax. On the contrary addition of too much sulphur can destroy the elastic properties completely and will produce hard rubber like which is used in storage batteries. (Chruma and Chapman, 1985)

REFERENCES

Braunsteiner, E. E., & Mark, H. F. (1974). Aromatic polymers. *Journal of Polymer Science: Macromolecular Reviews*, *9*(1), 83–126.

Fried, J. R. (1982). Polymer Technology. 1. The Polymers of Commercial Plastics. *Plastics Engineering*, *38*(6), 49–55.

Kim, H., Hong, J., Park, K. Y., Kim, H., Kim, S. W., & Kang, K. (2014). Aqueous rechargeable Li and Na ion batteries. *Chemical Reviews*, *114*(23), 11788–11827. doi:10.1021/cr500232y PMID:25211308

Manoravi, P., Selvaraj, I. I., Chandrasekhar, V., & Shahi, K. (1993). Conductivity studies of new polymer electrolytes based on the poly (ethylene glycol)/sodium iodide system. *Polymer*, *34*(6), 1339–1341. doi:10.1016/0032-3861(93)90799-G

Oh, W. C., & Chen, M. L. (2008). Synthesis and characterization of CNT/TiO$_2$ composites thermally derived from MWCNT and titanium (IV) n-butoxide. *Bulletin of the Korean Chemical Society, 29*(1), 159–164. doi:10.5012/bkcs.2008.29.1.159

Peacock, A. J., & Calhoun, A. (2012). *Polymer Chemistry: Properties and Application.* Carl Hanser Verlag GmbH Co KG.

Pradhan, D. K., Samantaray, B. K., Choudhary, R. N. P., & Thakur, A. K. (2005). Effect of plasticizer on microstructure and electrical properties of a sodium ion conducting composite polymer electrolyte. *Ionics, 11*(1-2), 95–102. doi:10.1007/BF02430407

Reiter, J., Velická, J., & Míka, M. (2008). Proton-conducting polymer electrolytes based on methacrylates. *Electrochimica Acta, 53*(26), 7769–7774. doi:10.1016/j.electacta.2008.05.066

Subban, R. H. Y., & Arof, A. K. (2003). Experimental investigations on PVC-LiCF$_3$SO$_3$-SiO$_2$ composite polymer electrolytes. *Journal of New Materials for Electrochemical Systems, 6*(3), 197–203.

Wieczorek, W., Florjanczyk, Z., & Stevens, J. R. (1995). Composite polyether based solid electrolytes. *Electrochimica Acta, 40*(13-14), 2251–2258. doi:10.1016/0013-4686(95)00172-B

Chapter 3
Reliability of CNTFET and NW–FET Devices

Sanjeet Kumar Sinha
Lovely Professional Univeersity, India

Sweta Chander
Lovely Professional University, India

ABSTRACT

The scaling of devices is a fundamental step for advancing technology in the semiconductor industry. The device scaling allows extra components as well as devices on a single chip, which provides large functionality and application for each integrated circuit (IC). The ultimate goal of device scaling is to make each IC smaller, faster, cheaper, and consumes low power. In today's nanoscale technology, the scaling has been continued and follows Moore's law in the initial phase of fabrication and also shows an exponential growth in ICs. The silicon-based semiconductor industry has reached its scaling limits due to tunneling and quantum-mechanical effects in deep nanometer level. The physics of such devices is not going to continue and hold true. This makes nanoelectronics the leading future of the semiconductor industry. The carbon nanotubes and nanowires are the most promising candidates to make illustrated devices.

INTRODUCTION

In next decade, the continuous scaling of integrated circuits increases the physical as well as technological problems. Gordon Moore was the first who observed that the constant shrinking of chip sizes over the last four decades. For scaling, the cost reduction and higher speed are the two most important driving forces, which is inherently depends on the area occupied by a transistor on the chip. International Technology Roadmap for Semiconductors (ITRS) identifies the necessary requirements and challenges to undergo the constant scaling of the device. The parameters such as trans-conductance, mobility and the current densities of the semiconductor material have to be enhanced. The scaling of silicon channels produces unwanted effects i.e. short channel effect. To reduce short channel effects, high doping

DOI: 10.4018/978-1-7998-1464-1.ch003

concentration is needed to get further scaling of the device. In metal interconnects, the scaling becomes tough for interconnect material because the physical size and electro migration results physical limits (Knoch, J., Riess, W., & Appenzeller, J. 2008). At high current densities, the electro migration occurs and this results interconnect failure. The current densities increase as the dimensions of the interconnect shrinks and this will more proclaim the interconnect failure.

The solution of the above problems needs some new promising devices such as carbon nanotubes field effect transistors (CNTFETs) (Appenzeller, J., 2008, Bachtold, A., Hadley, P., Nakanishi, T., & Dekker, C. 2001)-(Appenzeller, J., Lin, Y. M., Knoch, J., Chen, Z., & Avouris, P, 2005) and nanowire field effect transistors (Kim, R., & Lundstrom, M. S.,2008)-(Schulze, A., Hantschel, T., Eyben, P., Verhulst, A. S., Rooyackers, R., Vandooren, A., & Vandervorst, W. 2011). These devices are one dimensional self-assembled system with few unusual electronic properties. CNTs have metallic or semiconducting properties, which depending on their structure. The properties of CNTs provides potential for different nanoelectronics applications. This chapter deals with the details of the fabrication process and reliability related to the growth of CNTFET and Nanowire devices.

Fabrication of CNT and CNTFET

In compare to silicon/SiO_2 interface CNT with high current densities would be the ideal candidate to replace silicon in the future transistors. In semiconductor industry, it has already been demonstrated in laboratory that CNTFET (carbon nanotube field effect transistor) can have superior performance to the most advanced existing MOSFETs. In other aspects also, carbon nanotubes exhibit plenty number of device properties that make them interesting for semiconductor devices as well as for many other applications. Other than this at room temperature, CNTs exhibit thermal conductivities even higher than diamond.

Several number of fabrication methods has been developed to synthesize and characterized CNT. Arc Discharge Synthesis, Pulsed Laser Evaporation, and Catalytic Chemical Vapour Deposition are the three methods through which CNTs can be fabricated in laboratory.

Arc Discharge Synthesis (ADS)

Arc discharge. synthesis process is based on evaporation of carbon atoms into plasma by a high current. flowing between two graphite rods (Jinno, M., Bandow, S., & Ando, Y. 2004)- (Lange, H., Sioda, M., Huczko, A., Zhu, Y. Q., Kroto, H. W., & Walton, D. R. M. 2003). As shown in Figure 1, the direct .arc of the plasma reaches .temperatures around 3000°C. In order to .achieve an adequate .yield sufficient cooling is necessary .since nanotubes are prone to structural damage at such temperatures.

For the synthesis. of multi walled CNTs (MWCNT) no catalyst addition to the graphite rods is required, whereas for single walled CNTs (SWCNT) synthesis is only possible if small .amounts of catalysts. are added. The SWCNT .yield is usually less. than 20%. To make. electronic devices. out of SWCNT material, it has to. be cleaned, dispersed., cut, and finally. deposited on a substrate. An .oxidative process, using .nitric acid or .heating the material under oxygen, is usually used for cleaning. The dispersion and cutting is done by ultrasonication in appropriate solvents. Spinning on spraying on, or adsorption will then deposit the nanotubes on substrates. Unfortunately, the .cleaning and ultrasonication .might induce defects in the nanotube. and, thus, deteriorate the electronic properties.

Figure 1. Arc-discharge setup for CNTs

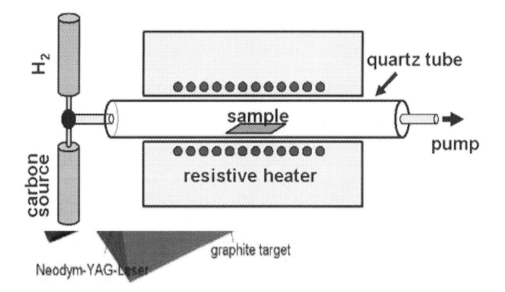

Pulsed Laser Evaporation (PLE)

In this. processes. a pulsed. laser evaporates. a target. within. a heated zone of a.. furnace at. 900°C to 1200°C as shown. in Figure 2 (Chrisey, D. B., & Hubler, G. K. (Eds.). 1994). The target. usually consists. of graphite with the addition of small amounts. of catalyst metals (Co/Ni). The yield can reach 70–90%. The material, therefore, still requires similar. cleaning and dispersion treatments as arc discharge synthesis.

Catalytic Chemical Vapour Deposition (CCVD)

The CCVD. growth of CNTs. might be the. most straightforward. approach for integration of nanotubes. in nanoelectronics. The PLE and the. arc discharge process are based on. the evaporation. of carbon atoms. at temperatures 3000 °C. These. Temperatures. are clearly incompatible. with the substrate based.

Figure 2. Schematic of pulsed laser setup

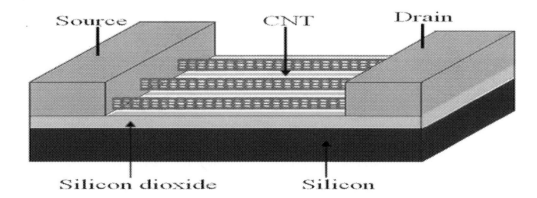

growth of CNT (Cantoro, M., Hofmann, S., Pisana, S., Scardaci, V., Parvez, A., Ducati, C.& Robertson, J, 2006). The CCVD synthesis of SWCNTs at much lower temperatures around 900 °C to 1000 °C. Compared. to all other methods. discussed earlier it has. the inherent. advantage that it allows. direct and patterned. growth of CNTs on a chip. The principal. mechanism of CCVD. is the decomposition of a carbon. source on metal. catalyst. The nanotubes obtained by CCVD. are usually very. clean. Therefore, the laborious. Purification. and dispersion treatment as described in ADS and PLE methods are not required. Damage of the structure and deterioration of the electronic properties can, thus, be widely excluded. CCVD growth of CNTs. is usually. performed in a furnace. at temperatures between. 600 °C to 1000 °C using an appropriate gas as the carbon source.

The growth. of CNTs have been. studied and briefed in. this thesis since. it is believed that CCVD. is the most straightforward. approach for integration. of CNTs into. electronic. devices.

CCVD. growth is based. on the deposition. of a suitable catalyst on a substrate. The catalyst. can be either deposited. as prefabricated catalyst. particles or by. deposition of a thin metal. or metal salt layer. If appropriate. particles have been formed, growth. of nanotubes will. occur under proper. conditions when. a carbon sources. is introduced.

Fabrication of CNTFET

The placement. of nanotubes. is one of the most. Important. challenges for their future integration. in devices. Nanotubes. will only be interesting. in nanoelectronics. if one can manage to. place a huge number. of nanotubes exactly. where they. are needed. with nano level. accuracy and extreme. reliability. There are two. Principal. Approaches. for the placement of nanotubes. The first one. is simply to grow. them where. they are needed.

This can only. be achieved by. Patterned. CCVD growth (Li, J., Zhang, Q., Yang, D. & Tian, J. 2004). The second .approach is based on. a post. growth assembly from solutions (Li, J., Zhang, Q., Yang, D., & Tian, J,2004).

Figure 3. Schematic of CCVD

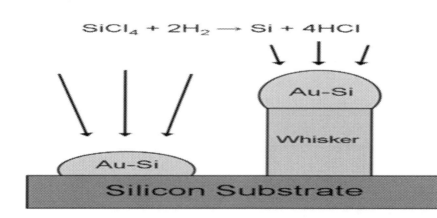

Figure 4. Schematic of CNTFET

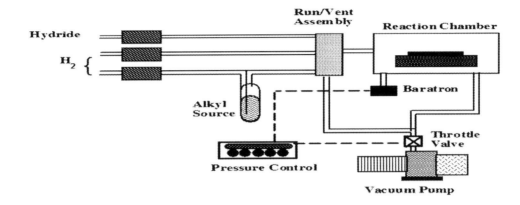

Patterned CCVD

CCVD allows the direct growth of carbon nanotubes on substrates. Therefore, patterning the catalyst can control the location of the nanotubes (Li, J., Zhang, Q., Yan, Y., Li, S., & Chen, L,2007). The perfect scenario for CCVD is to place a catalyst particle at a specific location and then to start the CNT growth. This requires precise control of the placement, the activity of the catalyst particle, the growth direction, and the suppression of bundle formation between adjacent CNTs. This methods work quite well for a low density of SWCNTs. However, for a high-density growth of SWCNTs these approaches are inappropriate since the nanotubes will stick together during growth due to thermal vibrations.

Post Growth Assembly

CNT material has been grown before in a separate process either by arc discharge synthesis, pulsed laser evaporation, or catalytic chemical vapour deposition to assemble and orientate CNTs on substrates.

Fabrication of NW-FET

Among the potential semiconducting nanostructures, NWs seem particularly attractive since they can, in principle, be incorporated into existing process flows and can be grown epitaxial at predefined locations (Schmid, H., Björk, M. T., Knoch, J., Riel, H., Riess, W., Rice, P., & Topuria, T, 2008). NWs represent key building blocks for a variety of electronic devices. An example of such a device with broad potential for applications is the NW field effect transistor (NW-FET). Studies of FETs enable evaluation of the performance level of NWs compared with corresponding planar devices. NW-FETs can exhibit performance comparable to the best reported for planar devices made from the same materials. Studies have also demonstrated the high electron mobility of NW-FETs.

Vapour Liquid Solid (VLS)

Growing nanowires (NWs) by the VLS technique has been found to be an attractive approach since VLS growth allows obtaining NWs with extremely small diameters [Wagner, R. S, 1965]. Moreover at the same time, there is no need for lithography and etching, which potentially leads to large surface-roughness-induced mobility degradation.

VLS is a mechanism for the growth of one-dimensional nanowires, from chemical vapor deposition. The growth of a crystal through direct adsorption of a gas phase on to a solid surface is generally very slow. The VLS mechanism circumvents this by introducing a catalytic liquid alloy phase which can rapidly adsorb a vapour to super saturation levels, and from which crystal growth can subsequently occur from nucleated seeds at the liquid–solid interface. The physical characteristics of nanowires grown in this manner depend, in a controllable way, upon the size and physical properties of the liquid alloy (Surawijaya, A., Anshori, I., Rohiman, A., & Idris, I, 2011). A thin 10 nm Au film is deposited onto a silicon wafer substrate by sputter deposition or thermal evaporation.Mixing Au with Si greatly reduces the melting temperature of the alloy as compared to the alloy constituents. The melting temperature of the Au-Si alloy reaches a minimum 363°C when the ratio of its constituents is 4:1 (Au:Si). Lithography techniques can also be used to controllably manipulate the diameter and position of the droplets. Nanowires are grown by physical vapor deposition process, which takes place in a vacuum deposition system. Au-Si droplets on the surface of the substrate act to lower the activation energy of normal vapor-solid growth. Au particles can form Au-Si droplets at temperatures above 363°C and adsorb Si from the vapor state until reaching a supersaturated state of Si in Au.

Metal Organic Chemical Vapour Deposition (MOCVD)

Metal Organic Chemical Vapour deposition (MOCVD) is one of the Chemical Vapour Depositionmethods which can also be used to produce nanowires (Nakamura, S., Harada, Y., & Seno, M, 1991). In MOCVD ultra pure gases are injected into a reactor and finely dosed to deposit a very thin layer of atoms onto a semiconductor wafer. Surface reaction of organic compounds or metal organics and hydrides containing the required chemical elements creates conditions for crystalline growth of materials and compound semiconductors. Unlike traditional silicon semiconductors, these semiconductors contain combinations of Group III and Group V elements (Appenzeller, J., Knoch, J., Tutuc, E., Reuter, M., & Guha, S, 2006).

Figure 5. Schematic of VLS growth

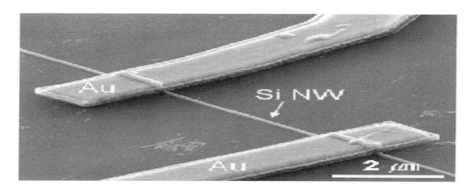

Figure 6. Schematic of MOCVD reactor

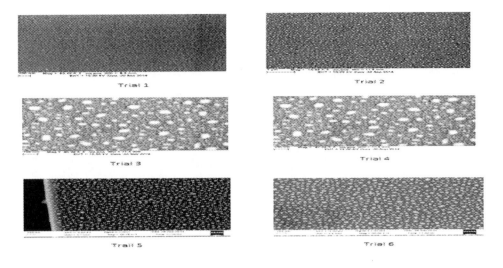

In MOCVD technique, reactant gases are combined at elevated temperatures in the reactor to cause a chemical interaction, resulting in the deposition of materials on the substrate. A reactor is a chamber made of a material that does not react with the chemicals being used and it must also withstand high temperature.

NW-FET with Contacts

Once nanowires have been grown they must be connected to the outside world to eventually become a part of an electronic circuit. The most widely used method to define contacts to nanowires is to remove the nanowires from the grown substrate and deposit them on a second substrate. The simplest approach to fabricate a NW-FET is by placing a NW on top of an oxidized silicon wafer and attaching metal contacts as source and drain electrodes, the silicon substrate then serves as a large-area back gate. During this process step, the resist mask is employed to define the contact areas and to protect the gate oxide outside the source/drain regions at the same time, in order to access the NW for source/drain formation, the gate oxide was removed using buffered oxide etch right before nickel deposition. During this process step, the resist mask is employed to define the contact areas and to protect the gate oxide outside the source/drain regions at the same time. After a standard lift off, a silicide is formed using rapid thermal processing. A rather low temperature of 280^0C is chosen since nickel tends to diffuse quickly into the SiNWs. In fact, the nickel diffusion strongly increases with temperature and decreasing NW diameter (Appenzeller, J., Knoch, J., Tutuc, E., Reuter, M., & Guha, S, 2006).

A field effect transistor requires a gate besides the source and drain contacts. It is predicted that the vertical geometry of nanowires, which makes it possible for a gate to surround the nanowire, will lead to superior electrostatic control of the conductivity in the FET channel. The important steps in forming a vertical gate is deposition of a dielectric layer followed by gate metal deposition and an etch back process (Cohen, G. M., Rooks, M. J., Chu, J. O., Laux, S. E., Solomon, P. M., Ott, J. A., & Haensch, W, 2007).

Figure 7. Schematic of NW-FET

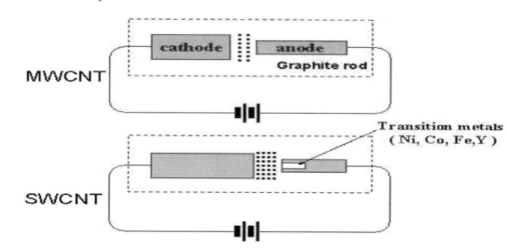

Nanowire Growth

Silicon nano wires (Si-NWs) are gaining great interest due to their electrical and optical properties. Si-NWs based nanodevices like, field effect transistors (FETs) have potential applications in electronics and biosciences. Vapour-Liquid-Solid (VLS) mechanism has been extensively investigated for the growth of Si-NWs. In VLS method, metal particles like Au act as catalyst for the growth and Si based gas sources like SiH_4 or $SiCl_4$ are used as precursors at temperatures typically around 500°C to 1000°C. The Au forms a eutectic with Si and forms molten Au/Si alloy. Si from gas source diffuses into liquid alloy and super saturates at the liquid/solid interface forming Si-NWs . Deposition of the Au catalyst could be carried out by several methods including sputtering, evaporation or solution based methods using Au nanocrystals .

Si-NWs are grown on p-type Si (100) and SiO_2 substrates using Au as catalyst layer. Au films are deposited by sputtering on samples of sizes 1cm x 1cm. A KI:I2 gold etchant solution is used to etch gold to reduce the thickness in the range of 20-30 nm. These samples are annealed at 450 °C for 60 minutes at forming gas annealing (FGA) furnace. The samples are then placed for LPCVD to grow Si-NWs at a temperature of 600°C and a pressure of 400 millitorr for 60 minutes. The samples are characterized using Scanning Electron Microscopy (SEM).

Si-NWs Growth by LPCVD

Au films of thickness 5 nm are sputtered on Si (100) samples. Au/Si eutectic droplets are formed by FGA annealing, with the droplets growing to diameters of 17-200 nm depending on the initial thickness of the film. NWs of diameter in the range of 30-390 nm are grown when these samples are used.

As seen from Figure 8., the diameter of Au droplets formed increases with higher thicknesses of Au catalyst layer. Also, inter droplet distance was seen to decrease drastically with the increasing thickness of the Au layer.

Figure 8. SEM image of Si/Au droplet at different trials

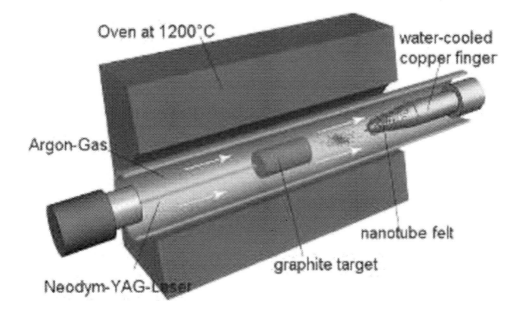

The diameter and the variation in diameter of the Si-NWs have increased with increasing thickness of the Au layer . This could be due to the higher variation in the Au droplets formed after FGA annealing for thicker Au catalyst layers. As shown in Figure 9, substrates with thicker Au layer produced shorter NWs, whereas less slanted and thinner NW's were obtained with a lower Au thickness.

Vapor Deposition, also known as sputtering, is mainly a physical process. In sputtering, heavy argon ions are electrically accelerated in high vacuum toward a pure metal. Upon impact, these ions sputter off the target material one by one. The atoms land on the wafer surface and form a solid metal layer. This layer can then be patterned and etched to form the conducting nanowires in a semiconductor device.

Figure 9. SEM image Si-NWs

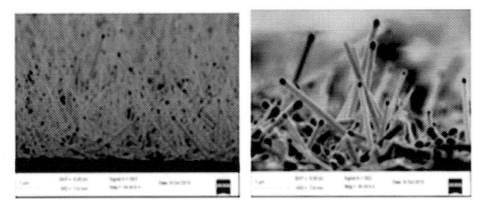

Si-NWs are grown in low pressure CVD (LPCVD) using Au layer deposited by FGA. Irrespective of the method of Au deposition, higher thickness of Au film resulted in the formation of wider Au droplets after annealing. This, in turn led to Si-NW's with shorter length and higher diameter, maybe due to the higher deposition of Si on Au droplets leading to a collapse of the growing NW.

CONCLUSION

In this dissertation we have mainly focused our efforts on simulation-based study on different MOSFET devices as well as on CNTFET and Nanowire FET devices. Then we compare and analyze the benefit of CNT and Nanowire FET as future semiconductor devices. However, the study is not complete until and unless it is not actually fabricated and characterized in a Fab lab.

In this chapter we have seen the state of art nano-fabrication techniques used worldwide, and nanowire growth, where a technology level up to 50 nm is possible. Further we have also grown Si-nanowires by sputtering method and taken SEM images to measure the dimensions. In the context of Indian process industry and R&D facilities, it is impossible to fabricate a device below 30 nm. Extracting a nanowire from the grown nanowires is a real challenge and for that we have to carry out the same processes in deep nanometer sufficiently below 30 nm. Under such circumstances a nanowire can be extracted and can be used as nanowire FET, which can be taken as future work.

REFERENCES

Appenzeller, J. (2008). Carbon nanotubes for high-performance electronics-Progress and prospect. *Proceedings of the IEEE, 96*(2), 201–211. doi:10.1109/JPROC.2007.911051

Appenzeller, J., Knoch, J., Tutuc, E., Reuter, M., & Guha, S. (2006, December). Dual-gate silicon nanowire transistors with nickel silicide contacts. In *2006 International Electron Devices Meeting* (pp. 1-4). IEEE. 10.1109/IEDM.2006.346842

Appenzeller, J., Lin, Y. M., Knoch, J., Chen, Z., & Avouris, P. (2005). Comparing carbon nanotube transistors-the ideal choice: A novel tunneling device design. *IEEE Transactions on Electron Devices, 52*(12), 2568–2576. doi:10.1109/TED.2005.859654

Bachtold, A., Hadley, P., Nakanishi, T., & Dekker, C. (2001). Logic circuits with carbon nanotube transistors. *Science, 294*(5545), 1317–1320. doi:10.1126cience.1065824 PMID:11588220

Baek, R. H., Baek, C. K., Jung, S. W., Yeoh, Y. Y., Kim, D. W., Lee, J. S., ... Jeong, Y. H. (2009). Characteristics of the Series Resistance Extracted From Si Nanowire FETs Using the $ Y $-Function Technique. *IEEE Transactions on Nanotechnology, 9*(2), 212–217. doi:10.1109/TNANO.2009.2028024

Bryllert, T., Wernersson, L. E., Froberg, L. E., & Samuelson, L. (2006). Vertical high-mobility wrap-gated InAs nanowire transistor. *IEEE Electron Device Letters, 27*(5), 323–325. doi:10.1109/LED.2006.873371

Cantoro, M., Hofmann, S., Pisana, S., Scardaci, V., Parvez, A., Ducati, C., & Robertson, J. (2006). Catalytic chemical vapor deposition of single-wall carbon nanotubes at low temperatures. *Nano Letters, 6*(6), 1107–1112. doi:10.1021/nl060068y PMID:16771562

Cantoro, M., Hofmann, S., Pisana, S., Scardaci, V., Parvez, A., Ducati, C., & Robertson, J. (2006). Catalytic chemical vapor deposition of single-wall carbon nanotubes at low temperatures. *Nano Letters*, 6(6), 1107–1112. doi:10.1021/nl060068y PMID:16771562

Chrisey, D. B., & Hubler, G. K. (Eds.). (1994). Pulsed laser deposition of thin films. Academic Press.

Clifford, J. P., John, D. L., Castro, L. C., & Pulfrey, D. L. (2004). Electrostatics of partially gated carbon nanotube FETs. *IEEE Transactions on Nanotechnology*, 3(2), 281–286. doi:10.1109/TNANO.2004.828539

Cohen, G. M., Rooks, M. J., Chu, J. O., Laux, S. E., Solomon, P. M., Ott, J. A., & Haensch, W. (2007). Nanowire metal-oxide-semiconductor field effect transistor with doped epitaxial contacts for source and drain. *Applied Physics Letters*, 90(23), 233110. doi:10.1063/1.2746946

Derycke, V., Martel, R., Appenzeller, J., & Avouris, P. (2002). Controlling doping and carrier injection in carbon nanotube transistors. *Applied Physics Letters*, 80(15), 2773–2775. doi:10.1063/1.1467702

Gnani, E., Gnudi, A., Reggiani, S., Luisier, M., & Baccarani, G. (2008). Band effects on the transport characteristics of ultrascaled snw-fets. *IEEE Transactions on Nanotechnology*, 7(6), 700–709. doi:10.1109/TNANO.2008.2005777

Jinno, M., Bandow, S., & Ando, Y. (2004). Multiwalled carbon nanotubes produced by direct current arc discharge in hydrogen gas. *Chemical Physics Letters*, 398(1-3), 256–259. doi:10.1016/j.cplett.2004.09.064

Khayer, M. A., & Lake, R. K. (2009). The quantum and classical capacitance limits of InSb and InAs nanowire FETs. *IEEE Transactions on Electron Devices*, 56(10), 2215–2223. doi:10.1109/TED.2009.2028401

Kim, R., & Lundstrom, M. S. (2008). Characteristic features of 1-D ballistic transport in nanowire MOSFETs. *IEEE Transactions on Nanotechnology*, 7(6), 787–794. doi:10.1109/TNANO.2008.920196

Kim, Y. A., Muramatsu, H., Hayashi, T., & Endo, M. (2012). Catalytic metal-free formation of multi-walled carbon nanotubes in atmospheric arc discharge. *Carbon*, 50(12), 4588–4595. doi:10.1016/j.carbon.2012.05.044

Knoch, J., Riess, W., & Appenzeller, J. (2008). Outperforming the conventional scaling rules in the quantum-capacitance limit. *IEEE Electron Device Letters*, 29(4), 372–374. doi:10.1109/LED.2008.917816

Lange, H., Sioda, M., Huczko, A., Zhu, Y. Q., Kroto, H. W., & Walton, D. R. M. (2003). Nanocarbon production by arc discharge in water. *Carbon*, 41(8), 1617–1623. doi:10.1016/S0008-6223(03)00111-8

Li, J., Zhang, Q., Yan, Y., Li, S., & Chen, L. (2007). Fabrication of carbon nanotube field-effect transistors by fluidic alignment technique. *IEEE Transactions on Nanotechnology*, 6(4), 481–484. doi:10.1109/TNANO.2007.897868

Li, J., Zhang, Q., Yang, D., & Tian, J. (2004). Fabrication of carbon nanotube field effect transistors by AC dielectrophoresis method. *Carbon*, 42(11), 2263–2267. doi:10.1016/j.carbon.2004.05.002

Nakamura, S., Harada, Y., & Seno, M. (1991). Novel metalorganic chemical vapor deposition system for GaN growth. *Applied Physics Letters*, 58(18), 2021–2023. doi:10.1063/1.105239

Neophytou, N., & Kosina, H. (2012). Numerical study of the thermoelectric power factor in ultra-thin Si nanowires. *Journal of Computational Electronics*, 11(1), 29–44. doi:10.100710825-012-0383-1

Razavieh, A., Singh, N., Paul, A., Klimeck, G., Janes, D., & Appenzeller, J. (2011, June). A new method to achieve RF linearity in SOI nanowire MOSFETs. In *2011 IEEE Radio Frequency Integrated Circuits Symposium* (pp. 1-4). IEEE. 10.1109/RFIC.2011.5940626

Schmid, H., Björk, M. T., Knoch, J., Riel, H., Riess, W., Rice, P., & Topuria, T. (2008). Patterned epitaxial vapor-liquid-solid growth of silicon nanowires on Si (111) using silane. *Journal of Applied Physics, 103*(2), 024304. doi:10.1063/1.2832760

Schulze, A., Hantschel, T., Eyben, P., Verhulst, A. S., Rooyackers, R., Vandooren, A., ... Vandervorst, W. (2011). Observation of diameter dependent carrier distribution in nanowire-based transistors. *Nanotechnology, 22*(18), 185701. doi:10.1088/0957-4484/22/18/185701 PMID:21415466

Stringfellow, G. B. (1985). Organometallic vapor-phase epitaxial growth of III–V semiconductors. *Semiconductors and Semimetals, 22*, 209–259. doi:10.1016/S0080-8784(08)62930-0

Surawijaya, A., Anshori, I., Rohiman, A., & Idris, I. (2011, July). Silicon nanowire (SiNW) growth using Vapor Liquid Solid method with gold nanoparticle (Au-np) catalyst. In *Proceedings of the 2011 International Conference on Electrical Engineering and Informatics* (pp. 1-3). IEEE. 10.1109/ICEEI.2011.6021750

Wagner, R. S. (1965). The vapor-liquid-solid mechanism of crystal growth and its application to silicon. *Transactions of the Metallurgical Society of AIME, 233*, 1053–1064.

Wind, S. J., Appenzeller, J., & Avouris, P. (2003). Lateral scaling in carbon-nanotube field-effect transistors. *Physical Review Letters, 91*(5), 058301. doi:10.1103/PhysRevLett.91.058301 PMID:12906636

Chapter 4
Traditional and Non–Traditional Optimization Techniques to Enhance Reliability in Process Industries

Ravinder Kumar
lovely Professional University, India

Hanumant P. Jagtap
Zeal College of Engineering and Research, India

Dipen Kumar Rajak
 https://orcid.org/0000-0003-4469-2654
Sandip Institute of Technology and Research Centre, India

Anand K. Bewoor
Cummins College of Engineering for Women, India

ABSTRACT

At present, optimization techniques are popular to solve typical engineering problems. It is the action of making the best or most effective use of a situation or resources. In order to survive in the competitive market, each organization has to follow some optimization technique depending on their requirement. In each optimization problem, there is an objective function to minimize or maximize under the given restrictions or constraints. All techniques have their own advantages and disadvantages. Traditional method starts with the initial solution and with each successive iteration converges to the optimal solution. This convergence depends on the selection of initial approximation. These methods are not suited for discontinuous objective function. So, the need of non-traditional method was felt. Some non-traditional methods are called nature-inspired methods. In this chapter, the authors give the description of the optimization techniques along with the comparison of the traditional and non-traditional techniques.

DOI: 10.4018/978-1-7998-1464-1.ch004

INTRODUCTION

Today's Reliability analysis of process industries is the main concern for long term profit in the organization. At present, optimization techniques are popular to solve these kinds of issues. It is the action of making the best or most effective use of a situation or resources. In order to survive in the competitive market, each organization has to follow some optimization technique depending on their requirement. In each optimization problem, there is an objective function to minimize or maximize under the given restrictions or constraints. These techniques help for reducing the effort required or maximizing the desired advantage in terms of cost. All optimization problems cannot be solved with the help of a common method, so different methods have been approached to solve technical issues for long term profit gain.

FORMULATION OF AN OPTIMIZATION PROBLEM

The process consists of the number of steps as given below:

- Selection of control and state variables
- Identification and formulation of constraints
- Set-up of variables at suitable limits
- Selection of a proper algorithm for the problem
- Solve the problem to obtain the optimum solution for the objective function.

OPTIMIZATION TECHNIQUES

These methods are classified into two groups, traditional and non-traditional techniques. Figure 1 shows traditional optimization techniques, and Figure 2 shows non-traditional optimization techniques. All techniques have their advantages and disadvantages. The traditional method starts with the initial solution and with each successive iteration converges to the optimal solution. This convergence depends on the selection of initial approximation. These methods are not suited for the discontinuous objective function. So, the need for the non-traditional method was felt. Some non-traditional methods are called nature inspired method. In this chapter, we aim to describe the optimization techniques along with the application and difference between traditional and non-traditional optimization techniques applied in process industries.

Traditional Optimization Techniques

The traditional (conventional) optimization methods help to obtain the optimal solution for continuous and differential functions. The techniques need complete information about the objective function for its dependency on every variable and type of nature function, which are the significant shortcomings of the techniques. Hence it does not guarantee the global optimum solution or may be trapped in local optima. Due to long execution time and generations of the weak number of non-dominated solutions, the non-traditional techniques have restricted scope in real life applications. Some of its applications are found in previous studies such as thermal power plant, nuclear power plant, wind power plant, manufacturing

Figure 1. Traditional optimization techniques

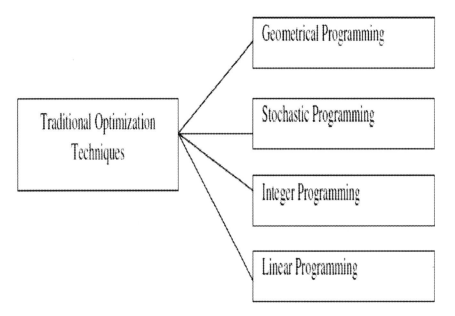

Figure 2. Non-Traditional and non-traditional optimization techniques

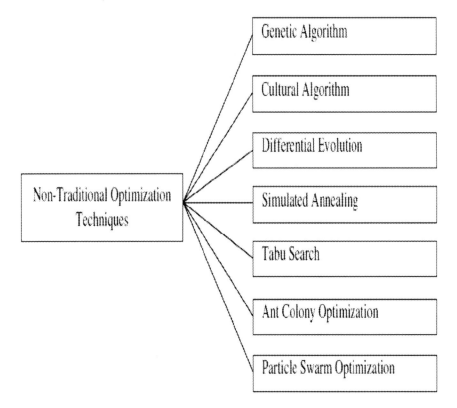

industries, and dairy plant. The objective function for enhancing the reliability may be maximizing system reliability or minimizing the cost of resources to be utilized for controlling the process parameters. The traditional optimization techniques consist of geometrical programming, stochastic programming, integer programming, and linear programming.

Geometrical Programming

The solution of the non-linear type of programming problem is obtained by geometric programming method by quantifying them into a mathematical optimization model. It used for minimizing the functions presented as polynomials and subjected to the constraint of a similar kind. The method initially determines the optimum values corresponding to the objective functions instead of finding the design variable first. The complexity optimization problem is decreased by having a set containing various simultaneous linear algebra equations.

Stochastic Programming

The stochastic programming method deals with the conditions that parameters are described by a stochastic variable rather than the deterministic. In this method, the stochastic type of problems is converted into an equivalent type of deterministic problem and further solved by using other techniques such as linear, geometric, and non-linear programming methods. (Duffuaa, S O, 1999) developed the model based on stochastic programming for scheduling of maintenance employees who work in the production industry. The model integrated stochastic and deterministic components for scheduling purpose. Their study revealed 10% enhancement in stochastic solution over the deterministic. The application of stochastic programming is also found in case of the thermal power plant. (Mohanta, Sadhu, & Chakrabarti, 2007) developed reliability analysis model based on the levelized risk method and implemented for the maintenance scheduling of the power plant. The study results revealed that the probabilistic methods are especially constituent as compared to a deterministic one.

Integer Programming

An integer programming optimization method contains the objective function and constraint as linear. In many situations, it is like to have fractional solutions, but the round-off solutions could provide a value which is distant from optimum. Such kinds of complexity are keeping away if the problem is solved using integer programming method. If the variables are constrained for integer value only, then the problem is known as an al-integer programming problem. Furthermore, if some variables are constrained for discrete integer values, then the type is a mixed-integer problem. (Hussin et al., 2018) provided the merits of an approach of mixed-integer linear programming for solving the maintenance scheduling problem of transmission and generation.

Linear Programming

Linear programming method allows for making optimal decisions in complex situations. In this method, the decision variables contain linear functions for the constraints and objective functions. (Khodr H.M., J.F. Gomez, L. Barnique, J.H. Vivas, P. Paiva, J.M. Yusta, 2002) presented a mathematical model for optimum selection schemes for power generation based on integer linear programming algorithm which considered from a reliability point of view. The problem was formulated by taking into account fuel cost, asset cost, maintenance cost, and operation cost. Under this conceptual framework, (Dotoli, Fanti, Mangini, & Ukovich, 2009) diagnosed the fault of the system using Petri Net framework A methodology adopted for storing the sequence of system events which facilitates to check that either the system behavior is stable.

Non-Traditional Optimization Techniques

In recent years, modern optimization techniques are widely used for complex engineering problems in various industries. The methods deal with various characteristics such a molecular, biological, swarm of insects, and neurobiological system. The non-traditional optimization techniques which include genetic algorithm, cultural algorithm, differential evolution, tabu search, simulated annealing, ant colony algorithm, and particle swarm optimization.

Genetic Algorithm

Genetic algorithm is one type of search algorithm based on the principles of natural genetics and natural selection. The fundamental elements are reproduction, crossover, and mutations. The genetic algorithm estimated the fitness for every string for directing the search as an alternative to the optimization function. It examines the search space for obtaining high performance. The problem describes the point population and utilizes the values of the corresponding objective function. The variables are presented as a string of binary variable. Therefore the method is valid for the integer as well as a discrete type of problems. (H. Peng & Ling, 2008) validated the applicability of genetic algorithm which combined with backpropagation neural network for optimizing the design of plate of fin heat exchanger application. (Lapa, Pereira, & De Barros, 2006) developed a model for scheduling preventive maintenance of a high-pressure injection system based on reliability and cost aspect. Besides, (Kaviri, Jaafar, & Lazim, 2012) studied and modeled the combined cycle power plant for energy optimization using a genetic algorithm. The study result shows that any change process parameters could lead to severe deviation in the objective function.

In the case of a nuclear power plant, improvement in the ability to detect and diagnose the faults using a genetic algorithm is highly essential. (Yangping, Bingquan, & Dongxin, 2000) developed a new method by combining genetic algorithm and classical probability with expert knowledge. The study results revealed that the proposed has relative adaptability to diagnose the faults. In addition, (Hao, Kefa, & Jianbo, 2001) optimized NOx emission of pulverized coal burner utility boiler with combined use of a genetic algorithm and neural network. In their study, an artificial neural network was used to build the neural network model. Besides, the optimum solution was obtained with suitable set points for a given operating condition. Based on the work of (Valdés, Durán, & Rovira, 2003), the genetic algorithm has used for optimizing the heat recovery steam generator by employing a genetic algorithm. For the study, thermodynamic parameters were considered for the optimizations which establish the configuration of the system. The two objective functions were established for the optimization purpose, such as minimizing the production cost and maximization of annual cash flow. The study result shows the achievable optimum strategy for the two functions and likely to find for the optimum solution for every design parameter.

The performance of gear fault is compared by (Samanta, 2004) using two techniques such as artificial neural network, and support vector machine. The vibration signals were proposed for extraction and provided as input for classifiers. The genetic algorithm is used for optimizing the features. The feature includes a number in a hidden layer, radial basic function kernel parameters along with a selection of input feature. The proposed methodology is validated experimentally for vibration signals of the gearbox. (Volkanovski, Mavko, Boševski, Čauševski, & Čepin, 2008) developed a method for scheduling the maintenance of the power system. In their study, loss of load expectation was taken for a measure of reliability, and maintenance was scheduled. The genetic algorithm utilized for obtaining the minimal value for loss of load expectation. An integrated combustion optimization approach proposed by (Si et al., 2009) for selective catalyst system of a coal-fired power plant. The regression model was developed for monitoring the coal quality and conditions of the plant. The genetic algorithm based optimization is carried out for a multi-objective optimization problem. The optimal operating setting for the boiler, selective catalyst reaction, and air preheated were obtained.

The performance evaluation and assessment are the two aspects of conventional power plant for management planning and control activities. (Azadeh, Saberi, Anvari, Azaron, & Mohammadi, 2011) studied non-parametric efficiency methods based on an adaptive network fuzzy interface system, and the performance evaluation power plant was carried out. While, (Kančev, Gjorgiev, & Čepin, 2011) used genetic algorithm and probabilistic safety assessment to optimize the surveillance test interval for the stand in the equipment of a nuclear power plant. Monte Carlo based methodology was employed for assessing the uncertainty, and the optimum test interval was obtained. The results used for obtaining the risk and importance by taking into account of aging data uncertainties for modeling of component aging. Moreover, (Kajal, Tewari, & Saini, 2013) developed a mathematical model for availability modeling for performance evaluation using a Markov approach for a coal handling system of the thermal power plant. In their study, a genetic algorithm based the availability optimization are studied. The steady state availability was compared for results obtained from the Markov approach and genetic algorithm.

A model based on the combined use of genetic algorithm and wavelet support machine for fault diagnosis of the gearbox is proposed by (F. Chen, Tang, & Chen, 2013). The collected vibration signals with its feature were processed using the empirical mode decomposition method. The study results revealed that the proposed method has a strong ability to fault diagnosis. Subsequently, (S. Peng & Vayenas, 2014) studied the different characteristics involved in predicting the maintainability of mining using a genetic algorithm. The field data was grouped for loads hand pump vehicle for maintainability analysis. On the other hand, (Qian, Wu, Duan, Kong, & Long, 2018) analyzed the multi-chromosome genetic algorithm, for optimizing parameters of support vector machine multi-classification problem.

Cultural Algorithm

The cultural algorithms are extensions of the genetic algorithm. The technique integrates the evolutionary-based populations affecting the search process for congregating the problem directly. The previous study reported that the cultural algorithm had been used for the functions of global optimization constraints as well as for scheduling of real-life applications (Bhattacharya, Mandal, & Chakravorty, 2012). In addition, (Yuan, 2006) developed and applied the cultural algorithm for scheduling the power generation of the hydrothermal power plant. The study result revealed that the developed algorithm has quick convergence solution with higher precision.

Differential Evolution

The technique makes use of real coding of floating numbers. The differential strategies are employed, which are based on the category of problem. (Saber, Yare, Member, Venayagamoorthy, & Member, 2009) implemented differential evolution technique for the maintenance scheduling for the generator of an Indonesian power plant. (Mohamed, Sabry, & Abd-Elaziz, 2013) used the differential evolution technique for real-life problems in the continuous domain. The best and worst individuals were considered for the entire population. A modified genetic algorithm and random mutation were combined for avoiding convergence and stagnation. The effectiveness and benefits of the developed algorithms have been investigated. The result obtained from the study revealed that the developed approach is not only effective but also robust.

An improved differential algorithm for solving the parametric and structural optimization is used by (Wang, Yang, Dong, Morosuk, & Tsatsaronis, 2014). The approach implemented for investigating the economically optimum design problem of supercritical coal-based power plants. The study deals with the investigation of feed water preheater, reheater, and turbine. Also, the study discussed in individual differences among the front of various plant and plant efficiency. Following the objective of improved algorithm, (Wang, Yang, Dong, Morosuk, & Tsatsaronis, 2014) proposed a modified differential algorithm for multi-objective type optimization problem of a coal-fired power plant. The various uncertainties related to cost function are analyzed for the sensitivity analysis.

Simulated Annealing

The technique searches the optimum solution by avoiding the trap in poor local optima. The large optimization problem is solved by this technique (S. Kirkpatrick, C.D. Gelatt, 1983). The technique deals with simulating thermal annealing of the heated solids. The technique simulates the cooling of molten for attaining the minimum function. The main strength of the simulated annealing method is that it optimizes the functions in non-linearity and boundary conditions, stochastic, and constraint. (E, Ames, & Ames, 2007) reported the simulated algorithm for finding the allocation of redundancy of a coherent system.

(K. Das, R.S. Lashkari, 2006) analyzed the problem for solving the multi-objective optimization problem of the manufacturing system. Furthermore, (Lapa et al., 2006) discussed the new method which combined genetic and simulated annealing technique and used for reliability based maintenance scheduling optimization problem. Besides, (Conradie, Morison, & Joubert, 2008) conducted a study for the optimization of scheduling the coal and blending processes. A genetic-based model was studied. In addition, simulated annealing metaheuristic was intended to make sure the acceptable superior solutions for great examples of the generic model. (Saraiva, Pereira, Mendes, & Sousa, 2011) reported the use of mixed integer technique for maintenance scheduling of generator. The optimization results are determined for the generation system. (R. Kumar, 2017) focused the research for identifying the redundancies for the subsystems of a coal-fired power plant through modeling, simulation, and optimization techniques. In the study, simulated annealing techniques have been utilized for profit optimization in term of the net present value.

Tabu Search

The method uses memory structures which describes the visited solutions. If a probable solution is visited before within a specified time, the algorithm will not use the possibility of time after time. It is a gradient search, and the memory saves the number of previously visited states. (Banerjee & Das, 2012) developed a Tabu search for solving the defect of searching neighborhood spaces and identifies the bottleneck arts of machines based on cost measures. In addition, (El-amin, Duffuaa, & Abbas, 2000) reported use of tabu search for scheduling the maintenance activity of the power generating system.

Ant Colony Optimization

The method deals with actions of real colonies of an ant who locate the short way from the nest to food. The solution candidate communicates with the other number of ant colony by providing pheromone to spot a pathway. Higher attentiveness of pheromones designated the positive path that other member needs to pursue getting the optimal solution. (Wai Kuan Fong, Holger Maier, 2008) developed the ant colony algorithm and implemented for scheduling the maintenance formation.

Particle Swarm Optimization

The method deals with actions of the colony of insects of the swarm, which is population-based (Pant, Anand, Kishor, & Singh, 2015). Every particle moves in a dispersed way with intelligence and collective information of the swarm. For any particle, if it finds out the best way the other warm needs to follow the same way. Every swarm has velocity and position. The particles communicate the information regarding good positions and adjust their velocities. (Suresh & Kumarappan, 2012) studied the probabilistic levelized method combined with particle swarm optimization for determining the risk associated with power generation planning, which has the potential to solve the maintenance scheduling problems. Besides, (Onwunalu & Durlofsky, 2010) used particle swarm optimization for determination of optimal well type and location. Also, (Ma, Yu, & Hu, 2013) developed the model for the life-cycle cost approach based on analyzing the behavior of the heating problem in the various system. The results revealed that the strategy is applied correctly and provides useful information while making maintenance planning decisions.

(Garg & Sharma, 2011) proposed an approach for analyzing the reliabilities indices by the use of fuzzy numbers allowing the experts opinion. Based on the approach, fuzzy based multi-objective type optimization problem was generated by taking the account for the preference of decision maker. The developed approach was used for solving the multi-objective series and parallel system of urea plant. (Zhao, Ru, Chang, & Li, 2015) applied particle swarm optimization combined the effect with first order reliability method for reliability enhancement. In their study, the chaotic particle is combined with swarm optimization. The reliability analysis of circular tunnel was carried out.

(Roy, Mahapatra, & Dey, 2017) developed a dynamic weighted model which is based on an artificial neural network used for software reliability prediction. Also, a neighborhood-based adaptive particle swarm optimization is studied for finding the global optimum weights. The obtained empirical results show that the proposed model has more predicting capability and presents fairly accurate fitting for software reliability models. (Kundu, Rossini, & Portioli-staudacher, 2019) discussed various characteristics

Table 1. Review of optimization techniques used in various process industries

Sr. No.	Optimization technique used	Identified Process Industry	References
1.	Stochastic Programming	Production Industry	(Duffuaa, S O, 1999)
		Thermal Power Plant	(Mohanta et al., 2007)
2.	Integer Programming	Manufacturing Industry	(Taylor, Ashayeri, Teelen, & Selenj, 1996)
		Power generation and transmission industry	(Hussin et al., 2018)
3	Linear Programming	Thermal Power Plant	(Khodr H.M., J.F. Gomez, L. Barnique, J.H. Vivas, P. Paiva, J.M. Yusta, 2002)
4	Genetic Algorithm	Thermal power plant	(Hao et al., 2001; Valdés et al., 2003) (Azadeh et al., 2011; Si et al., 2009; Volkanovski et al., 2008) (Kajal et al., 2013)
		Nuclear power plant	(Yangping et al., 2000)(Kančev et al., 2011)
		Manufacturing Industry	(Samanta, 2004)(F. Chen et al., 2013) (Sortrakul, Nachtmann, & Cassady, 2005)
		Paper Industry	(Informa & Marrocos, 1999)
		Chemical Process Plants	(Nguyen & Bagajewicz, 2010)
		Aerospace (Rocket) Industry	(T. Chen, Li, Jin, & Cai, 2013)
		Mining Industry	(S. Peng & Vayenas, 2014)
		Fermentation processes	(L. Chen, Kiong, Dong, & Mei, 2004)
		Tennessee Eastman Plant	(Nguyen & Bagajewicz, 2008)
5	Cultural Algorithm	Hydrothermal power plant	(Yuan, 2006)
6	Differential Evolution	Thermal power plant	(Wang et al., 2014b)(Wang et al., 2014)
		Generator maintenance scheduling	(Saber et al., 2009)
7	Simulated Annealing	Thermal power plant	(Conradie et al., 2008)(R. Kumar, 2017)
		Manufacturing Industry	(K. Das, R.S. Lashkari, 2006)
		Generator maintenance scheduling	(Saraiva et al., 2011)
		Coherent system	(E et al., 2007)
8	Tabu Search	Electric generating unit.`	(El-amin et al., 2000)
9	Ant Colony Optimization	Thermal Power Plant	(Wai Kuan Fong, Holger Maier, 2008)
10	Particle Swarm Optimization	Thermal Power Plant	(Onwunalu & Durlofsky, 2010) (Ma et al., 2013)
		Beverage plant	(P. Kumar & Tewari, 2017)
		Fertilizer Plant	(Garg & Sharma, 2011)
		Tunnel	(Zhao et al., 2015)
		Manufacturing industry	(Kundu et al., 2019)

of line feeding and proposed a mathematical model for simulation using particle swarm optimization technique. The developed model is utilized for the practical situation. The study results are obtained for showing the kanban influence of the assembly line feeding system.

Applications of Optimization in Process Industries

The earlier researchers have implemented the various optimization techniques for enhancing the availability level and reliability level of various process industries. Table 1 shows an overview of such optimization techniques in various applications.

CONCLUSION

This chapter reported the necessary information about the traditional and non-traditional optimization techniques which are used to enhance the system reliability in process industries. Overview of traditional optimization techniques viz. geometrical programming, stochastic programming, integer programming, and linear programming are discussed. Also, the use and application of non-traditional optimization techniques viz. genetic algorithm, cultural algorithm, differential evolution, tabu search, simulated annealing, ant colony optimization, and particle swarm optimization method are discussed in detail.

ACKNOWLEDGMENT

The contributors of this chapter would like to recognize Dr. Mohammad Hossein Ahmadi, Shahrood University of Technology, for his contribution in the research and writing of this chapter.

REFERENCES

Azadeh, A., Saberi, M., Anvari, M., Azaron, A., & Mohammadi, M. (2011). An adaptive-network-based fuzzy inference system-genetic algorithm clustering ensemble algorithm for performance assessment and improvement of conventional power plants. *Expert Systems with Applications*, 38(3), 2224–2234. doi:10.1016/j.eswa.2010.08.010

Banerjee, I., & Das, P. (2012). Group technology based adaptive cell formation using a predator-prey genetic algorithm. *Applied Soft Computing*, 12(1), 559–572. doi:10.1016/j.asoc.2011.07.021

Bhattacharya, B., Mandal, K. K., & Chakravorty, N. (2012). Cultural Algorithm Based Constrained Optimization for Economic Load Dispatch of Units Considering Different Effects. *International Journal of Soft Computing and Engineering*, 2(2), 45–50.

Chen, F., Tang, B., & Chen, R. (2013). A novel fault diagnosis model for gearbox based on wavelet support vector machine with immune genetic algorithm. *Measurement: Journal of the International Measurement Confederation*, 46(1), 220–232. doi:10.1016/j.measurement.2012.06.009

Chen, L., Kiong, S., Dong, X., & Mei, X. (2004). Modeling and optimization of fed-batch fermentation processes using dynamic neural networks and genetic algorithms. *Biochemical Engineering Journal, 22*(1), 51–61. doi:10.1016/j.bej.2004.07.012

Chen, T., Li, J., Jin, P., & Cai, G. (2013). Reusable rocket engine preventive maintenance scheduling using a genetic algorithm. *Reliability Engineering & System Safety, 114,* 52–60. doi:10.1016/j.ress.2012.12.020

Conradie, D. G., Morison, L. E., & Joubert, J. W. (2008). Scheduling at coal handling facilities using Simulated Annealing. *Mathematical Methods of Operations Research, 68*(2), 277–293. doi:10.100700186-008-0221-1

Das, K., Lashkari, R. S., & Sengupta, S. (2006). Reliability considerations in the design of cellular manufacturing systems. *International Journal of Quality & Reliability Management, 23*(7), 880–904. doi:10.1108/02656710610679851

De Informa, C., & De Marrocos, P. (1999). Global optimization of energy and production in process industries : A genetic algorithm application. *Control Engineering Practice, 7*(4), 549–554. doi:10.1016/S0967-0661(98)00194-4

Dotoli, M., Fanti, M. P., Mangini, A. M., & Ukovich, W. (2009). On-line fault detection in discrete event by petri nuts and integer linear programming. *Autamatica, 45*(11), 2665–2672. doi:10.1016/j.automatica.2009.07.021

Duffuaa, S. O., & Sultan, K. S. (1999). A stochastic programming model for scheduling maintenance personnel. *Applied Mathematical Modelling, 25*(5), 385–397. doi:10.1016/S0307-904X(98)10009-4

E, J., Ames, K., & Ames, K. (2007). A simulated annealing algorithm for system cost minimization subject to reliability constraints. *Communications in Statistics-Simulation and Computation,* 37–41.

El-Amin, I., Duffuaa, S., & Abbas, M. (2000). A tabu search algorithm for maintenance scheduling of generating units. *Electric Power Systems Research, 54*(2), 91–99. doi:10.1016/S0378-7796(99)00079-6

Fong & Maier. (2008). Power plant maintenance scheduling using Ant Colony Optimization – An improved formulation. *Engineering Optimization, 404*(4), 309–329.

Garg, H., & Sharma, S. P. (2011). Multi-Objective Optimization of Crystallization Unit in a Fertilizer Plant Using Particle Swarm Optimization. *International Journal of Applied Science and Engineering, 9*(4), 261–276.

Hao, Z., Kefa, C., & Jianbo, M. (2001). Combining neural network and genetic algorithms to optimize low NOx pulverized coal combustion. *Fuel, 80*(15), 2163–2169. doi:10.1016/S0016-2361(01)00104-1

Hussin, S. M., Hassan, M. Y., Wu, L., Abdullah, M. P., Rosmin, N., & Ahmad, M. A. (2018). Mixed Integer Linear Programming for Maintenance Scheduling in Power System Planning. *Indonesian Journal of Electrical Engineering and Computer Science, 11*(2), 607–613. doi:10.11591/ijeecs.v11.i2.pp607-613

Kajal, S., Tewari, P. C., & Saini, P. (2013). Availability optimization for coal handling system using a genetic algorithm. *International Journal of Performability Engineering, 9*(1), 109–116.

Kančev, D., Gjorgiev, B., & Čepin, M. (2011). Optimization of test interval for aging equipment: A multi-objective genetic algorithm approach. *Journal of Loss Prevention in the Process Industries*, *24*(4), 397–404. doi:10.1016/j.jlp.2011.02.003

Kaviri, A. G., Jaafar, M. N. M., & Lazim, T. M. (2012). Modeling and multi-objective exergy-based optimization of a combined cycle power plant using a genetic algorithm. *Energy Conversion and Management*, *58*, 94–103. doi:10.1016/j.enconman.2012.01.002

Khodr, H. M., Gomez, J. F., Barnique, L., Vivas, J. H., Paiva, P., Yusta, J. M., & Urdaneta, A. J. (2002). A Linear Programming Methodology for the Optimization of Electric Power – Generation Schemes. *IEEE Transactions on Power Systems*, *17*(3), 864–869. doi:10.1109/TPWRS.2002.800982

Kirkpatrick, S., & Gelatt, C.D., M. P. V. (1983). Optimization by Simulated Annealing. *Science*, *220*(4598), 671–680. PMID:17813860

Kumar, P., & Tewari, P. C. (2017). Performance analysis and optimization for CSDGB filling system of a beverage plant using particle swarm optimization. *International Journal of Industrial Engineering Computations*, *8*, 303–314. doi:10.5267/j.ijiec.2017.1.002

Kumar, R. (2017). Redundancy optimization of a coal-fired power plant using a simulated annealing technique. *International Journal of Intelligent Enterprise*, *4*(3), 191–203. doi:10.1504/IJIE.2017.087625

Kundu, K., Rossini, M., & Portioli-staudacher, A. (2019). A study of kanban assembly line feeding system through integration of simulation and particle swarm optimization. *International Journal of Industrial Engineering and Computations*, *10*, 421–442. doi:10.5267/j.ijiec.2018.12.001

Lapa, C. M. F., Pereira, C. M. N. A., & De Barros, M. P. (2006). A model for preventive maintenance planning by genetic algorithms based on cost and reliability. *Reliability Engineering & System Safety*, *91*(2), 233–240. doi:10.1016/j.ress.2005.01.004

Ma, R., Yu, N., & Hu, J. (2013). Application of Particle Swarm Optimization Algorithm in the Heating System Planning Problem. *The Scientific World Journal*, 1–13. PMID:23935429

Mohamed, A. W., Sabry, H. Z., & Abd-Elaziz, T. (2013). Real parameter optimization by an effective differential evolution algorithm. *Egyptian Informatics Journal*, *14*(1), 37–53. doi:10.1016/j.eij.2013.01.001

Mohanta, D. K., Sadhu, P. K., & Chakrabarti, R. (2007). A deterministic and stochastic approach for safety and reliability optimization of captive power plant maintenance scheduling using GA/SA-based hybrid techniques: A comparison of results. *Reliability Engineering & System Safety*, *92*(2), 187–199. doi:10.1016/j.ress.2005.11.062

Nguyen, D., & Bagajewicz, M. (2008). Optimization of Preventive Maintenance Scheduling in Processing Plants. *European Symposium on Computer Aided Process Engineering*, 319–324. 10.1016/S1570-7946(08)80058-2

Nguyen, D., & Bagajewicz, M. (2010). Optimization of Preventive Maintenance in Chemical Process Plants. *Industrial & Engineering Chemistry Research*, *49*(9), 4329–4339. doi:10.1021/ie901433b

Onwunalu, J. E., & Durlofsky, L. J. (2010). Application of a particle swarm optimization algorithm for determining optimum well location and type. *Computers & Geosciences*, *14*(1), 183–198. doi:10.100710596-009-9142-1

Pant, S., Anand, D., Kishor, A., & Singh, B. (2015). A Particle Swarm Algorithm for Optimization of Complex System Reliability. *International Journal of Performability Engineering*, *11*(1), 33–42.

Peng, H., & Ling, X. (2008). Optimal design approach for the plate-fin heat exchangers using neural networks cooperated with genetic algorithms. *Applied Thermal Engineering*, *28*(5–6), 642–650. doi:10.1016/j.applthermaleng.2007.03.032

Peng, S., & Vayenas, N. (2014). *Maintainability Analysis of Underground Mining Equipment Using Genetic Algorithms: Case Studies with an LHD Vehicle*. Academic Press.

Qian, R., Wu, Y., Duan, X., Kong, G., & Long, H. (2018). SVM multi-classification optimization research based on multi-chromosome genetic algorithm. *International Journal of Performability Engineering*, *14*(4), 631–638.

Roy, P., Mahapatra, G. S., & Dey, K. N. (2017). An Efficient Particle Swarm Optimization-Based Neural Network Approach for Software Reliability Assessment. *International Journal of Reliability Quality and Safety Engineering*, *24*(4), 1–24. doi:10.1142/S021853931750019X

Saber, A. Y., Yare, Y., Member, S., Venayagamoorthy, G. K., & Member, S. (2009). Economic Dispatch of a Differential Evolution Based Generator Maintenance Scheduling of a Power System. *Proceedings of IEE Power and Energy Society General Meeting Conference*, 1–8.

Samanta, B. (2004). Gear fault detection using artificial neural networks and support vector machines with genetic algorithms. *Mechanical Systems and Signal Processing*, *18*(3), 625–644. doi:10.1016/S0888-3270(03)00020-7

Saraiva, J. T., Pereira, M. L., Mendes, V. T., & Sousa, J. C. (2011). A Simulated Annealing based approach to solve the generator maintenance scheduling problem. *Electric Power Systems Research*, *81*(7), 1283–1291. doi:10.1016/j.epsr.2011.01.013

Si, F., Romero, C. E., Yao, Z., Schuster, E., Xu, Z., Morey, R. L., & Liebowitz, B. N. (2009). Optimization of coal-fired boiler SCRs based on modified support vector machine models and genetic algorithms. *Fuel*, *88*(5), 806–816. doi:10.1016/j.fuel.2008.10.038

Sortrakul, N., Nachtmann, H. L., & Cassady, C. R. (2005). Genetic algorithms for integrated preventive maintenance planning and production scheduling for a single machine. *Computers in Industry*, *56*(2), 161–168. doi:10.1016/j.compind.2004.06.005

Suresh, K., & Kumarappan, N. (2012). Particle swarm optimization based generation maintenance scheduling using probabilistic approach. *Procedia Engineering*, *30*, 1146–1154. doi:10.1016/j.proeng.2012.01.974

Taylor, P., Ashayeri, J., Teelen, A., & Selenj, W. (1996). A production and maintenance planning model for the process industry. *International Journal of Production Research*, *34*(12), 37–41.

Valdés, M., Durán, M. D., & Rovira, A. (2003). Thermoeconomic optimization of combined cycle gas turbine power plants using genetic algorithms. *Applied Thermal Engineering, 23*(17), 2169–2182. doi:10.1016/S1359-4311(03)00203-5

Volkanovski, A., Mavko, B., Boševski, T., Čauševski, A., & Čepin, M. (2008). Genetic algorithm optimisation of the maintenance scheduling of generating units in a power system. *Reliability Engineering & System Safety, 93*(6), 779–789. doi:10.1016/j.ress.2007.03.027

Wang, L., Yang, Y., Dong, C., Morosuk, T., & Tsatsaronis, G. (2014). Multi-objective optimization of coal-fired power plants using differential evolution. *Applied Energy, 115*, 254–264. doi:10.1016/j.apenergy.2013.11.005

Wang, L., Yang, Y., Dong, C., Morosuk, T., & Tsatsaronis, G. (2014). Parametric optimization of supercritical coal-fired power plants by MINLP and differential evolution. *Energy Conversion and Management, 85*, 828–838. doi:10.1016/j.enconman.2014.01.006

Yangping, Z., Bingquan, Z., & Dongxin, W. (2000). *Application of genetic algorithms to fault diagnosis in nuclear power plants*. Academic Press.

Yuan, X., & Yuan, Y. (2006). Application of cultural algorithm to generation scheduling of hydrothermal systems. *Energy Conversion and Management, 47*(15-16), 2192–2201. doi:10.1016/j.enconman.2005.12.006

Zhao, H., Ru, Z., Chang, X., & Li, S. (2015). Reliability Analysis Using Chaotic Particle Swarm Optimization. *Quality and Reliability Engineering International, 31*(8), 1537–1552. doi:10.1002/qre.1689

Chapter 5
Residual Life Estimation of Humidity Sensor DHT11 Using Artificial Neural Networks

Pardeep Kumar Sharma
Lovely Professional University, India

Cherry Bhargava
Lovely Professional University, India

ABSTRACT

Electronic systems have become an integral part of our daily lives. From toy to radar, system is dependent on electronics. The health conditions of humidity sensor need to be monitored regularly. Temperature can be taken as a quality parameter for electronics systems, which work under variable conditions. Using various environmental testing techniques, the performance of DHT11 has been analysed. The failure of humidity sensor has been detected using accelerated life testing, and an expert system is modelled using various artificial intelligence techniques (i.e., Artificial Neural Network, Fuzzy Inference System, and Adaptive Neuro-Fuzzy Inference System). A comparison has been made between the response of actual and prediction techniques, which enable us to choose the best technique on the basis of minimum error and maximum accuracy. ANFIS is proven to be the best technique with minimum error for developing intelligent models.

INTRODUCTION TO RELIABILITY AND LIFE ESTIMATION

Across the globe, every industry is trying to lure the customers for electronics and electrical items and for that they need to provide better performance, high quality and low cost. Another factor which comes into picture is the Time To Market (TTM) for these products(Barnes, 1971). As competition is increasing day by day every sector or industry tries to launch their product as soon as possible because there may be chances that likewise product may get launched and industry may face great loss economically. Another factor which is most important now-a-days is the "Reliability" of any system. All the big brands,

DOI: 10.4018/978-1-7998-1464-1.ch005

Figure 1. Bath Tub Curve for Product Life

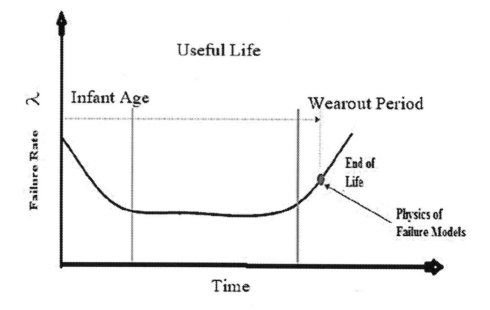

big industry with repute are moving forward to develop as reliable system as possible to contribute towards customer's safety as well as to maintain the quality. Higher the brand, higher the cost, high is the Reliability(Neri, Allen, & Anderson, 1979). Generally, a big trade-off is they're between the Reliability, Cost and Time. The overall Life of any product has been depicted by Bath-Curve shown in Figure 1,

"Reliability" is the general term in our day-to-day life which means how much we can rely over that particular thing. Talking about electronics world than now-a-days this word has been a great buzz. This term is becoming so vast that we have one entire engineering stream with this name i.e. "Reliability Engineering". Now what is the meaning of term for an electronics Engineer. "It is the degree which tells how reliably a particular electronic system or component will work as it is expected to, in the specific or desired duration."

Now operational Reliability includes three parts:

1.   Reliability, whether system is giving desired output or not.
2.   Reliability, whether it's working fine under different environment.
3.   Reliability, whether different conditions are not fluctuating the output.

If we go deeper in the Reliability then comes the term "Life Estimation or Prediction" Normally Prediction and Estimation are two different terms. Prediction is based on the historical observation or data available but Estimation is about the real time data available for that particular system. The data which we are using here is the data about that particular electronic component only that's why the topic recited as LIFE ESTIMATION. The meaning of this term is to estimate or observed the remaining useful Life (RUL) of that particular system under the various stressors we provide(Siddiqui & Çağlar, 1994).

Stressors are parameter, whose variation beyond the range could make the system fail. There are many categories of stressors i.e., Environmental stressors (Thermal Stressor), Electrical stressor (Voltage), Mechanical stressor (shock or vibration). More the stressor we select, more is the accuracy about failure or Life Estimation we get. If we talk about practical ground than it's very complicated to design the observation or experimental system to examine any component and more the parameters we add, more it get complicated. Mostly the big manufacturing companies only developed this kind of experiment setup because they want their selling product to be highly reliable and it is also the demand of the growing competition of the market. MTBF stands for Mean Time Between Failure, is the time duration for which the system is in operational state. In case of Repairable system, it is the total time component was active excluding the time for which it was under failure or under Maintenance and in case of Non-Repairable system, it is MTTF i.e., Mean Time to Failure as here we are not repairing. It is the time until which component meets its failure. 'Tool Life' or simple 'Life' of the component is the MTBF we calculate. Failure Rate is defined as the reciprocal of MTBF.

PARAMETERS AFFECTING RELIABILITY

There can be many parameters practically which tends to vary the Reliability graph. Factors affecting Reliability are mentioned below,

DESIGN FACTOR

Design plays major role in efficient operation of any system. Many time failures are observed not due to external changes but due to design faults such as voltage, current etc., these failures don't come under any prediction model and mostly it can't be handled at basic level since it needs changes in the design. Hence, careful designing must be done prior so to avoid any fault at design level(C Bhargava, Banga, & Singh, 2018).

COMPLEXITY

More complex the system less is its reliability as the operation becomes tough. A complete unit is a combination of many components, more the number of components, more complexes it becomes. Also, the reliability factor adds with increase in numbers of elements.

STRESS

One of the most prominent factors which affect the life of the component is stress. Factors for stress can be rise in current which tends to affect the temperature, vibration, voltage etc. Thermal stress is the parameter where special derating is needed as it tends to complete failure of the system as well. It this problem is quite prominent; coolers or fan should be used.

GENERIC (INHERENT) RELIABILITY

It involves selecting that component individually for which failure rate is very less. As the entire system is formed by individual components, it they would be having less failure rate, overall failure rate would reduce for the system(Lu & Christou, 2017). In case if we select high failure rate components, the overall reliability would decrease(Huston & Clarke, 1992).

DHT11 HUMIDITY SENSOR

The electronic component we have taken for observation is fabricated Humidity sensor cum Thermistor which is available in the market as DHT11. It's a resistive humidity sensor whose dielectric is made up of polymer. When the moisture level increase, the concentration of ions in the dielectric and in turn it decreases the resistance, which effect the voltage and hence these variation gives us relative Humidity and Temperature reading(Cherry Bhargava, Vijay kumar Banga, & Singh, 2018).

DHT11 is an Arduino compatible sensor hence we have used Arduino board for the experimental reading. The stressor parameter used here is temperature. The normal range if DHT11, is Humidity-[20-90] %RH, Temperature-[0-50] ^{0}celsius. That means this is the working range for this particular sensor(Korotcenkov, 2013). For checking the response, different temperatures have been provided to see the response as well the condition of the sensor until and unless it gets completely disfunction. Application of Humidity Sensor can be observed in many practical areas like soil-humidity detection, cold store monitoring, maintaining green-house, lie detector etc(Fei, Jiang, Liu, & Zhang, 2014).

CALCULATION OF TOOL LIFE

Calculation of DHT11 tool life is the main task we need to do. For calculating, Reliability, Failure Rate or MTBF we have many approaches shown in Figure 2,

EMPIRICAL METHOD

These methods are the one based on the analytical data collected from manufacturer, fields, Test laboratories etc. The most prominent set of such data is in MIL-HDBK-217, Bellcore, Prism etc. these are examples of Empirical method widely used by the Military purpose in twentieth century

ANALYTICAL METHOD

Many numerical Models were developed for the prediction of Reliability or Failure rate which provides better results and this numerical Model comes under Analytical Methods for example FEM stands for Finite Element Methods where we look for the faulty characteristics of an element. Another one is Fault Tree Analysis (FTA), where in hierarchical ways we check for the faulty units in the entire system.

Figure 2. Methods for Reliability Prediction

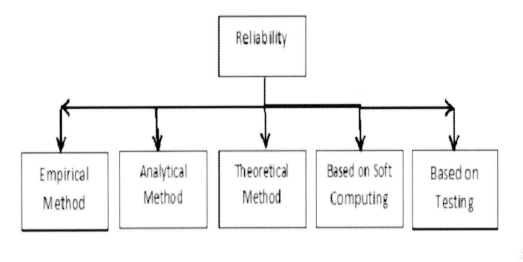

THEORETICAL MODEL

Normally, determining the failure of any model is tough and costly that's why we go for Physics of Failure approach which can deduce the reason of failure of any part by applying various models of physics to the data(Huston & Clarke, 1992). In real time application such as aeronautics where topology and environment changes abruptly, finding exact failure is very complexes hence there we can use theoretical Models.

APPROACH BASED ON SOFT COMPUTING MODELS

Here we used different Models or Intelligent Modelling techniques based on Artificial Intelligence like Artificial Neural Network, Adaptive Neuro-Fuzzy Inference System, Fuzzy Inference System Genetic Programming etc. Apart from this basic approach we have highly efficient approach like Monte Carlo Method, Naïve-Bayes etc. Along with this we can also apply hybrid approach which is a combination of two or more approaches(Gokulachandran & Mohandas, 2015).

BASED ON TESTING

This method is mainly used by big manufacturing unit where large number of samples or electrical units is subjected under tests. These tests can be environmental, Electrical, thermal and so on. For having reliable commercial product, better understanding of the failure mechanism is highly needed. Hence, thereby testing we get to know the exact failure reason or Model for any component. One of the examples of such testing is Highly Accelerated Life Testing (HALT).

For calculating we took Empirical Method which includes MIL-HDBK 217F, Bellcore, NTT Procedure etc. These all are Reliability book having methods to calculate reliability as well as directed reliability values are given for many active and passive component. Here we are considering MIL-HDBK217 F as a reference to calculate Tool Life. This book has been published by Washington D.C Military in year 1990

January, 2. They observed the data provided various manufacturers about the failure of the component and use that in predicting Tool Life or failure rate of many electrical and electronic component(Solomon, Sandborn, & Pecht, 2000).

For preparing an Intelligent Model we took Soft-computing approach using the various Artificial Intelligence techniques. Here we have taken Artificial Neural Network, Fuzzy Inference System and Adaptive Neuro-Fuzzy Inference System (ANFIS). Data set has been generated and feed to these techniques as an input and hence we get to know which technique is the most efficient(Wang, 2009).

NEED OF INTELLIGENT MODELLING

In today's world we need everything smart. Be it our vehicles, gadgets, home or surroundings(Venet, Perisse, El-Husseini, & Rojat, 2002). We want every system to be autonomous so that we could remove human intervention from their operation. This comfort leads to the basis of Intelligent Modelling, where we tend to developed intelligent model for every system(Wang, 2009). The basic terminology we required to understand for this id Artificial Intelligence. "Artificial Intelligence is the method or way where we try to provide intelligence or human like thinking capabilities to the machine or system through learning or any other method."

The approach to implement Artificial Intelligence is through Soft Computing where we have various algorithms that tend to make the system adaptive or intelligent or developed "Intelligent Model". The basic implementation approach one can take is Artificial Neural Network (ANN), Fuzzy Inference System (FIS), Adaptive Neuro-Fuzzy Inference System (ANFIS), Genetic Programming (GPs), Support Vector Machine (SVM) etc. In this work we have taken first three approaches to implement the Intelligent Model for our designed sensor(Chen, Zhang, Vachtsevanos, & Orchard, 2011).

MODELLING OF INTELLIGENT SYSTEM USING ARTIFICIAL NEURAL NETWORK (ANN)

As the world is getting modernized, we are moving more towards intelligent system which could work same way as human use to do. We are searching for method which could remove or reject human intervention by developing intelligent system. Here, our first method is Artificial Neural Network (ANN). Artificial Neural Network is an analogous system of human neural network which tries to mimic the functioning of actual brain. Input data along with target data has been fed to the network. Activation function has been provided to start the process where system learns by itself how output is coming(Cherry Bhargava & Handa, 2018). The system gets train with the number of epochs we specified. The system will train itself and reduce the error after every epoch and hence after specific number of epochs we get the best result(Rajeev, Dinakaran, & Singh, 2017). The ANN based figure is shown below,

MODELLING THROUGH FUZZY INFERENCE SYSTEM

Fuzzy Inference System or Fuzzy Logic is used to handle Ambiguity and Uncertainty in Data(Lefik, 2013). As the complexity increases, we can't make exact statement about the behaviour of the system as

Figure 3. Artificial Neural Network

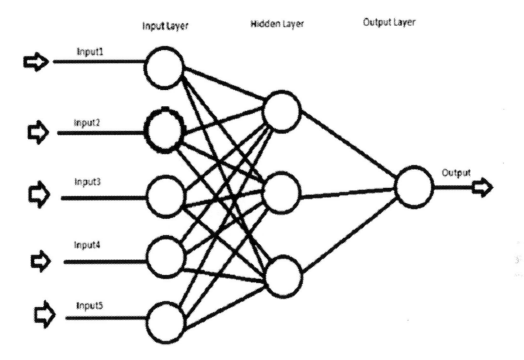

in the traditional method we were using Binary Logic which says 0 or 1, i.e., YES or NO, but the real-world problems are beyond as it can't be TRUE or FALSE only(Hayward & Davidson, 2003). Taking water Problem than the possible answer could be hot, cold, slightly cold, slightly hot, extremely cold, extremely hot etc.(Hayward & Davidson, 2003)

For this purpose, we deal with Linguistic variables in fuzzy which are user understandable. Entire input set is known as Crisp Set which after fuzzification converts into fuzzy sets(Jiao, Lei, Pei, & Lee, 2004). Here we use the concept of Membership function, it defines the membership of particular input value in the fuzzy sets, and its range is from 0 to 1. If input value has complete membership, it is 1 otherwise it can be any value in this range(Kirby & Chen, 2007). If we have fuzzy set A than considering the Universal set, X can be defined as, $A = \{(x, \mu_A(x))|x \in X\}$, where μ_A is known as A's Membership Function. In FIS we defined certain rules for fuzzification to defines crisp relation into Fuzzy relation in IF, THEN, ELSE format e.g.

IF (f is x_1, x_2.....a_n) THEN (g is y_1, y_2....y_n) ELSE (g is z_1, z_2....z_n)

This fuzzified data goes to decision-making unit which decides about the membership function and hence attached the related linguistic Variable for that particular value(Aronson, Liang, & Turban, 2005). The fuzzy output from this block directly goes to defuzzifier Interface unit, which is reverse of Fuzzifier. And hence after this block we get proper output in Crisp set form as defuzzifier convert fuzzy set back to crisp set(Virk, Muhammad, & Martinez-Enriquez, 2008). Fuzzification as well as defuzzification unit are assisted by knowledge base which has design base as well as Rule base for making rules and modifying data(Chen et al., 2011). The Block Diagram of FIS is given below in Figure 4,

Figure 4. Block Diagram of Fuzzy Inference System

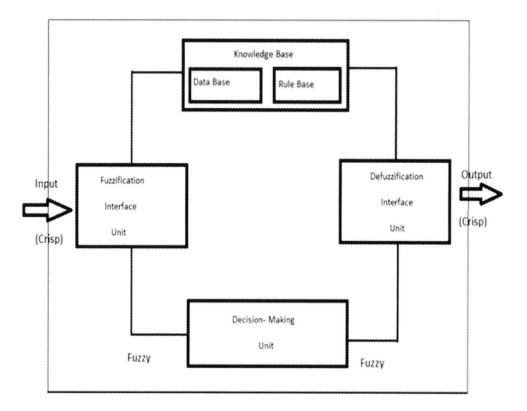

MODELLING THROUGH ADAPTIVE NEURO-FUZZY INFERENCE SYSTEM (ANFIS)

ANFIS is a hybrid technique comprises both ANN as well as Fuzzy Tool. It has advantage of both the technique as ANN has this self-learning mechanism but it doesn't know how the hidden process is following to reach the particular target and the disadvantage is that the output is not that user understandable also, need very precise and accurate(Manogaran, Varatharajan, & Priyan, 2018). It can't handle ambiguity(Chen et al., 2011). On the other hand, the advantage with Fuzzy logic is that it can handle uncertain data and also, we use linguistic variable to have better understanding but no self-learning is there(Jiao et al., 2004). Hence to omit each other advantages, these two techniques have been combined to formed third technique that is ANFIS (Adaptive neuro-fuzzy Inference system). Here the rules needed by fuzzy get self-updated through the self-learning mechanism possessed by ANN(Parler, 1999). That's why a smaller number of errors is shown by the predicted Data of ANFIS. The basic structure of ANFIS is shown in the figure 5 below,

Figure 5. Adaptive Neuro-Fuzzy Inference System

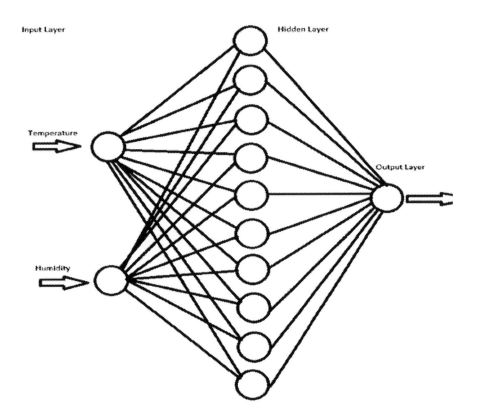

COMPONENTS AND TOOLS USED

Components Used

In this section, all the electrical equipment's, Chemicals, Lab instruments and components and software tools used for this work are mentioned,

TOOLS USED

Matlab

MATLAB stands for Matrix Laboratory, tools comprise Simulator as well as programming window, used for computation, integration, designing graphical user Interface(Simulink & Natick, 1993). It is highly used in the field like Communication, Machine Learning, Reliability Prediction, Digital Signal Processing etc. MATLAB Simulink is an in-built Window comprises many Interfaces like Neural Network, Fuzzy logic, Neuro- Fuzzy etc. The programming Interface has Editor Window to write the program, Command History stores all the related variables, matrix etc. Command Window is used for running the programming and Workspace is used load all the necessary matrix, xls. file etc. Here, are working on both Interfaces. ANN and ANFIS on MATLAB SIMULINK and MATLAB Programming Interface for Naïve-Bayes Classifier Algorithm.

ARDUINO 1.6.13

Arduino 1.6.13 is an Integrated Development Environment Tool with Java as basic for programming which is used to write basic programme related to the operation and to load it on Arduino-Board. Our digital sensor DHT11 is an Arduino compatible sensor. We can run this software on Windows OSX, Mac and Linux. We have used DHT library for programming to have compatible function to run the coding.

ELECTRICAL COMPONENTS AND CHEMICAL EQUIPMENT USED

Electrical components used for this work are mentioned in the table 1 and table 2 as below,

Table 1. Electrical Component Used

Sr. No.	Component Name	Numbers	Company
1.	DHT-11 sensor	3	D-robotics
2.	LCD	1	Blaze Display
3.	Resistor (2.2 K)	2	Reckon Electronics
4.	Connecting Wires	14	----------
5.	Arduino-Board	1	Bharthi Electronics
6.	PC	1	Personal computer

Table 2. Chemicals and Equipment Used

Sr. No.	Chemical or Equipment
1.	Zinc-Oxide
2.	Carbon Black
3.	Pottasium Bromide (KBr)
4.	Mortar Pestle
5.	Hydraulic Pelletizer
6.	Spatula
7.	Petri-Dish
8.	Hot air Oven

EXPERIMENTAL SETUP FOR DHT11

The experimental set up for calculating Data set of DHT11 comprises one computer system which has Arduino Library as well as software installed in it, an Arduino board, Bread –board, DHT11 sensor, LCD-Display, 2.2K resistors and connecting wires. For giving the overview of connection an image is shown below(Aggarwal & Bhargava, 2016),

According to DHT11 commercial manual range its normal operating range is 0 to 50 degrees and for Humidity it is 20 to 90%RH, that's why for failure mechanism it was necessary to provide temperature more than this range. Behaviour of sensor has been keenly observed for entire range until and unless it's been immediately failed at 60%. Temperature has been increased through hot dryer and sample of sand whose temperature has been increased every time up to specified limit and their temperature has been measured through Hygrometer(Cherry Bhargava et al., 2018). In spite of having range up to 50-degree, sensor was working normally uptill 63 degree. It started showing deflection at 64 and completely failed at 68 degree. For checking the sensor at lower temperature, it has been kept in ice box keeping in mind that it didn't get drenched because of that.

Talking about Humidity which has been observed by keeping sensor in air-tight box along with calcium carbonate powder. More the concentration of the powder, more the humidity it absorbed. Likewise, Humidity is increased through keeping the sensor in air-tight box having glass of ho6t water in it which increases the humidity level uptill 99%RH. Talking about effect of Humidity then only long exposure to high humidity can affect the health of sensor. If momentarily exposure is considered than it may not cause any harm unlike temperature. Long exposure can cause rusting to the contact pin of the sensor(Nazari, Kashanian, Moradipour, & Maleki, 2018).

Figure 6. DHT11 Experiment

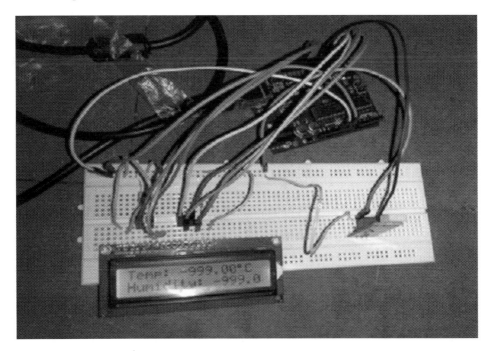

TOOL LIFE CALCULATION

For Tool Life Calculation, Military Handbook i.e. MIL-HDBK217 F has been used to provide numerical model which gives equation for failure rate i.e.

$$\Lambda_p = \lambda_b \pi_T \pi_A \pi_Q \pi_E, \text{ Failure}/10^6 \text{ hours}$$

where, Λ_p = Overall Failure rate
λ_b = Base Failure Rate.
π_T = Temperature Factor.
π_Q = Quality Factor.
π_E = Environment factor.

None of the value has been given directly. Each of the above parameter involve numerical analysis and formula to predict exact value or tool Life at that particular Temperature(Kötz, Hahn, & Gallay, 2006).

INTELLIGENT MODELLING OF SENSOR

Since the era of Artificial Intelligence has started, every other observed system is there on its way to become smart or analogous to Human. Intelligent Modelling has become the base of every new upcoming technology. With Reliability also Manufacturers are moving towards developing an intelligent system that could sense its need, behaviour and it should be able to perform all the analysis required. Here also we are trying to utilize different Artificial Intelligence technique to develop successful model which could estimate the Life for the sensor we designed.

A Guide How To Use Artificial Neural Network (ANN)

Let us take a case of DHT 11 where input parameters are five, output parameter is single(Seth, 2014). An expert system is designed using following steps:

1.   Right click on workspace---new----name it

Three files will be created in workspace, one for input, one for sample (only one column of input), one for output
Click on name, a workspace will be opened
Paste the excel data (transpose form) into the worksheet

2.   Function--->nntool
3.   Click on input data---import
4.   Click on input---import data (for sample)
5.   Click on target—import data (output)
6.   Click on networks--new
7.   Click on network1---open

8.   Train the sample
9.   Simulate
      Input---sample
      Change name of output
      Then simulate
10.   Click on network1_output
      Export—select all
11.   Export and it will be in workspace

A GUIDE HOW TO USE FUZZY LOGIC (FL)

Let us take a case of DHT11 where input parameters are five, output parameter is single. An expert system is designed using fuzzy logic as per following steps:

1.   Open MATLAB Toolbox
2.   Open fuzzy using "Fuzzy" command in the command window of MATLAB toolbox.
3.   Fuzzy editor window will get open.
4.   Now, add the number of inputs as per the number of input variables using "Edit" menu. In the edit menu, go to "Add variable" and then select "input or output according to number of inputs and output.

The selected number of inputs will be available on FIS editor window. Rename the inputs by changing the name in dialog box beyond the "Name" that is shown below the Fuzzy structure.
In the same way, name of another input and output can be changed.

5.   In the next step, the input data and the output data are loaded in the input variables and output variables in the form of linguistic variables by selecting the particular variable. The variable will be selected by double clicking that particular variable and the membership function editor window will get open(Manogaran et al., 2018).

The membership function can be of any types such as triangular, Gaussian or trap etc and one of all these types will be selected according to the range of loaded data. The name of membership functions can also be changed using "Add MFs" from the edit menu.
The name and range of membership functions is changed by selecting each membership function. The range of the input variables is also changed by editing the range in the edit box named as "Range" as shown below.
In the same way, the range and name of the membership functions of all the inputs and output will be changed as per the range of data.

6.   After editing all the input and output variables, the rules are defined using edit menu and then rules.

A rule editor window gets open after selecting the rules by right click.

Rules are defined for each combination of inputs with the output such as when both inputs (time and temperature) are at low range, the output (life) will be high.

Similarly, rules will be defined for every combination of inputs and output such as if the input have 3 membership functions low, medium and high & if the output also have 3 membership functions the 3*3 = 9 rules will be formed.

7.   Once all the rules are defined, the output is verified by viewing the rule viewer using view menu and then select "Rules".

Rule viewer window gets open and from this window output is obtained by giving the input data ranges in the input dialog box(Mansouri, Gholampour, Kisi, & Ozbakkaloglu, 2018).

A GUIDE FOR ADAPTIVE NEURO FUZZY INFERENCE SYSTEM (ANFIS)

Let us take a case of DHT11 where input parameters are five, output parameter is single. An expert system is designed using ANFIS as per following steps:

1.   Open MATLAB and ANFIS window by command anfisedit
2.   Right side, workspace is there. Right click---new----name
     Double click- open excel
     Copy paste own data
     Close it.
3.   After writing >>anfisedit
4.   Load data----training--- wrksp--- click on load data
5.   Generate FIS
     Select membership function ---gaussian
     MF type-linear
6.   Train now---epochs-50
7.   Test FIS
8.   Click on Structure
9.   Edit—FIS editor (to change name and range)
10. Edit—membership function
11. Edit---rules
12. View-surface
13. View-rule (for output values)

You can get new output, by changing input. Using intelligent modelling techniques, the failure of DHT11 can be very well pre-accessed and necessary replacement of the components can be done before the actual destruction of the whole system.

REFERENCES

Aggarwal, J., & Bhargava, C. (2016). Reliability Prediction of Soil Humidity Sensor using Parts Count Analysis Method. *Indian Journal of Science and Technology*, *9*(47).

Aronson, J. E., Liang, T.-P., & Turban, E. (2005). *Decision support systems and intelligent systems*. Pearson Prentice-Hall.

Barnes, F. (1971). *Component Reliability*. Springer.

Bhargava, Banga, & Singh. (2018). Failure prediction of humidity sensor DHT11 using various environmental testing techniques. *Journal of Materials and Environmental Sciences, 9*(7), 243-252.

Bhargava, C., Banga, V., & Singh, Y. (2018). Mathematical Modelling and Residual Life Prediction of an Aluminium Electrolytic Capacitor. *Pertanika Journal of Science & Technology*, *26*(2), 785–798.

Bhargava, C., & Handa, M. (2018). An Intelligent Reliability Assessment technique for Bipolar Junction Transistor using Artificial Intelligence Techniques. *Pertanika Journal of Science & Technology*, *26*(4).

Chen, C., Zhang, B., Vachtsevanos, G., & Orchard, M. (2011). Machine condition prediction based on adaptive neuro–fuzzy and high-order particle filtering. *IEEE Transactions on Industrial Electronics*, *58*(9), 4353–4364. doi:10.1109/TIE.2010.2098369

Fei, T., Jiang, K., Liu, S., & Zhang, T. (2014). Humidity sensors based on Li-loaded nanoporous polymers. *Sensors and Actuators. B, Chemical*, *190*, 523–528. doi:10.1016/j.snb.2013.09.013

Gokulachandran, J., & Mohandas, K. (2015). Comparative study of two soft computing techniques for the prediction of remaining useful life of cutting tools. *Journal of Intelligent Manufacturing*, *26*(2), 255–268. doi:10.100710845-013-0778-2

Hayward, G., & Davidson, V. (2003). Fuzzy logic applications. *Analyst (London)*, *128*(11), 1304–1306. doi:10.1039/b312701j PMID:14700220

Huston, H. H., & Clarke, C. P. (1992). *Reliability defect detection and screening during processing-theory and implementation*. Paper presented at the IEEE 30th Annual Symposium on International Reliability Physics, San Diego, CA.

Jiao, Y., Lei, S., Pei, Z., & Lee, E. (2004). Fuzzy adaptive networks in machining process modeling: Surface roughness prediction for turning operations. *International Journal of Machine Tools & Manufacture*, *44*(15), 1643–1651. doi:10.1016/j.ijmachtools.2004.06.004

Kirby, E. D., & Chen, J. C. (2007). Development of a fuzzy-nets-based surface roughness prediction system in turning operations. *Computers & Industrial Engineering*, *53*(1), 30–42. doi:10.1016/j.cie.2006.06.018

Korotcenkov, G. (2013). Chemical Sensors: Simulation and Modeling: Vol. 5. *Electrochemical Sensors*. Momentum Press.

Kötz, R., Hahn, M., & Gallay, R. (2006). Temperature behavior and impedance fundamentals of supercapacitors. *Journal of Power Sources*, *154*(2), 550–555. doi:10.1016/j.jpowsour.2005.10.048

Lefik, M. (2013). Some aspects of application of artificial neural network for numerical modeling in civil engineering. *Bulletin of the Polish Academy of Sciences. Technical Sciences, 61*(1), 39–50. doi:10.2478/bpasts-2013-0003

Lu, Y., & Christou, A. (2017). Lifetime Estimation of Insulated Gate Bipolar Transistor Modules Using Two-step Bayesian Estimation. *IEEE Transactions on Device and Materials Reliability, 17*(2), 414–421. doi:10.1109/TDMR.2017.2694158

Manogaran, G., Varatharajan, R., & Priyan, M. (2018). Hybrid recommendation system for heart disease diagnosis based on multiple kernel learning with adaptive neuro-fuzzy inference system. *Multimedia Tools and Applications, 77*(4), 4379–4399. doi:10.100711042-017-5515-y

Mansouri, I., Gholampour, A., Kisi, O., & Ozbakkaloglu, T. (2018). Evaluation of peak and residual conditions of actively confined concrete using neuro-fuzzy and neural computing techniques. *Neural Computing & Applications, 29*(3), 873–888. doi:10.100700521-016-2492-4

Nazari, M., Kashanian, S., Moradipour, P., & Maleki, N. (2018). A novel fabrication of sensor using ZnO-Al2O3 ceramic nanofibers to simultaneously detect catechol and hydroquinone. *Journal of Electroanalytical Chemistry, 812*, 122–131. doi:10.1016/j.jelechem.2018.01.058

Neri, L., Allen, V., & Anderson, R. (1979). Reliability based quality (RBQ) technique for evaluating the degradation of reliability during manufacturing. *Microelectronics and Reliability, 19*(1-2), 117–126. doi:10.1016/0026-2714(79)90369-X

Parler, S. G. (1999). *Thermal modeling of aluminum electrolytic capacitors.* Paper presented at the IEEE 34th Annual Meeting on Industry Applications Conference, Phoenix, AZ. 10.1109/IAS.1999.799180

Rajeev, D., Dinakaran, D., & Singh, S. (2017). Artificial neural network based tool wear estimation on dry hard turning processes of AISI4140 steel using coated carbide tool. *Bulletin of the Polish Academy of Sciences. Technical Sciences, 65*(4), 553–559. doi:10.1515/bpasts-2017-0060

Seth, S. (2014). MExS A Fuzzy Rule Based Medical Expert System To Diagnose The Diseases. *IOSR Journal of Engineering, 4*(7), 57–62. doi:10.9790/3021-04735762

Siddiqui, M., & Çağlar, M. (1994). Residual lifetime distribution and its applications. *Microelectronics and Reliability, 34*(2), 211–227. doi:10.1016/0026-2714(94)90104-X

Solomon, R., Sandborn, P. A., & Pecht, M. G. (2000). Electronic part life cycle concepts and obsolescence forecasting. *IEEE Transactions on Components and Packaging Technologies, 23*(4), 707–717. doi:10.1109/6144.888857

Venet, P., Perisse, F., El-Husseini, M., & Rojat, G. (2002). Realization of a smart electrolytic capacitor circuit. *IEEE Industry Applications Magazine, 8*(1), 16–20. doi:10.1109/2943.974353

Virk, S. M., Muhammad, A., & Martinez-Enriquez, A. (2008). *Fault prediction using artificial neural network and fuzzy logic.* Paper presented at the IEEE Seventh Mexican International Conference on Artificial Intelligence (MICAI'08), Atizapan de Zaragoza, Mexico. 10.1109/MICAI.2008.38

Wang, X. (2009). *Intelligent modeling and predicting surface roughness in end milling.* Paper presented at the IEEE Fifth International Conference on Natural Computation (ICNC'09), Tianjin, China.

Chapter 6
Nanocomposite–Based Humidity Sensor:
Reliability Prediction Using Artificial Intelligence Techniques

Pardeep Kumar Sharma
Lovely Professional University, India

Cherry Bhargava
Lovely Professional University, India

ABSTRACT

A humidity sensor detects, measures, and reports the content of moisture in the air. Using low cost composite materials, a humidity sensor has been fabricated. The characterization has been done using various techniques to prove its surface morphology and working. The fabricated sensor detects relative humidity in the range of 15% to 65%. The life of the sensor has been calculated using different experimental and statistical methods. An expert system has been modeled using different artificial intelligence techniques that predict failure of the sensor. The failure prediction of fabricated sensor using Fuzzy Logic, ANN, and ANFIS are 81.4%, 97.4%, and 98.2% accurate, respectively. ANFIS technique proves to be the most accurate technique for prediction of reliability.

INTRODUCTION

As electronic gadgets are becoming ever more popular in day by day activities, such as mobile devices, so do their operating conditions become harsher, means using electronic components at the threshold of their functional parameters and very close to or even exceeding their maximum operating temperature which leads to a dramatic decrease in component lifetimes. Under these extreme circumstances, classical failure prediction methodologies are perturbed, giving unreliable results. This generated the need to establish new methods of failure prediction.

DOI: 10.4018/978-1-7998-1464-1.ch006

BACKGROUND

Salah-Al-Zubaidi predicts as well as generate a comparison about the tool life in End Milling of Ti-6A1-4V Alloy by taking multiple regression technique along with artificial intelligence (Artificial Neural Network) technique (Al-Zubaidi, Ghani, & Haron, 2011). It has been discussed that the traditional methods for prediction could not handle the non-linearity in data which acts as drawbacks and we move towards using AI Techniques which draws the non-linearity very well. Regression model is being generated using SPSS software which given us the mathematical equation showing the link between multiple inputs we are providing with the output. We get linked beta coefficient for relation through which we can determine our tool life. Talking about AI techniques than we have taken ANN model using MAT-LAB neural network simulink. It is a supervised learning method which is feed with input and output both and the system is trained to give desire output by training or learning. As a result, we infer that the ANN model is better than RM model as the accuracy for former is 92% and latter is 86.53%. We take 3-7-1 hidden layer for ANN as it gives 0.1967 average mean square errors. Dealing with linear data is the prime advantage of this technique.

K. Raj Kumar used the copper-graphite tri-biological composite for the determination, efficiency and its reliability to be used as electrical sliding contacts. Copper is taken as it has considerably high Thermal as well as electrical conductivity. Its conductivity is checked by taking SEM (Scanning Electron Microscope) Test. Since the normal testing method consumer long duration which won't be feasible so accelerated wear testing has been used to determine the tool life(Antony, Bardhan Anand, Kumar, & Tiwari, 2006). In this test pressure has been taken as a parameter within the range 1250-5000kpa at constant sliding velocity 1.88m/s and out of entire range four parameters value has been taken. Reliability model was built using Weibull software under Reliasoft for Weibull distribution corresponding to different stressor also IPL (Inverse Power Law) model is used to check the effect of stress over the life result through Weibull. Mainly it is taken for non-thermal stress like pressure. The graphical method is used to represent the result which says initially under normal pressure the life span was high but it gradually starts decreasing as we increased the pressure up till 5000kpa likewise probability of failure also decreases for the same.

According to Cherry Bhargava, the need of reliability has been discussed that why it has become one of the ace research engineering area now a days. For electronics market low cost and high performance are the most desirable characteristics and they are somewhat highly dependent reliability for that particular system(Cherry Bhargava, Vijay kumar Banga, & Singh, 2014). Various techniques has been discussed for failure prediction like empirical method (based on MIL-HDBK-217, Bell core etc.), analytical methods (based on Fault-Tree Analysis, Finite Element Method etc.), theoretical methods (based on Physics of Failure model, Real time Prediction Model etc.), based on soft computing (based on Neural network, Fuzzy Logic, Genetic etc.) and analysis based on testing (Electrical, Thermal, Monte Carlo analysis etc.) are made. Brief introduction of all these techniques has been given along with examples. Conclusion has been made that empirical method MIL-HDBK still has been considered as a standard but they consider failure model under constant hazard rate only but component load profile is not considered.

Salah Al-Zubaidi discuss the intelligent modelling has been done on milling tool Ti6Al4V using adaptive neuro-fuzzy inference system(Al-Zubaidi, Ghani, & Haron, 2013). Practical data of 14 sets has been extracted about cutting speed, feed rate and depth of cut. Surface roughness has been taken as output parameter to develop intelligent model. Different assumptions are made when ANFIS parameters are taken for PVD coated carbide and another for uncoated carbide. Also, different number of epochs is taken to determine which will give best result based on their implementation. Obtained result says the model 1 with 100 numbers of epochs has given best result.

P Yadav discuss the intelligent greenhouse monitories set is proposed using GSM. It's an effective and smart way to avoid or abolish manual labour from the field, which sometime become challenging due to large area(Yadav, Singh, Chinnam, & Goel, 2003). The parameters like temperature, humidity, light and CO_2 has been taken for monitoring through sensors (temperature sensor-LM35, humidity sensor-HSM-20G, light dependent resistor (LDR) etc.) and its equivalent digital is taken through analog to digital converter and fed to microcontroller as input. Microcontroller is programmed to generate predefined specific output in terms switching ON fan through relay 1 if temperature is high, glowing of bulb/LED through relay2 if light intensity is low, starting sprinklers if humidity or moisture level is less amongst the plants through relay 3 and switching ON ventilators if CO_2 level is high through relay4. Also, one of the inputs is connected to GSM module; it needs SIM card to generate an SMS for the user. Entire system is feed with predefined range of working parameter beyond which they raise the alarm.

Zhigang Tian discuss a new approach that has been taken to implement ANN model for finding RUL (Remaining useful Life) of any system(Tian, Wong, & Safaei, 2010). The data implemented is taken from pump bearings in the site. Traditionally only failure data was taken but here suspension data of any system is also taken as it greatly affects the operational life. Suspension data is the log which is been created when system is un-operational for maintenance and actually not fail. The real challenge in this paper was to generate data log for suspension data. The artificial neural network has been used to train the system with failure data as input and the percentage of RUL as output. The two hidden layers with 2 neurons each have been used. The parameter for performance is not MSE (Mean square error) but validation data set is used to do so. In the entire data set two –third is used for training and one-third is used for simulation as sample (validation-set). ANN network used is resilient back propagation network and for better performance, it has been trained 30 times.

Seung Wu Lee discussed about the reliability for electronic component based on various aspects and failure models. Reliability is not only the function of components, parts, or subsystem's parameters but also on the working or operational environment(Wu, Ho, & Lee, 2004). So far, many methods have come into limelight for predicting the reliability, but the main factor is still the failure model we are considering. On general purpose we have MIL-HDBK217 N2, PRISM, Telcordia, NTT procedure etc. MIL-HDBK217 is highly used for electronic components. Telcordia is used for the serially oriented blocks to predict reliability and PRISM mainly for military purpose equipment. Here, three methods have been taken for comparison over their result for MTBF i.e., Mean time between failures. The first when is web system which is having huge information feed related to time, environment etc. It provides the reliability result to the user without any need to use any applied program just by having access to internet. Another is through using numerical method based on the model provided by the MIL-HDBK217, which is having genre like microcircuit, diodes etc. out of which GAAs MMIC from micro sub circuits category is taken. The third one is through software named Relex has been taken which is a commercial link available for reliability. The result through developed web, MIL-HDBK217 and Relex were almost same i.e., MTBF=0.169 failures/million hours on an average(Handbook, 1995). Overall, the developed web system is expected to contribute more because of the consideration of small and medium sub circuit or component which is not possible in case of general-purpose system due to financial aspect.

Amit Sachdeva et al., discuss a composite material that has been prepared using fly ash and potash alum(sinteza & galun-lete, 2013). Fly ash is a harmful element for environment as it's produced in abundant through factories where coal is burnt that's why utilizing this as a raw material is highly environment friendly. This composite material has high electrical conductivity than normal material when taken the best composite ratio which is 65:35 (fly ash /potash alum) having conductivity 1.5×10^{-5}. After this proportion the conductivity decreases. It has application as a humidity sensor which gives good result up till 50^0C of temperature as after that the number of H^+ ions reduced. Various test has been performed over the sample which yields to this result such as Infrared Spectroscopy (IR) which confirms about the proportion that could yield best result in terms of electronic conductivity, XRD (X-ray diffraction) for conforming the composite nature of electrolyte and Scanning electron microscope(SEM) confirms the overall blending of the potash alum and fly ash. After results using aluminium paste and copper wire the humidity sensor has been fabricated and keen observation is done under air tight chamber over the change in humidity level through sensor. Good results have been taken.

Aronson discuss the characterisation of lead titanate and calcium oxide composite is done for its nano structural behaviour and based on that it's intelligent modelling is done using Artificial Neural Network and Adaptive neuro fuzzy inference system(Aronson, Liang, & Turban, 2005). The sol-gel spin coating technique used to prepare composite where calcium oxide acts as dopant. Lead titanate is one of the multiple oxides of titanate oxide. It's a cost-effective method also yield better homogeneity for the sample. The composite has ability to detect variation in humidity sensor. After fabrication characterisation of the composite is done using XRD (X-ray diffraction), its output is the crystalline size and phase of the fabricated nano sensor. For intelligent modelling, the input is dopant ratio, heat rate and temperature and based on that tetragonality of the humidity sensor is the output. While implementation of models it was observed that both artificial Intelligent techniques provide good result but artificial neural networks gave comparatively better result and these can also be used to detect faults as well as for fabrication efficiency.

Aggarwal discuss the two different reliability prediction model that has been used for determining MTTF (Mean Time to Failure) for instrumentation amplifier and BJT Transistor and their comparison has been done(Agarwal, Paul, Zhang, & Mitra, 2007). Also, the analysis has been used to determine the estimate cost of reliability prediction using both methods. For traditional approach, RIAC 217+ has been used to determine reliability using numerical method and here temperature is the considerable parameter. All the constant values have been taken through this Handbook only. Same method has been done for BJT Transistor also. Another method is Physics of Failure (POF) method where analysis of stressor parameter is needed to determine exact failure result. Testing is done at physical level (accelerated stress testing) to record the minimum or maximum value which could lead to failure. In comparison, the better result is achieved using physics of failure method for both the components. For cost analysis, the parameter depends on the cost of reliability handbook for traditional method and the circuit and setup cost for physics of failure method(Gu & Pecht, 2008). It's a variable parameter depending on the characteristics of the system.

Huang discuss the reliability of printed circuit board that has been examine under Environmental condition(Huang et al., 2017). If we will observe than working conditions for practical ground is very different from what has been specified theoretical. Proper analysis of printed circuit board is strictly needed in early ages as if it's been fabricated nothing can be undoing. Environmental temperature is one of the factors which highly deviates the operation of any practical system. Reliability analysis done here is through Weibull analysis and MIL-HDBK 217 F for predicting exact failure model for future uses also. Final prediction is done using Monte Carlo analysis which gives error of 22%, for final reliability model of printed circuit board.

BK Klass discuss that one of the soft computing techniques termed as "Genetic Programming" has been taken to draw intelligent model for monitoring online failure system for electronics system (Klass BK, 2006). Stressor is the name given to the parameter whose variability tends to change the reliability or behaviour of the system or may even failure. Susceptibility is the property of the system that how susceptible it is for stressor parameter. In this paper, stressor- susceptibility relation has been used to draw the data set for empirical model. Also, experimental value has been compared with genetic programming predicted values.

Youmin Rong discuss the Laser Brazing i.e., LB has an advantage with very high speed, less heat rate and high quality. The aim of this paper is to optimize seam shape in this process using Back propagation Network (BPNN) and Genetic Algorithm (GA). After performing experiment through Taguchi method, the included parameter for input is welding speed, gap and welding speed rate(Antony et al., 2006). The experimental data set is fed to BPNN for predicting the output and that output is fed to genetic programming so that through optimization the error percentage may get reduce. The final output of this BPNN-GA technique is very effective.

Prabhakar V. Varde discuss the life extension methods using early failure and early maintenance warning. This early warning method also tend to increase the life cycle of the products as if before failure warning has been generated than necessary measures could be taken to extend the life like repair or replacement of any component(Varde, 2010). Approaches used here involve sensing the performance, operational environment, remaining useful life (RUL), health diagnosis has been used. This kind of technique is very useful for manufacturing industries as it develops an intelligent as well as reliable system for maintenance.

Yang Zhao discuss the quantitative analysis for reliability prediction and assessment of electronic System. Here, analysis on reliability is done to implement better fault testing, health management and maintenance system(Zhao, Chen, & Xu, 2009). During analysis knowledge of hardware as well as software system is needed. Quantitative analysis is mainly a prognostic health analysis for reliability assessment of electronic systems.

Malgorzata Kutylowska discuss two artificial intelligent modelling; artificial neural network using radial basis function and artificial neural network using multilayer perceptron. Data include connectivity of home, water mains and distribution pipe. Record of year 1990-2013 has been taken for training and of 2014 for testing the system(Yang, Ge, & Xu, 2008). There were seven input signals to the system and 3 output neurons. Amongst the two the best performance has been shown by artificial neural network using Multi-Layer perceptron.

Tomohiro Takagi discuss a mathematical model that has been implement to model of a fuzzy system. Here all the fuzzy reasoning and fuzzy implications has been used. Also, fuzzy rules have been developed along with membership function. Using this mathematical model, two applications have also been implemented named one in steel making process and other in water cleaning process(Srinivasan & Weidner, 1999).

SYNTHESIS OF ZINC-OXIDE AND CARBON COMPOSITE

Nanotechnology has got great advancement in last decades. Numerous new materials have been formed through nanotubes, nanowires, and nanoparticles etc., which possess high efficiency then their parent material(Shanthi et al., 2018). These nanoparticles materials are ruling over the traditional matter we are using. Now, why we are switching towards nanoparticles? Traditional materials we are having one or other drawback, which posed limitation to their use for example, non-environment friendly, complex fabrication, non-abundant in nature, low covering range etc. Hence, we are looking for new materials which are effective in one or the other way. Moreover Nano-particles are considered as one-dimensional particle with the measurement of 100 nm. When we grind them, their properties get changed because now they are having more surface area per weight, which tends them to react more with other molecules and hence with rough surface they possess great capability as a sensing material. As an individual also, Zinc-oxide has capability to shield woods, textiles, materials from UV ray's exposure.

For this work Zinc-oxide has been prepared by using Sol-Gel method and carbon black is a pollutant which has been collected through fumes in a closed chamber. Zinc-oxide and Carbon both are insulator but when combine them in a definite ratio, it can have conductivity. In another word, we can say by combining two insulators we are getting a material with semiconductor property which can be used to sense the Humidity in the environment. The ZnO-Carbon composite we have got has the conductivity property which has been proved by series of test we did over the best sample we have got to confirm its conductivity, morphology, Humidity sensing capability etc.

ZINC OXIDE

Zinc oxide commonly known as ZnO comes under inorganic compound having molar mass (81.408g/mol), density (5.606g/cm³) and odourless in nature. It's a visible white colour powder which is almost insoluble in H_2O (Water) but can be soluble in alkalis or substance which are acidic in nature. Natural ZnO is present in earth's crust in abundance in the form of mineral zincate. Synthetically we can prepare ZnO with Sol-Gel process. In medical application it is utilised in skin ointments which are used to cure skin allergies. As a compound it is insulating in nature. Its colour remains same even when it is exposed to ultraviolet rays. Considered as most important content in rubber industry. They also have commendable anti-corrosion, anti-bacterial ultraviolet filtering and catalytic properties.

The sol-gel method which is used to form ZnO is carried out using zinc acetate dehydrate and Sodium hydroxide and ethanol is used as solvent for the process along with distilled water to act as a medium. Over all the zinc content is 55.38% and oxygen content remain 44.62%. ZnO used is shown in the Figure 1.

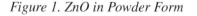
Figure 1. ZnO in Powder Form

CARBON BLACK

Carbon black is a black colour material generate after combustion of any fuel, petroleum, oil etc. Mostly it is considered as environment pollutant. Its para crystalline form of carbon having high surface area to the volume ratio. Mainly used in tyres and rubber products also as an ink due to its colour property. It has oil absorption property. For extracting any functional group out of it we need to trigger it by keeping it at high temperature. Normally its conductivity is very low or we can consider it as an insulator. It has the tendency to absorb ultraviolet rays. Most of its application is in automobile industry.

COMPLEX IMPEDANCE SPECTROSCOPY (CIS)

CIS stands for Complex Impedance Spectroscopy is used for having characterization of any sample, material, dielectric, ferroelectric ceramics, composites etc. Traditionally frequency-based characterization was used to know dielectric characteristic for the microstructure of the material but it wasn't giving enough information about the conductivity hence then these frequencies relying electrical-dielectric characteristics were determined using complex impedance spectroscopy. This approach was first proposed by Cole-Cole, who plots the real and imaginary data of complex permittivity (Є) and this plot is known as Cole-Cole plot.

In this work we are plotting real and imagine axis with the electric Impedance to determine the electrical conductivity over complex plane plots. CIS is a beneficial technique for having characterization of electrochemical and electrical traits of the sample with respect to the microstructure of the composite materials. Apart from impedance and permittivity properties like dielectric relaxation, charge transport, diffusion of charge can also be determined using CIS. It gives good analyses of grain and grain boundary along with electrode interface properties.

Figure 2. Carbon Black in Powder Form

Chamber or environment where complex impedance spectroscopy is performed, there need is wide temperature (500^0 – room temperature) and frequency in the range of 1 to 1000 KHz, that too in air atmosphere.

X-RAY DIFFRACTION (XRD)

In 1913 Sir W. H. Bragg along with his son Sir W.L. Bragg, works for a statement that the crystal lattice of any compounds tends to diffract the X-ray beam with an angle of incidence theta (θ). Considering d as the gap between the two layers of the atoms in the crystal lattice, λ as the considered wavelength of the beam we incident and n, as an integer gives an expression,

$$N \lambda = 2 d \sin \theta \qquad (1)$$

This is Röntgenstrahlinterferenzen expression, commonly known as X-ray diffraction (XRD) and also Bragg's law. Although diffraction is made through X-ray but the crystal lattice properties or the information about the structure of the matter can also be demonstrated using any other beam like, electrons, protons, ions etc. having wavelength directly proportional to the gap between atoms in the structure. XRD is mainly done to calculate the optimum gaps between the layer and rows of atoms in the structure. If we talk about single grain, where the orientation is uniform in one domain, we can determine its orientation. XRD over all tell the structure, size, shape and stress in any lattice structure and also helps in confirming the presence of any unknown material by drawing its internal characteristics.

Figure 3. X-ray Diffraction

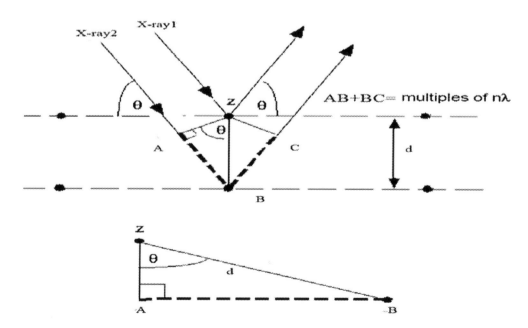

Mainly the incident beam of X-rays tends to reflect through the atomic places of the material's crystal, which makes the beam to diffract and interfere with one another as they penetrate the crystal. This diffraction takes place only when the condition for Bragg's law is satisfied. In all the cases θ must be directly proportional to 'd' that means if d is changing θ must change. Diffraction through different planes of the atoms in the crystal lattice generate different diffraction pattern, and these patterns carries the information of the atomic arrangement of any material with in the crystal.

FOURIER TRANSFORM INFRARED SPECTROSCOPY (FTIR)

In the Electromagnetic Spectrum of light, Infrared radiation exists between the visible and microwave region, which means its wavelength (λ) falls longer the visible region and shorter of microwave and follows opposite in case of frequency. Infrared radiation is directly proportional to thermal radiation. More the thermal heat is produced by the object due to the motion of atoms or molecules in the material more is the Infrared radiation from that particular body. Through the principle theory of Infrared Radiation, if applied Infrared radiation frequency is equal to the normal or natural or normal frequency of vibration of that particular material, the molecules vibrates due to the absorption of IR. Each and every bond and the different functional group present in that particular molecule need different frequencies to absorb hence, we can observe the characteristics peak for different functional group present in any molecule.

For having the Infrared Spectroscopy for any material, we go for fourier transform infrared spectroscopy. The material in the form of pallet is pass through infrared, where a portion of it is absorb and remaining is transmitted through it. The spectrum or characteristic peak we get after that depict the different functional group present in that particular material. It's like a fingerprint claiming that no two composites can have similar fingerprints hence it helps in determining the property or functional group. Mainly it provides the presence of different materials in the composite, presence of an unknown material if it is there, quality or any unwanted mixing if there etc.

FTIR also known as interferometry infrared spectroscopy due to the Michelson Interferometer it used. Previously dispersive spectroscopy was used for this purpose but it has got various disadvantage related to sensitivity, wavelength accuracy and speed also there was Infrared loss while it tends to pass through a sample. FTIR is quite fast as compared to this technique along with better sensitivity and easy mechanical Implementation as advantage. FTIR operational instrument is a highly calibrated box with no user Interference with the user except the initial basic starting step. FTIR set up used is shown in Figure 4.

SCANNING ELECTRON MICROSCOPE (SEM)

SEM i.e., Scanning Electron Microscope, development started few years later after the TEM came into picture in 1931 through RUSKA, but overall scenario becomes clear after 30 years when it started getting commercialised. It is a technique that exploit very high energy electron drawn in a focused beam to produce a variety of signals over the surface of solid structure. It is mainly used to obtain physical information about the sample like surface morphology, crystalline structure, orientation of the sample, chemical composition etc. Data is focused over small area of sample surface and for that 2-dimensional image is obtained that shows the surface roughness or spatial variations. The range of area scanned could

Figure 4. FTIR Set-up

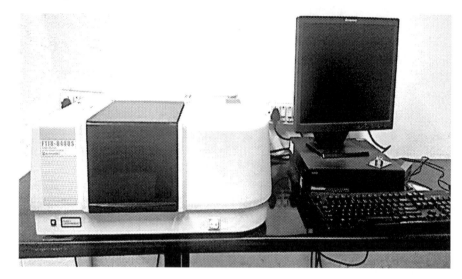

be between 1 centimetre to 5 microns approximately in width using this technique. Also, we can target the specific area of the sample for analyses in case of requirement.

The highly accelerated electrons tend to produce SEM images, also we have secondary electrons which tends to show the morphology and orientation of the sample. We also have X-ray involvement in SEM but that's "Non-Destructive" in nature, means it won't make any damage or loss in the sample hence making the same material analyse repeated manner. SEM machine has one detector for secondary electron.

TRANSFERENCE NUMBER

Transference number is a dimensionless entity determines to know the contribution of ionic species in the sample to the overall concentration. It helps in knowing the ionic conductivity of the sample also; we can determine the sensitivity of the sample to ensure its quality. It is also known as transport number defined as proportion of current carried through ions in a sample. This number is shows high fluctuation if we vary the temperature. Temperature could make cat ion and anion number to reach 0.5. Another factor is mobility of ion. Method used to calculate this is Hittorf's method and another one is emf method.

HUMIDITY VS. VOLTAGE GRAPH

For knowing the Humidity sensing range of the raw sensor fabricated through the ZnO-Carbon nanoparticles composite experiment is performed. In a closed, air tight chamber the sensor is placed and the humidity is varied using moisture, calcium carbonate etc. and supply is provided to the chamber and a potentiometer is connected outside to observe that whether change in humidity is varying the voltage or not. Likewise, temperature is also provided to see the behaviour of the sensor that how much it can withstand and what is the upper and lower limit of the temperature it can survive. Based on that a data set has been generated which will confirm the range of humidity sensor which can be fabricated using the ZnO-Carbon composite nanoparticles we synthesised and characterised. On the basis of that comparison will be made with the already fabricated sensor.

NEED OF INTELLIGENT MODELLING

In today's world we need everything smart. Be it our vehicles, gadgets, home or surroundings. We want every system to be autonomous so that we could remove human intervention from their operation. This comfort leads to the basis of intelligent modelling, where we tend to developed intelligent model for every system. The basic terminology we required to understand for this id artificial intelligence. "Artificial Intelligence is the method or way where we try to provide intelligence or human like thinking capabilities to the machine or system through learning or any other method."

The approach to implement artificial intelligence is through soft computing where we have various algorithms that tend to make the system adaptive or intelligent or developed "Intelligent Model". The basic implementation approach one can take is Artificial Neural Network (ANN), Fuzzy Inference System (FIS), Adaptive Neuro-Fuzzy Inference System (ANFIS), Genetic Programming (GPs), Support Vector Machine (SVM) etc. In this work we have taken first three approaches to implement the Intelligent Model for our designed sensor.

MODELLING OF INTELLIGENT SYSTEM USING ARTIFICIAL NEURAL NETWORK (ANN)

As the world is getting modernized, we are moving more towards intelligent system which could work same way as human use to do. We are searching for method which could remove or reject human intervention by developing intelligent system. Here, our first method is Artificial Neural Network (ANN). Artificial Neural Network is an analogous system of human neural network which tries to mimic the functioning of actual brain. Input data along with target data has been fed to the network. Activation function has been provided to start the process where system learns by itself how output is coming. The system gets train with the number of epochs we specified. The system will train itself and reduce the error after every epoch. And hence after specific number of epochs we get the best result. The ANN Figure 5 is shown.

MODELLING THROUGH FUZZY INFERENCE SYSTEM

Fuzzy Inference System or Fuzzy Logic is used to handle ambiguity and uncertainty in data. As the complexity increases, we can't make exact statement about the behaviour of the system as in the traditional method we were using binary logic which says 0 or 1, i.e., yes or no, but the real-world problems are beyond as it can't be true or false only. Taking water problem than the possible answer could be hot, cold, slightly cold, slightly hot, extremely cold, extremely hot etc.

For this purpose, we deal with linguistic variables in fuzzy which are user understandable. Entire input set is known as Crisp Set which after fuzzification converts into fuzzy sets. Here we use the concept of membership function, it defines the membership of particular input value in the fuzzy sets, and

Figure 5. Artificial Neural Network

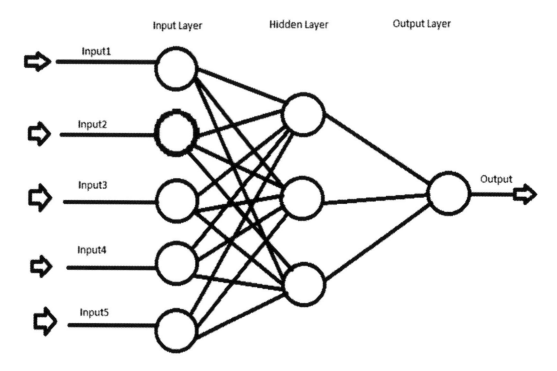

its range is from 0 to 1. If input value has complete membership, it is 1 otherwise it can be any value in this range. If we have fuzzy set A than considering the Universal set, X can be defined as,

A= {(x, μ_A(x))|x$\in$ X},

where μ_A is known as A's membership function. In FIS, we defined certain rules for fuzzification to defines crisp relation into Fuzzy relation in IF, THEN, ELSE format e.g.

IF (f is x_1, x_2.....a_n) THEN (g is y_1, y_2....y_n) ELSE (g is z_1, z_2....z_n)

This fuzzified data goes to decision-making unit which decides about the membership function and hence attached the related Linguistic Variable for that particular value. The fuzzy output from this block directly goes to defuzzifier Interface unit, which is reverse of Fuzzifier. And hence after this block we get proper output in Crisp set form as defuzzifier convert fuzzy set back to crisp set(Macin, Tormos, Sala, & Ramirez, 2006). Fuzzification as well as defuzzification unit are assisted by knowledge base which has design base as well as rule base for making rules and modifying data(Manogaran, Varatharajan, & Priyan, 2018). The block diagram of FIS is given in Figure 6.

Figure 6. Block Diagram of Fuzzy Inference System

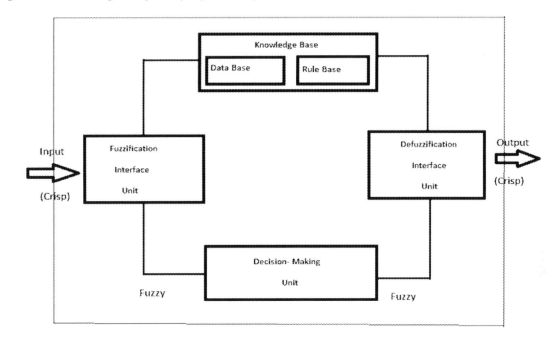

MODELLING THROUGH ADAPTIVE NEURO-FUZZY INFERENCE SYSTEM (ANFIS)

ANFIS is a hybrid technique comprises both ANN as well as Fuzzy Tool. It has advantage of both the technique as ANN has this self-learning mechanism but it doesn't know how the hidden process is following to reach the particular target and the disadvantage is that the output is not that user understandable also, we need very precise and accurate. It can't handle ambiguity(Cherry Bhargava, Vijay kumar Banga, & Singh, 2017). On the other hand, the advantage with fuzzy logic is that it can handle uncertain data and also, we use linguistic variable to have better understanding but no self-learning is there(Virk, Muhammad, & Martinez-Enriquez, 2008). Hence to omit each other advantages, these two techniques have been combined to formed third technique that is ANFIS (Adaptive neuro-fuzzy Inference system). Here the rules needed by fuzzy get self-updated through the self-learning mechanism possessed by ANN(Bradley & Gupta, 2003). That's why a smaller number of errors is shown by the predicted Data of ANFIS(Xu & Wang, 2012). The basic structure of ANFIS is shown in the figure 7.

FLOW CHART

A flow chart of completed work can be seen in figure 8.

Figure 7. Adaptive Neuro-Fuzzy Inference System

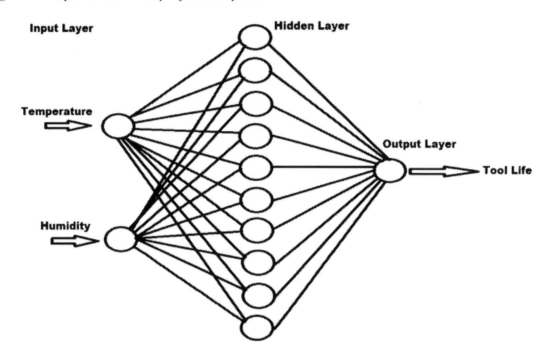

SYNTHESIS AND CHARACTERISATION OF ZNO-CARBON COMPOSITE NANO-PARTICLES

For the characterisation of the ZnO-Carbon composite nanoparticle, series of tests has been performed over 11 samples (9 composites, 1 ZnO and 1 Carbon) to know about its conductivity, Impedance, Morphology, Orientation, particle length, conductivity, range etc. The obtained results are shown in the forthcoming sections below,

COMPLEX IMPEDANCE SPECTROSCOPY

CIS result is obtained in the form of data which has been further modified and converted into graphical forms to get the overall conductivity of the 11 samples which undergone this test. Origin 7.1 software is used for this purpose. Acquired graph is shown below in the increasing order of ratio of ZnO and Carbon. Here X-axis represents real impedance and Y-axis represent imaginary impedance.

- **ZnO-Carbon (20/80):**

The sample was prepared for ZnO-Carbon Composite ratio (10/90), but while making pallet the sample was very unstable. Even after attempting number of times pallet was not formed. The conductivity measure for all these samples are mentioned in the Table 1,

Out of the result obtained for the conductivity, it is obvious that the ZnO-Carbon Composite ratio (20/80) has maximum conductivity. Hence rest all the test are performed over this sample.

Figure 8. Flow Chart of Work Done

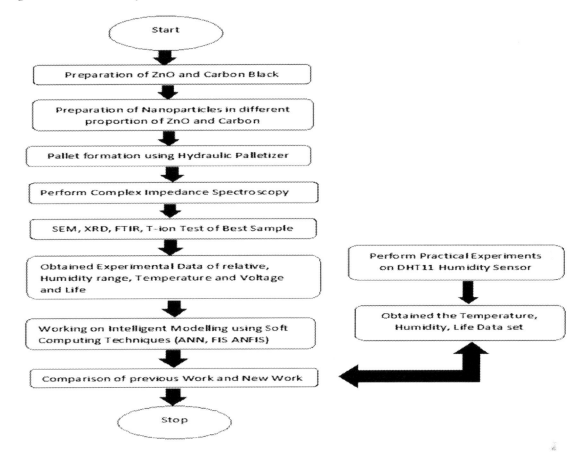

Figure 9. Cole-Cole Plot for (20/80) Ratio of ZnO-Carbon Composite

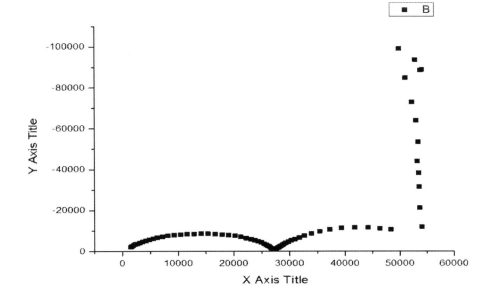

Table 1. Conductivity values of Samples

Sr. No.	ZnO-Carbon Ratio	Conductivity value ($\mho^{-1}$)
1.	20-80	0.0064
2.	30-70	0.000125
3.	40-60	0.000241
4.	50-50	0.000111
5.	60-40	0.00025
6.	70-30	0.000388
7.	80-20	0.0007
8.	90-10	0.000128
9.	Pure Carbon	0.000109
10.	Pure ZnO	0.000005833

FOURIER TRANSFORM INFRARED SPECTROSCOPY

For FTIR test, the maximum conductivity sample i.e., ZnO-Carbon Composite (20/80) is used to form a pallet whose weight was 10mg. To this weight 100mg of Potassium Bromide is being and Pallet is formed using Hydraulic Palletiser with a pressure of 2-3 ton.

Through the observation on the IR graphs, It can be followed that,

- All the spikes and peaks in ZnO and Carbon is present in ZnO-Carbon Composite.
- Curve for the composite is covering all the slopes in the waveform which is there for is respective parent material.
- All the peaks of the composite can be deduced either for ZnO or Carbon.
- ZnO waveform is showing two extra peak that's because of the presence of little bit impurities in it as sample can get contaminate even while performing the test.

As per the observation the peaks for the graph of ZnO and Carbon is showing certain values related to their properties. The point at which we are getting these peaks are the point where we can incur the functional group for ZnO and Carbon which will confirm their existence.

For ZnO graph we received first peak at 3449.81^{-1} which confirms that stretching mode found in hydroxyl group. Another peak is observed at 2045.96^{-1}, which depicts the (C-H) stretching vibration present in the alkaline group. Third peak observed at 1626.91^{-1} shows the (COO⁻), Carboxylate group. For next two peaks i.e, (1513.88^{-1} and $1423,92^{-1}$) we can say it is C=O of acetate group. For 909.9 to covering 725.95^{-1}, it is the organic coating which is encapsulating the nanoparticles. For 537.47^{-1}, oxygen deficiency or oxygen defect is observed in ZnO and for 419.91^{-1} we can say E^2 mode of the hexagonal, ZnO-Validation or we can say pure ZnO is observed.

For carbon we have observed peak at 6 places and corresponding wave pattern is also observed in composite material's waveform. First peak is at 3733.57^{-1}, depicting phenolic hydroxyl group with O-H stretch vibration. The same group is decreasing when we move further covering 2988.56^{-1} and 1925.55^{-1} where the effect of this group is slightest. Next peak is at 1529.55^{-1} holding C=O stretching frequency of the COOH group and for 1213.55 also same group followed but with C-O stretching of COOH group(Nazari, Kashanian, Moradipour, & Maleki, 2018).

We can also say that apart from the peaks of the parent material ZnO and Carbon, the composite doesn't have any other peak which confirms that it is purely a composite of ZnO and Carbon and exhibit the properties of ZnO and Carbon both. The FTIR image for ZnO, Carbon and composite is shown in the Figures 10, 11 and 12.

X-RAY DIFFRACTION

X-Ray Diffraction test is performed to confirm the crystalline nature of the composite material of ZnO-Carbon From the result it is revealed that Carbon is showing no spikes in it's resultant graph which confirms it's amorphous nature. XRD of Carbon is shown in the Figure 13

Spikes of the Zinc-Oxide are very sharp as compared to Carbon and all the peaks which are available in Carbon as well as ZnO is present in the graph of Zinc-Oxide and Carbon composite which confirms that there are no impurities present in the composite and its composition is purely Carbon and Zinc-Oxide. Also, it confirms the homogenous nature of the composite. XRD graph of Zinc-Oxide confirms the particles size of the nanoparticles in the material for that Scherer Formula is used,

Nano-Particle Size= $(0.9 \times \lambda)/ (d \cos\theta)$,

where $\lambda= 1.540$ A
D= half of the width of the maximum intensity peak or spike.

Figure 10. FTIR waveform of Carbon

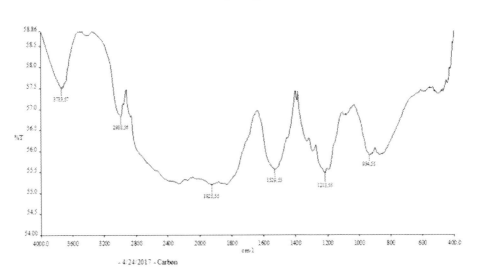

Figure 11. FTIR Waveform of ZnO-Carbon Composite

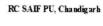

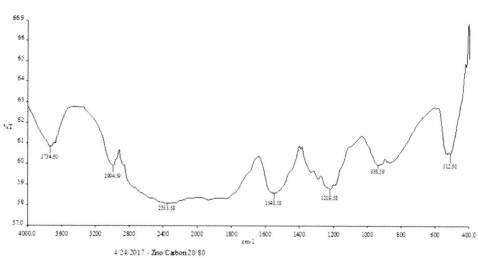

4/24/2017 - Zno/Carbon 20/80

Figure 12. FTIR Waveform of ZnO

RC SAIF PU, Chandigarh

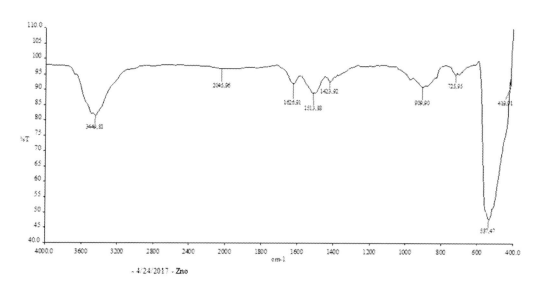

- 4/24/2017 - Zno

Figure 13. XRD of Carbon

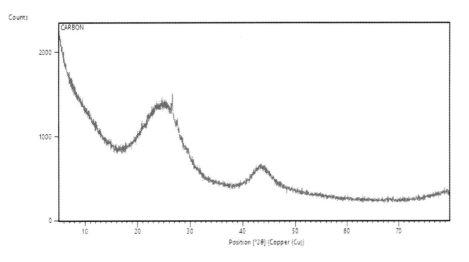

Using the spikes available in the graph XRD result confirms that the Nano-Particle size of Carbon is nearly 12.6μm, which is nearly equal to the result acquired through SEM and Zinc-oxide it is 3.2 μm, almost equal of what SEM result has given.

SCANNING ELECTRON MICROSCOPE

This test is done mainly to acquire surface images of the material. Hence determining the morphology, orientation, surface roughness of the composite and parent material. Also, it confirms the presence of both the material at micro-level in the composite. For SEM micrograph for Carbon, the minimum size of the particle is 1 μm and the maximum size is 25 μm. Hence average size of the Carbon nano-particle is,

Average Size = (Minimum Size + Maximum Size)/2,

For Carbon it is 13μm. SEM Micrograph for ZnO, the minimum size is 0.5 μm and the maximum size is 6μm. Hence average size of the ZnO Nano-particle is 3.25μm. The SEM micrograph for ZnO, composite and Carbon is shown below,

Observation of SEM Micrographs

- Long rigid flakes of carbon are observed or visible from the SEM micrograph.
- Carbon appears little bit opaque as compared to the image of the zinc-oxide nano-particles.
- From the composite Micrograph it is clearly visible that long flakes of carbon have completely mixed homogenously with ZnO nano-particles.

Figure 14. XRD of Zinc-Oxide

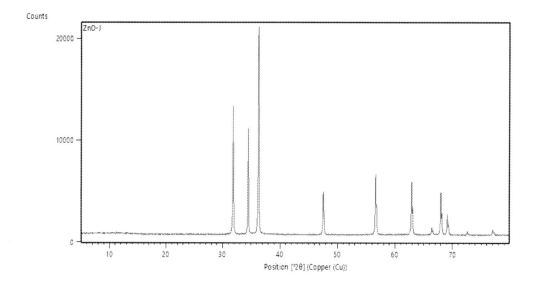

Figure 15. XRD of Zinc-Carbon Composite

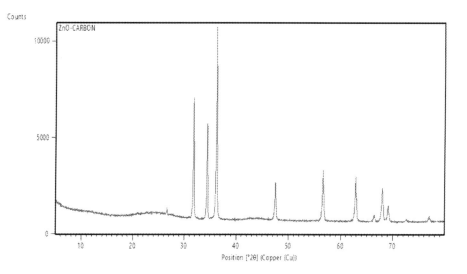

The Carbon has not been crushed properly which makes variable dimension flakes of black Carbon visible in the SEM micrograph

- Surface morphology appears to be rough rather than smooth which is perfect condition for absorption of water molecules.

Figure 16. Combined Image of Carbon, Composite and ZnO Micrograph

TRANSFERENCE NUMBER TEST

Transference number test is performed to know the ionic conductivity of the sample. Through the minimum and maximum, we find out the average ionic conductivity of the sample. More the conductivity, better the sample is. The result is shown in the Figure 17.

Current and conductivity shares directly proportional relationship. Also, we know that conductivity is inversely proportional with the resistivity. Through the Figure 16, we can say that with respect to time the value of conductivity was high in initial hours and in gradually decrease to lowest in at the end of 2nd hour. Which means the value of resistivity was low in initial stage but it gradually increases with time. It confirms that it is happening due to the rise in humidity. In initial stage the humidity was low hence the resistivity was low and ions were moving freely with high conductivity but as the humidity rise, the overall conductivity decreases to support increase in the resistance. To determine the conductivity we have relation,

Figure 17. T-ion Test result for Ionic Conductivity

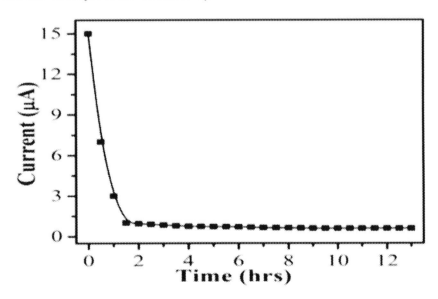

Ionic Conductivity = (Upper Limit-Lower Limit)/Upper Limit.

Hence here the ionic conductivity of the composite in terms of current or in proportion to current is can say, is 0.9333 µA.

Humidity vs. Voltage

Practical experiment is performed over the sensor made of best ratio i.e., (20/80) composite nano-particle made out of Zinc-Oxide and Carbon. The purpose of experiment is to know the entire range the sensor sense for humidity plus its relationship correspond to the Voltage change(Fei, Jiang, Liu, & Zhang, 2014). The overall range for this humidity sensor if from 30% RH to 95%RH, after which it will stop sensing. This range is extract and interpreted in terms of voltage which is beneficial for further requirement where we need to find its life. Figure 17 shows the result obtained from the experiment in the form of graph(Jing & Zhan, 2008).

In the graph X-axis depicts humidity value and Y-axis shows voltage value. From this graph it is clear that with the increase in humidity voltage also increasing but at one point it is becoming constant with respect to humidity(Jose et al., 2017). At this point the slope will become almost zero. Through this graph we can find the sensitivity of this sensor we fabricate. Table 2 will give the sensitivity value at different point and hence we can easily calculate overall sensitivity from this point.

From the table we can make following conclusion,

- In starting range, the sensitivity of the sensor is better with respect to change in the environmental humidity.
- From the range of (70-95) %RH, the slope is becoming stagnant or constant almost. This depicts the observation that from range (30- 70), sensitivity is decreasing at very high rate.

Table 2. Sensitivity Table

Sr. No.	Humidity Variation (%RH)	Slope Value
1.	30-35	0.48
2.	35-40	0.52
3.	40-45	0.44
4.	45-50	0.64
5.	50-55	0.42
6.	55-60	0.28
7.	60-65	0.34
8.	65-70	0.28
9.	70-75	0.04
10.	75-80	0.04
11.	80-85	0.02
12.	85-90	0.02
13.	90-95	0.02

- Beyond this limit sensor would not sense any change in the humidity. Best result is between 45-50%RH for sensitivity.
- From initial point to the top we can observe from the graph that it's showing full range of humidity sensing capability of the sensor which is 30 to 95%RH(Jose et al., 2017).
- The overall sensitivity of the sensor is 99.7% which claims to better efficiency.

The image of the fabricated sensor is shown in Figure 18.

INTELLIGENT MODELLING

To develop Intelligent Model for the fabricated Humidity Sensor we are considering three techniques of soft computing approach such as Artificial Neural Network (ANN), Fuzzy Inference System (FIS) and Adaptive Neuro-Fuzzy Inference System (ANFIS)(Samanta & Al-Balushi, 2003). Implementation is done using MATLAB simulink 2014a Software(Simulink & Natick, 1993).

ARTIFICIAL NEURAL NETWORK

The topology of the network used here is 3-10-1 i.e., 3 input, 1 output and 10 neurons in hidden layer. These three inputs indicate humidity, voltage and temperature and output is life estimation(Mazhar, Kara, & Kaebernick, 2007). Neural Network formed is shown in the Figure 19.

We have 14 samples for training and 10 has been used for testing. The number of epochs taken are 1000 and best validation performance 0.35 observed at iteration 6. After training with the feed data, the predicted value given by neural network is shown in the Table 3.

Figure 18. Image for Fabricated Sensor

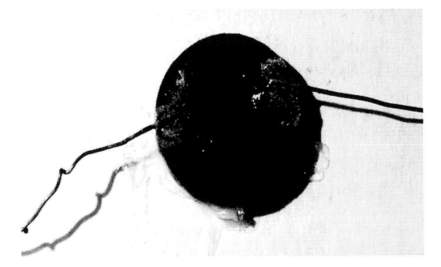

Figure 19. ANN formed

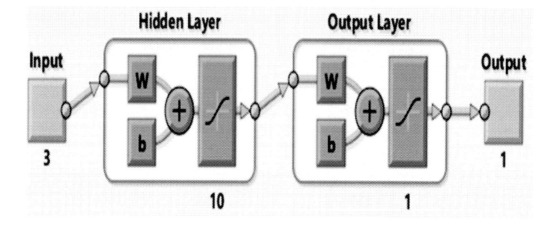

Table 3. Artificial Neural Network Life Estimation Predicted Data

Sr. No.	Humidity (%RH)	Voltage (V)	Temperature (Celsius)	Analytical Method Life	Predicted Life (ANN)
1.	30	2.4	0	225.3	212.32
2.	35	4.6	5	132.1	154.45
3.	40	7.2	10	79.01	81.23
4.	45	9.4	15	48.07	39.23
5.	50	12.6	20	29.74	19.56
6.	55	14.7	25	18.7	18.5
7.	60	16.1	30	11.94	13.99
8.	65	17.8	35	7.74	7.74
9.	70	19.2	40	5.08	4.98
10.	75	19.4	45	3.38	3.36
11.	80	19.6	50	2.28	2.24
12.	85	19.7	55	1.55	1.33
13.	90	19.6	60	1.07	0.9
14.	95	19.7	65	0.75	0.75

CONCLUSION

This chapter signifies the fabrication and characterisation of humidity sensor using nanocomposite materials. An intelligent model is designed which interacts with user and update the live health condition of humidity sensor. The prediction or Intelligent Model formed by Artificial Neural Network has accuracy of 90.67 with Root Mean Square error of 9.33%. The user can replace the faulty component well before the actual destruction of the system.

REFERENCES

Agarwal, M., Paul, B. C., Zhang, M., & Mitra, S. (2007). *Circuit failure prediction and its application to transistor aging.* Paper presented at the 25th IEEE VLSI Test Symposium, Berkeley, CA. 10.1109/VTS.2007.22

Al-Zubaidi, S., Ghani, J. A., & Haron, C. H. C. (2011). Application of ANN in milling process: A review. *Modelling and Simulation in Engineering, 2011,* 9. doi:10.1155/2011/696275

Al-Zubaidi, S., Ghani, J. A., & Haron, C. H. C. (2013). Prediction of tool life in end milling of Ti-6Al-4V alloy using artificial neural network and multiple regression models. *Sains Malaysiana, 42*(12), 1735–1741.

Antony, J., Bardhan Anand, R., Kumar, M., & Tiwari, M. (2006). Multiple response optimization using Taguchi methodology and neuro-fuzzy based model. *Journal of Manufacturing Technology Management, 17*(7), 908–925. doi:10.1108/17410380610688232

Aronson, J. E., Liang, T.-P., & Turban, E. (2005). *Decision support systems and intelligent systems.* Pearson Prentice-Hall.

Bhargava, Banga, & Singh. (2014). *Failure prediction and health prognostics of electronic components: A review.* Paper presented at the IEEE Conference on Recent Advances in Engineering and Computational Sciences (RAECS), Chandigarh, India.

Bhargava, Banga, & Singh. (2017). Fabrication and Failure Prediction of Carbon-alum solid composite electrolyte based humidity sensor using ANN. *Science and Engineering of Composite Materials.* doi:10.1515ecm-2016-0272

Bradley, D. M., & Gupta, R. C. (2003). Limiting behaviour of the mean residual life. *Annals of the Institute of Statistical Mathematics, 55*(1), 217–226. doi:10.1007/BF02530495

Fei, T., Jiang, K., Liu, S., & Zhang, T. (2014). Humidity sensors based on Li-loaded nanoporous polymers. *Sensors and Actuators. B, Chemical, 190,* 523–528. doi:10.1016/j.snb.2013.09.013

Gu, J., & Pecht, M. (2008). *Prognostics and health management using physics-of-failure.* Paper presented at the IEEE Annual Symposium on Reliability and Maintainability Symposium (RAMS 2008), Las Vegas, NV 10.1109/RAMS.2008.4925843

Handbook, M. S. (1995). MIL-HDBK-217F. Department of Defense, US.

Huang, X., Denprasert, P. M., Zhou, L., Vest, A. N., Kohan, S., & Loeb, G. E. (2017). Accelerated life-test methods and results for implantable electronic devices with adhesive encapsulation. *Biomedical Microdevices, 19*(3), 46. doi:10.100710544-017-0189-9 PMID:28536859

Jing, Z., & Zhan, J. (2008). Fabrication and gas-sensing properties of porous ZnO nanoplates. *Advanced Materials, 20*(23), 4547–4551. doi:10.1002/adma.200800243

Jose, S., Voogt, F., van der Schaar, C., Nath, S., Nenadović, N., Vanhelmont, F., . . . Šakić, A. (2017). *Reliability tests for modelling of relative humidity sensor drifts.* Paper presented at the IEEE International Symposium on Reliability Physics (IRPS'17), Monterey, CA

Klass, B. K. J. C., & Van, P. (2006). System Reliability: Concepts and Applications. Edward Arnold.

Macin, V., Tormos, B., Sala, A., & Ramirez, J. (2006). Fuzzy logic-based expert system for diesel engine oil analysis diagnosis. *Insight-Non-Destructive Testing and Condition Monitoring*, *48*(8), 462–469. doi:10.1784/insi.2006.48.8.462

Manogaran, G., Varatharajan, R., & Priyan, M. (2018). Hybrid recommendation system for heart disease diagnosis based on multiple kernel learning with adaptive neuro-fuzzy inference system. *Multimedia Tools and Applications*, *77*(4), 4379–4399. doi:10.100711042-017-5515-y

Mazhar, M., Kara, S., & Kaebernick, H. (2007). Remaining life estimation of used components in consumer products: Life cycle data analysis by Weibull and artificial neural networks. *Journal of Operations Management*, *25*(6), 1184–1193. doi:10.1016/j.jom.2007.01.021

Nazari, M., Kashanian, S., Moradipour, P., & Maleki, N. (2018). A novel fabrication of sensor using ZnO-Al2O3 ceramic nanofibers to simultaneously detect catechol and hydroquinone. *Journal of Electroanalytical Chemistry*, *812*, 122–131. doi:10.1016/j.jelechem.2018.01.058

Samanta, B., & Al-Balushi, K. (2003). Artificial neural network based fault diagnostics of rolling element bearings using time-domain features. *Mechanical Systems and Signal Processing*, *17*(2), 317–328. doi:10.1006/mssp.2001.1462

Shanthi, S., Poovaragan, S., Arularasu, M., Nithya, S., Sundaram, R., Magdalane, C. M., ... Maaza, M. (2018). Optical, Magnetic and Photocatalytic Activity Studies of Li, Mg and Sr Doped and Undoped Zinc Oxide Nanoparticles. *Journal of Nanoscience and Nanotechnology*, *18*(8), 5441–5447. doi:10.1166/jnn.2018.15442 PMID:29458596

Sinteza, K. I. M., & Galun-Lete, N. (2013). Synthesis, Characterization and Sensing Application of a Solid Alum/Fly Ash Composite Electrolyte. *Materiali in Tehnologije*, *47*(4), 467–471.

Srinivasan, V., & Weidner, J. W. (1999). Mathematical modeling of electrochemical capacitors. *Journal of the Electrochemical Society*, *146*(5), 1650–1658. doi:10.1149/1.1391821

Tian, Z., Wong, L., & Safaei, N. (2010). A neural network approach for remaining useful life prediction utilizing both failure and suspension histories. *Mechanical Systems and Signal Processing*, *24*(5), 1542–1555. doi:10.1016/j.ymssp.2009.11.005

Varde, P. (2010). Physics-of-failure based approach for predicting life and reliability of electronics components. *Barc Newsletter, 313*.

Virk, S. M., Muhammad, A., & Martinez-Enriquez, A. (2008). *Fault prediction using artificial neural network and fuzzy logic.* Paper presented at the IEEE Seventh Mexican International Conference on Artificial Intelligence (MICAI'08), Atizapan de Zaragoza, Mexico. 10.1109/MICAI.2008.38

Wu, C.-H., Ho, J.-M., & Lee, D.-T. (2004). Travel-time prediction with support vector regression. *IEEE Transactions on Intelligent Transportation Systems*, *5*(4), 276–281. doi:10.1109/TITS.2004.837813

Xu, W., & Wang, W. (2012). *An adaptive gamma process based model for residual useful life prediction.* Paper presented at the IEEE Conference on Prognostics and System Health Management (PHM 2012), Beijing, China.

Yadav, O. P., Singh, N., Chinnam, R. B., & Goel, P. S. (2003). A fuzzy logic based approach to reliability improvement estimation during product development. *Reliability Engineering & System Safety, 80*(1), 63–74. doi:10.1016/S0951-8320(02)00268-5

Yang, Y.-m., Ge, Z.-x., & Xu, Y.-c. (2008). *Fault diagnosis of complex systems based on multi-sensor and multi-domain knowledge information fusion.* Paper presented at the IEEE International Conference on Networking, Sensing and Control (ICNSC 2008), Sanya, China. 10.1109/ICNSC.2008.4525374

Zhao, F., Chen, J., & Xu, W. (2009). Condition prediction based on wavelet packet transform and least squares support vector machine methods. *Journal of Process Mechanical Engineering, 223*(2), 71–79. doi:10.1243/09544089JPME220

Chapter 7
Role of Artificial Neural Network for Prediction of Gait Parameters and Patterns

Kamalpreet Sandhu

School of Design II, Product and Industrial Design, Lovely Professional University, India

Vikram Kumar Kamboj

School of Electronics and Electrical Engineering, Lovely Professional University, India

ABSTRACT

Walking is very important exercise. Walking is characterized by gait. Gait defines the bipedal and forward propulsion of center of gravity of the human body. This chapter describes the role of artificial neural network (ANN) for prediction of gait parameters and patterns for human locomotion. The artificial neural network is a mathematical model. It is computational system inspired by the structure, processing method, and learning ability of a biological brain. According to bio-mechanics perspective, the neural system is utilized to check the non-direct connections between datasets. Also, ANN model in gait application is more desired than bio-mechanics strategies or statistical methods. It produces models of gait patterns, predicts horizontal ground reactions forces (GRF), vertical GRF, recognizes examples of stand, and predicts incline speed and distance of walking.

INTRODUCTION

Ergonomics is the study of individuals at work. This field got its name in the mid-year of 1949 when a group of intrigued people gathered in Oxford, England to discuss human performance and its limits. Ergonomics comes from Greek word 'Ergo' and 'Nomics'. 'Ergo' implies work and 'Nomics' implies study. A few specialists define the objective of ergonomics and as designing machines to fit the human operator requirements. However, it is also necessary to fit operations to machine in the form of personnel channel selection and training. It is probably more accurate to describe this field as the study of human machine systems, with an emphasis on the human aspects. Ergonomists deals with the fact that people

DOI: 10.4018/978-1-7998-1464-1.ch007

come in different sizes and shapes, varying greatly in their strength, endurance and work capacity. A basic understanding of human anatomy, physiology and psychology can help ergonomists to find solutions that deal with these issues and help to prevent problems that can cause injury to workers. Scope of this problem, can be studied by considering some of the components system terms by the human body. Walking is very important exercise associated with human being. There are different types of walking i.e. Brisk walking, Treadmill waking, Interval walking, Water walking and Hill walking. Human want to walk for the purpose of the health benefits also. The human being has a stride length of 640.08 to 742mm, it take over 2000 steps to walk one mile. On an average a person covers 10,000 steps a day. Human foot movement is based on the gait cycle. Walking is characterized by Gait. Locomotion produced through the movement of human limbs. Gait defines the bipedal and forward propulsion of center of gravity (COG) of the human body. Natural gaits are of two types i.e. Walk and skip. The healthy person up to 6 years follows walking while under age of the 4 to 5 years called as skipping by children. The various stages of the gait cycle show in Figure1 are Heel contact, Flat foot, Mid-stance, Heel off and Toe off.

- **Heel Contact:** The starting of the stance phase is called the heel-strike because there is an impact between the heel and the ground.
- **Foot Flat:** After heel contact, the rest of the foot comes in contact with the ambulatory surface at foot flat. This generally occurs at about 8% of the gait cycle, just before toe off of the opposite leg. During the interval between heel strike and foot flat the GRF increase rapidly in magnitude.
- **Mid Stance:** The period of time between foot flat and heel off which occurs at about 30% of the gait cycle. At this point, the swing phase leg passes the stance phase leg.
- **Heel Off:** When the heel begins to lift from the walking surface. It occurs between 40-50% of the gait cycle.
- **Toe Off:** This occurs at about 60% of the gait cycle when the stance phase ends and swing phase begins. This phase is also commonly called the push off phase.
- **Mid Swing:** This is the opposite of mid-stance as the mid-swing on one leg corresponds to the mid-stance of the other. It is the time when the swinging leg passes the stance leg.

Figure 1. Various stages of the gait cycles (Marco, Augusto, Fabio, Giovanni, Carlo, Stefano, & Antonella, 2013)

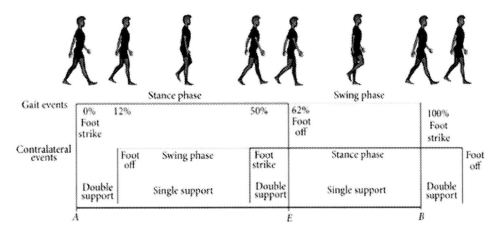

Methods of Measuring Gait

Different types of methods used to measure the gait analysis. The current technologies used for measure gait are: Motion capture camera, Force plates, Electromyography, Inertial measurement systems, Foot scan insole/Plantar pressure distribution insoles.

- **Motion Capture Camera:** The motion capture camera is basically used for measuring motion of objects or peoples. This system consider as gold standardize system and it require heavy equipment's and tight clothes for placement which may cause patients. Although wearing of heavy equipment's effect in natural gait of the human being. In addition this system is good for measuring the joint motions of lower extremity. The possibility of use this instrument in only laboratory and hard to use in daily living.
- **Force Plate:** The first commercial force plate discovered in 1969 (R Baker, 2007). Force plate consists of piezoelectric sensors mounted between two plates. The task of force plate is basically to calculate the ground reaction forces. Force plate is also used along with the motion capture camera. The disadvantage of force plate is basically it can be used in the laboratory.
- **Electromyography:** Electromyography in gait analysis used to measure the electrical activity of the muscle during contraction. Sensors are placed on the skin or the fine wire inserted into the muscle of interest.
- **Foot Scan Insole:** Foot scan insole is also known as plantar pressure distribution instrument. Basically used to measure the pressure on feet at different pressure points i.e. forefoot, mid-foot, hind-foot, heel, lateral, medial and overall foot. This system consists of thousands of pressure sensors which gives the pressure applied by the foot.
- **Sensor Based Insole:** It consists of different types of sensors. The typically used sensors are force sensitive resistors, PVDF, air pressure sensors, bi-directional sensors, bend sensor, flex sensors and electric field height sensors. This type of insole embeds into the shoes to collect pressure.

Figure 2. Current Gait measurement technologies

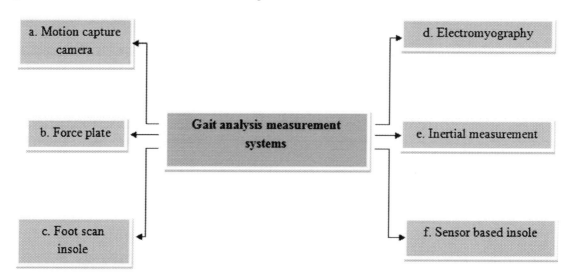

BACKGROUND

It is the use of man-made reasoning using for predicting the exact outcomes. It can get input information and used statistical analysis to predict the output value within an acceptable range.

Prediction using Machine Learning Models

Machine Learning comprises of two types supervised and unsupervised. Learning has different models: decision tree, random forest, ada boost, support vector machine, linear model and artificial neural network.

Random Forest

Random Forest is a machine learning method mainly used in regression and classification problems. After training, it generates multiple numbers of decision trees. Random forest work when different subset of training data are selected and with replacement to train each tree. The remaining data used for error calculations and importance of variables. Numbers of votes from the entire tree select the class alignment.

Decision Tree

Decision tree is a machine learning method using a tree like graph of decision. It used a set of binary rules and calculates the target values. It can be used as classification as well as regression problems. To determine the best node different types of algorithm was used.

Ada Boost

Ada boost is the machine learning method plays a good rule in linear classification problems. This type of method basically used in combination with many classification algorithms for improve the performance of the model.

Support Vector Machine (SVM)

The support vector machine is supervised learning model. The SVM is basically like a one layer and multiple neural networks. All machine learning method working on linear regression but the support vector machine working on the Non-linear regression problems. It uses the quadratic optimization problems.

Linear Model

The linear model is machine learning model, it the least complex model among all models. It will be nearer to the linear regression model. It shows the connection between the dependent and independent variable. Only one independent variable, it called as a simple regression model. At least two independent variables, it called multiple linear regression models.

Artificial Neural Networks

Artificial neural networks used in supervised, unsupervised and reinforced learning. This model looks like a neurons working in the mind. Neurons passed signal to another signal till the output is given by the body. The fundamental condition of the neural systems is given beneath:

Input $= W_1I_1 + W_2I_2 + ...W_nI_n + b$ 1.1
W= Weights
I= Inputs
b= bias
Output $= f(WI+b)$ 1.2

It has been observed that force plate was expected to give us the GRF step by step (Winter, 1990) yet a few specialists express that the current strategies don't precisely reflect individual physical activity levels, and further techniques does not precisely respond individual physical action levels, and further strategies advancement of physical movement apparatuses should be high need of research (Goran, M.I, Sun, M, 1998). It has been discovered that the utilization of ANN model in gait application is more desired than biomechanics strategies or different methods (Sepulveda, F., Wells, D.M, Vaughan, C.L, 1993). Various researchers observed that ANN is good model for discovering complex connections between examples of various signals (Sepulveda, F., Wells, D.M, Vaughan, C.L, 1993), (Breit, G.A., Whalen, R.T., 1997). For a human locomotion, the ANN can be utilized to predict gait parameters (Su, F.C., Wu, W.L., 2000). And gait patterns [Sepulveda, F., Wells, D.M, Vaughan, C.L, 1993), (Srinivasan, S, Gander, R.E., Wood, H.C, 1992). It produces models of gait patterns (Savelberg, H.H.C.M, de Lange, A.L.H, 1999) predicts horizontal GRF (Gioftsos, G, Grieve, D.W., 1996), vertically GRF, recognizes examples of stand (Aminian, K, Robert, P, Jequier, E, Schutz, Y, 1995) and predicts incline speed and distance of walking (Crowe, A, Samson, M.M, Hoitsma, M.J, van Ginkel, A.A, 1996). GRF has been utilized to decide the walking pattern (Goran, M.I, Sun, M, 1998), (Bertani, A, Cappello, A, Benedetti, M.G, Simoncini, L, Catani, F, 1999), (Giakas, G, Baltzopoulos, V, 1997) and balance of gait and posture (Bertani, A, Cappello, A, Benedetti, M.G, Simoncini, L, Catani, F, 1999), (Andriacchi, T.P, Ogle, J.A, Galante, J.O, 1977). A study demonstrated relationship between the variety of walk forces parameters, speed, and power, which vary with foot ground contact time (Cavagna, G.A, Kaneko, M, 1977), (Cavanagh, P.R, Lafortune, M.A, 1980), (Grasso, R, Bianchi, L, Lacquaniti, F, 1998), (Munro, C.F, Miller, D.I, Fuglevand, A.J, 1987). When human being increase the walking speed results in higher vertically force, horizontal force and decrease in the foot ground contact time (Keller, T.S., Weisberger, A.M., Ray, J.L, Hassan, S.S, Shiavi, R.G,Spengler, D.M,1996). The Pictorial view of the ANN model.

The Objective of this chapter is to describe the role and importance of ANN for predicting the gait parameters and patterns.

MAIN FOCUS AREA OF CHAPTER

The focus area of this chapter is to describe the role, importance and utilization of ANN for predicting the gait parameters and patterns by different researchers. Wang *et.al* (Wang, C.S, Wang, C.C. and Chang, T.R, 2013) researched the fitting of the human foot in various shoe insoles. Grey relational approach

Figure 3. Pictorial view of ANN

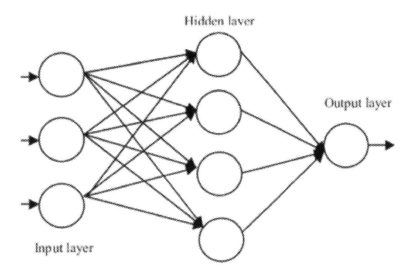

was utilized to evaluate the foot shapes and shoe insoles. ANN (Back propagation neural networks) was utilized to predict the most proper insoles for the foot. Results from plantar pressure experiment were analyzed with grey relational calculations and found the best insole for the human foot. Joo *et.al* (Joo, S.B, Oh, S.E, Sim, T, Kim, H, Choi, C.H, Koo, H. and Mun, J.H, 2014) predicted gait speed in stance and swing stage in reference to foot scan insole data gained by the plantar pressure measuring device. Information was gathered from 20 adults (10 guys and 10 females) having age 24.5±.3years and height 1.68±0.08m. The data collection was filtered with cut off frequency of 100 Hz. ANN was utilized to predict the gait speed. 99 weight sensors were used to create input information for the model. Robustness of models was checked by five k-fold validation. The model was evaluated for gait speed in three unique conditions normal walking, slow walking and fast walking. The correlation coefficient (r) was calculated and found 0.963 for normal walking, 0.998 for slow walking and 0.95 for fast walking.

Favre *el.al* (Favre, J, Hayoz, M, Erhart-Hledik, J.C. and Andriacchi, T.P, 2012) used ANN to predict knee adduction moments during walking based on the Ground Reaction Forces (GRF) and Anthropo-metric measurements. Force plate was used to measure forces and human link locomotion was capture by motion capture camera. The model was evaluated by using a correlation coefficients and mean absolute error. Correlation coefficients(r) were calculated for slow speed (mid-stance peak-0.74, terminal stance peak-0.55, mid stance angular impulse-0.78 and terminal stance angular impulse-0.63), normal speed (mid-stance peak-0.76, terminal stance peak-0.64, mid stance angular impulse-0.77 and terminal stance angular impulse-0.74) and fast speed (mid-stance peak-0.78, terminal stance peak-0.64, mid stance angular impulse-0.76 and terminal stance angular impulse-0.69).

Zhang *et.al* (Zhang, K., Sun, M., Lester, D.K, Pi-Sunyer, F.X., Boozer, C.N. and Longman, R.W, 2005) suggested new method for measuring human locomotion. Portable insole system was used to measure the GRF. Data from 40 participated was collected in different conditions i.e. walking, running, ascending and descending stairs in slow, normal and fast speed. ANN was used to predict the type of walking and identify the human gait patterns. The model was evaluated on the basis of accuracy. Accuracy recorded 98.77%, 98.3% for 97.3% and 97.2% for walking, running, ascending and descending respectively.

Figure 4. Pictorial view of the joo et.al (Joo, S.B, Oh, S.E, Sim, T, Kim, H, Choi, C.H, Koo, H. and Mun, J.H, 2014) research efforts

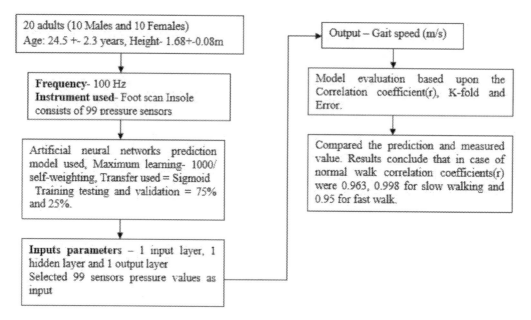

Figure 5. ANN architecture selected as prediction model (Joo, S.B, Oh, S.E, Sim, T, Kim, H, Choi, C.H, Koo, H. and Mun, J.H, 2014)

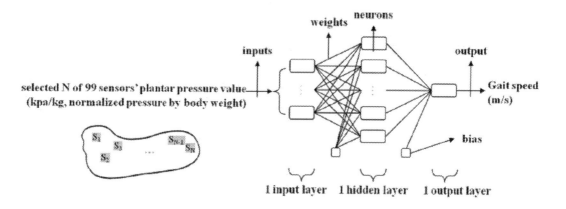

OH *el.al* (Oh, S.E, Choi, A. and Mun, J.H, 2013) described the prediction of Ground Reaction Forces during gait, based on kinematics. In this study Newtonian mechanics and ANN was used. The data was filtered with the cut off frequency of 120 Hz and walking time of 10 minute. The characteristics of ANN used were self-weighting and bipolar sigmoid. The input of the model was gait cycle and output was forces and moments in X, Y and Z directions. Ten k-fold cross validation was used to check the robustness of the model. The model was evaluated by using correlation coefficient and root mean square error (RMSE). Correlation coefficients(r) found for forces in X, Y and Z directions were 0.91, 0.98 and 0.99. A recorded value of moments in X, Y and Z directions were 0.98, 0.89 and 0.86.

Figure 6. Pictorial view of the Favre et.al (Favre, J, Hayoz, M, Erhart-Hledik, J.C. and Andriacchi, T.P, 2012) research efforts

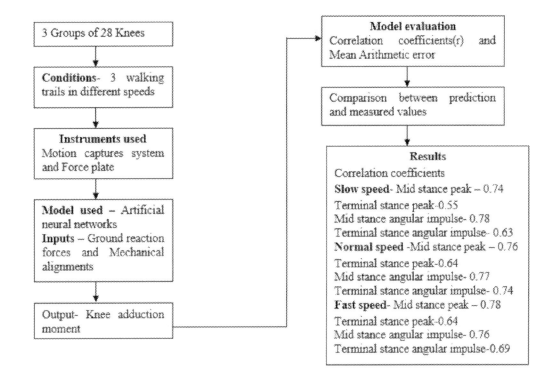

Figure 7. Pictorial view of the Zhang et.al (Zhang, K., Sun, M., Lester, D.K., Pi-Sunyer, F.X., Boozer, C.N. and Longman, R.W, 2005) research efforts

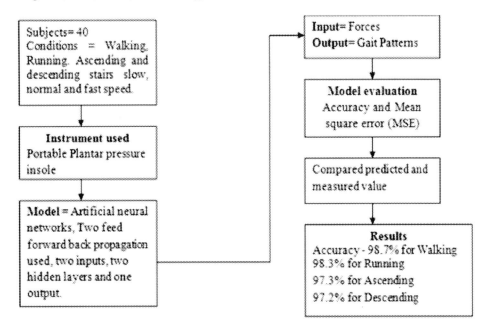

Figure 8. Pictorial view of the OH et.al (Oh, S.E, Choi, A. and Mun, J.H, 2013) research efforts

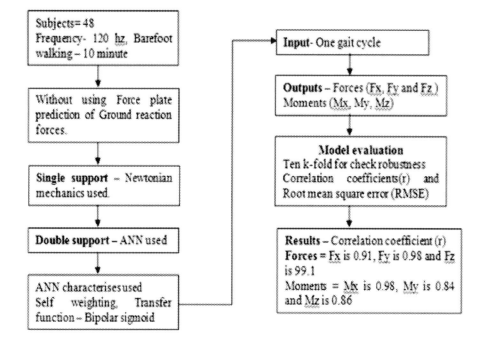

Figure 9. Process of predicting the GRF (Oh, S.E., Choi, A. and Mun, J.H., 2013)

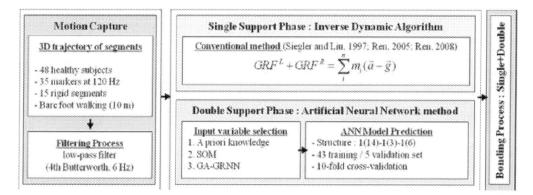

Toso *el.al* (Toso, M.A. and Gomes, H.M, 2014) calculated vertical force (Fz) on force plate system. It calibrated using artificial neural networks to reduce uncertainties. Math-works (2011) was used in this work. Model was runs on default settings. Each cell data calculated from force plate was used as input for the ANN. ANN model predicted force (Fz) accurately. Ruperez *et.al* (Rupérez, M.J, Martín-Guerrero, J.D, Monserrat, C. and Alcañiz, M, 2012) predicted of pressure on the foot surface using artificial neural networks model using Multilayer perception. Inputs used for ANN model was characteristics of materials and data obtained from 14 sensors. The prediction values were compared with the measured values. The value of correlation coefficients(r) was found to be close to 0.9.

FUTURE RESEARCH DIRECTIONS

- Experiments with a larger domain of parameters can be performed to achieve better results.
- From the Literature survey it has been found that most of research efforts are made on short duration of time and single foot consider. For future scope of work, it may consider long step duration of time two foot consider and perform ANN to predict gait parameters.
- Technique with muscle activity as intrinsic feedback to obtain accurate end-point limb movements could form the basis of neural prosthesis interface for allowing people with SB and other paralyzed patients to independently control voluntary movements of prosthetic limbs. Looking ahead, we will develop a self-organizing adaptive controller with low-power and high-performance computing technology for neural prostheses to enhance independent movement for people with disability.
- Also, in Literature survey it has been found that mostly predicting of the gait parameters by using insole sensors but for future scope consider human age, height, weight for predict force and plantar pressure between the foot.
- In the future it will be necessary to develop a model that can Predict gait speed for abnormal gait by using patients' data and to study a new prediction method to overcome the problems of the training process in artificial neural networks.
- Since the gait speed prediction method through ANN suggested in this study is not affected by the limitation of experiment location and procedure, it is expected to contribute to the future studies of related research by being used as basic material in studies that require data on both plantar pressure and gait speed.

CONCLUSION

In this chapter effort has been made on role of artificial neural network for prediction of Gait parameters and patterns. It has been found that lot of research are pending in predicting the gait parameters and patterns. This chapter include that outcomes of previous research were tested with further subjects and results were found to be in close proximity.

REFERENCES

Aminian, K., Robert, P., Jéquier, E., & Schutz, Y. (1995). Incline, speed, and distance assessment during unconstrained walking. *Medicine and Science in Sports and Exercise*, 27(2), 226–234. doi:10.1249/00005768-199502000-00012 PMID:7723646

Andriacchi, T. P., Ogle, J. A., & Galante, J. O. (1977). Walking speed as a basis for normal and abnormal gait measurements. *Journal of Biomechanics*, 10(4), 261–268. doi:10.1016/0021-9290(77)90049-5 PMID:858732

Baker, R. (2007). The history of gait analysis before the advent of modern computers. *Gait & Posture*, 26(3), 331–342. doi:10.1016/j.gaitpost.2006.10.014 PMID:17306979

Bertani, A., Cappello, A., Benedetti, M. G., Simoncini, L., & Catani, F. (1999). Flat foot functional evaluation using pattern recognition of ground reaction data. *Clinical Biomechanics (Bristol, Avon)*, *14*(7), 484–493. doi:10.1016/S0268-0033(98)90099-7 PMID:10521632

Breit, G. A., & Whalen, R. T. (1997). Prediction of human gait parameters from temporal measures of foot-ground contact. *Medicine and Science in Sports and Exercise*, *29*(4), 540–547. doi:10.1097/00005768-199704000-00017 PMID:9107638

Cavagna, G. A., & Kaneko, M. (1977). Mechanical work and efficiency in level walking and running. *The Journal of Physiology*, *268*(2), 467–481. doi:10.1113/jphysiol.1977.sp011866 PMID:874922

Cavanagh, P. R., & Lafortune, M. A. (1980). Ground reaction forces in distance running. *Journal of Biomechanics*, *13*(5), 397–406. doi:10.1016/0021-9290(80)90033-0 PMID:7400169

Crowe, A., Samson, M. M., Hoitsma, M. J., & van Ginkel, A. A. (1996). The influence of walking speed on parameters of gait symmetry determined from ground reaction forces. *Human Movement Science*, *15*(3), 347–367. doi:10.1016/0167-9457(96)00005-X

Favre, J., Hayoz, M., Erhart-Hledik, J. C., & Andriacchi, T. P. (2012). A neural network model to predict knee adduction moment during walking based on ground reaction force and anthropometric measurements. *Journal of Biomechanics*, *45*(4), 692–698. doi:10.1016/j.jbiomech.2011.11.057 PMID:22257888

Giakas, G., & Baltzopoulos, V. (1997). Time and frequency domain analysis of ground reaction forces during walking: An investigation of variability and symmetry. *Gait & Posture*, *5*(3), 189–197. doi:10.1016/S0966-6362(96)01083-1

Gioftsos, G., & Grieve, D. W. (1996). The use of artificial neural networks to identify patients with chronic low-back pain conditions from patterns of sit-to-stand manoeuvres. *Clinical Biomechanics (Bristol, Avon)*, *11*(5), 275–280. doi:10.1016/0268-0033(96)00013-7 PMID:11415632

Goran, M. I., & Sun, M. (1998). Total energy expenditure and physical activity in prepubertal children: Recent advances based on the application of the doubly labeled water method. *The American Journal of Clinical Nutrition*, *68*(4), 944S–949S. doi:10.1093/ajcn/68.4.944S PMID:9771877

Grasso, R., Bianchi, L., & Lacquaniti, F. (1998). Motor patterns for human gait: Backward versus forward locomotion. *Journal of Neurophysiology*, *80*(4), 1868–1885. doi:10.1152/jn.1998.80.4.1868 PMID:9772246

Joo, S. B., Oh, S. E., Sim, T., Kim, H., Choi, C. H., Koo, H., & Mun, J. H. (2014). Prediction of gait speed from plantar pressure using artificial neural networks. *Expert Systems with Applications*, *41*(16), 7398–7405. doi:10.1016/j.eswa.2014.06.002

Keller, T. S., Weisberger, A. M., Ray, J. L., Hasan, S. S., Shiavi, R. G., & Spengler, D. M. (1996). Relationship between vertical ground reaction force and speed during walking, slow jogging, and running. *Clinical Biomechanics (Bristol, Avon)*, *11*(5), 253–259. doi:10.1016/0268-0033(95)00068-2 PMID:11415629

Kram, R., & Taylor, C. R. (1990). Energetics of running: A new perspective. *Nature*, *346*(6281), 265–267. doi:10.1038/346265a0 PMID:2374590

McMahon, T. A., Valiant, G., & Frederick, E. C. (1987). Groucho running. *Journal of Applied Physiology, 62*(6), 2326–2337. doi:10.1152/jappl.1987.62.6.2326 PMID:3610929

Munro, C. F., Miller, D. I., & Fuglevand, A. J. (1987). Ground reaction forces in running: A reexamination. *Journal of Biomechanics, 20*(2), 147–155. doi:10.1016/0021-9290(87)90306-X PMID:3571295

Oh, S. E., Choi, A., & Mun, J. H. (2013). Prediction of ground reaction forces during gait based on kinematics and a neural network model. *Journal of Biomechanics, 46*(14), 2372–2380. doi:10.1016/j.jbiomech.2013.07.036 PMID:23962528

ResearchGate. (n.d.). Retrieved from https://www.researchgate.net/figure/The-gait-cycle-A-schematic-representation-of-gait-cycle-with-stance-red-and-swing_fig1_249968026

Rupérez, M. J., Martín-Guerrero, J. D., Monserrat, C., & Alcañiz, M. (2012). Artificial neural networks for predicting dorsal pressures on the foot surface while walking. *Expert Systems with Applications, 39*(5), 5349–5357. doi:10.1016/j.eswa.2011.11.050

Savelberg, H. H. C. M., & De Lange, A. L. H. (1999). Assessment of the horizontal, fore-aft component of the ground reaction force from insole pressure patterns by using artificial neural networks. *Clinical Biomechanics (Bristol, Avon), 14*(8), 585–592. doi:10.1016/S0268-0033(99)00036-4 PMID:10521642

Sepulveda, F., Wells, D. M., & Vaughan, C. L. (1993). A neural network representation of electromyography and joint dynamics in human gait. *Journal of Biomechanics, 26*(2), 101–109. doi:10.1016/0021-9290(93)90041-C PMID:8429053

Srinivasan, S., Gander, R. E., & Wood, H. C. (1992). A movement pattern generator model using artificial neural networks. *IEEE Transactions on Biomedical Engineering, 39*(7), 716–722. doi:10.1109/10.142646 PMID:1516938

Su, F. C., & Wu, W. L. (2000). Design and testing of a genetic algorithm neural network in the assessment of gait patterns. *Medical Engineering & Physics, 22*(1), 67–74. doi:10.1016/S1350-4533(00)00011-4 PMID:10817950

Toso, M. A., & Gomes, H. M. (2014). Vertical force calibration of smart force platform using artificial neural networks. *Revista Brasileira de Engenharia Biomédica, 30*(4), 406–411. doi:10.1590/1517-3151.0569

Wang, C. S., Wang, C. C., & Chang, T. R. (2013, July). Neural network evaluation for shoe insoles fitness. In *2013 Ninth International Conference on Natural Computation (ICNC)* (pp. 157-162). IEEE. 10.1109/ICNC.2013.6817962

Winter, D. A. (1990). Biomechanics and motor control of human movement Wiley. New York: Academic Press.

Zhang, K., Sun, M., Lester, D. K., Pi-Sunyer, F. X., Boozer, C. N., & Longman, R. W. (2005). Assessment of human locomotion by using an insole measurement system and artificial neural networks. *Journal of Biomechanics, 38*(11), 2276–2287. doi:10.1016/j.jbiomech.2004.07.036 PMID:16154415

136

Chapter 8
Modelling Analysis and Simulation for Reliability Prediction for Thermal Power System

Vikram Kumar Kamboj

School of Electronics and Electrical Engineering, Lovely Professional University, India

Kamalpreet Sandhu

School of Design II, Product and Industrial Design, Lovely Professional University, India

Shamik Chatterjee

School of Electronics and Electrical Engineering, Lovely Professional University, India

ABSTRACT

The size of the power system is growing exponentially due to heavy demand of power in all the sectors (e.g., agricultural, industrial, and commercial). Due to this, the chance of failure of individual units leading to practical or complete collapse of power supply is common to be encountered. The reliability of power system is therefore the most important feature to be maintained above some acceptable threshold value. Furthermore, the maintenance of individual units can also be planned and implemented once the level of reliability for given instant of time is known. The proposed research therefore aims at determining the threshold reliability of generation system. The generation system consists of boiler, water, blade angle in turbine, shaft coupling, excitation system, generator winding, circuit breaker, and relay. This chapter presents the mathematical model of reliability of individual components and equivalent reliability of the entire generation system. It suggests the approach to determine the critical reliability of both individual and equivalent reliability of the generation system.

DOI: 10.4018/978-1-7998-1464-1.ch008

INTRODUCTION

The size of the power system is growing exponentially due to heavy demand of power in all the sectors viz. agricultural, industrial, residential and commercial ones. As such the chance of failure of individual units leading to practical or complete collapse of power supply is common to be encountered. Also a most successful power system is one which works with minimum interruptions. The reliability of power system is therefore most important feature to be maintained above some acceptable threshold value. Further the maintenance of individual units can also be planned and implemented once the level of reliability for given instant of time is known. The research proposal therefore aims at determining the threshold reliability of generation system. The generation system consists of boiler, water, blade angle in turbine, shaft coupling, excitation system, generator winding, circuit breaker and relay. It is therefore the reliability of generation system shall be effected even when any one of the component's reliability is at stake. The Generation System is basically the heart of any power system. The collapse of generation system leads to the biggest collapse, the recovery of which is not only time consuming but tougher than any other system such as Transmission system and Distribution system. The failure of Generation System leads to prolonged interruption period besides being uneconomical. Thus the reliable operation of Generation System needs a thorough care and periodic servicing of its constituent parts such as boiler, turbine, generator and circuit breaker. Our industries, agriculture and software companies require consistent and uninterrupted power supply for toing the needs of continuous customer demand. It therefore drew the attention of researcher to determine the condition when reliability of generation system would be at stake. Further due to complexity of generation system, the determination of its reliability is filled with risk of skipping of some important factors which are responsible for loss of reliability. Thus, it needed a trusted and proof model to deal with the reliability of Generation System. Moved by these innovative ideas the research for assessing the reliability of reliability of Generation System has been undertaken. The reliability is based on the incremental reliability of its parts, which includes boiler, turbine, generator and circuit breaker. The chapter therefore presents a mathematical model for determining the reliability of generation system expressed as equivalent reliability by taking each of its component all together. In order to validate the result a model of ANN has been developed to obtain the reliability of Generation System. It could be found that the results obtained by two methods agree with each other. It is therefore possible to estimate the threshold value R_{Th} of equivalent reliability upto which the Generation System can be taken to be reliable. It however loses reliability beyond threshold value of equivalent reliability.

Problem Formulation

It is aimed to assess the reliability of consider a Generation System consisting of:

1. Mechanical Unit which again consists of boiler and turbine.
2. Electrical Unit which again consists of Generator and Circuit Beaker as shown in Figure1.
3. Obtain Mathematical model of equivalent reliability, Req of Mechanical and Electrical Units.
4. Obtain validation of reliability assessment made by mathematical model by way of ANN model.

Figure 1. Proposed Generation System Model

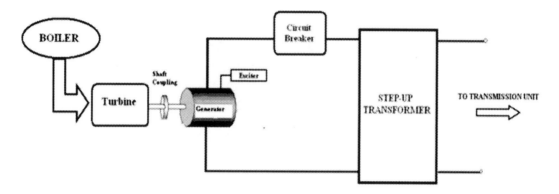

Proposed Solution

It is required to develop

1. Mathematical Model of reliability for individual components as well as for Generation System as a whole.
2. ANN platform for determining the reliability of Generation System.
3. Comparison and conclusion of study made by Mathematical and ANN approaches.

Real Out Come

Both Mathematical Model and ANN model for assessment of reliability of Generation System have been developed. The outcomes of one model supported the outcomes of another model. Illustration gave a satisfactory support and thus it could be possible to assess the reliability of generation system by both the approaches. The outcome of threshold value of reliability of generation system is an appreciable outcome of the work. The results of reliability of constituent parts of generation system has been plotted against time. Also, the threshold reliability of generation system could be determined by graphical solution.

IMPORTANCE OF RELIABILITY IN POWER SYSTEM

Reliability is, in a general sense, a measure of the ability of any system to perform its intended function without failure under certain conditions for a stated period of time. In the case of a generation system, reliability is usually defined in a quite different way. A Generation system is designed to perform its function for a relatively longer period of time. A Generation system can have failures at different points in the system at certain time, but these failures can always be repaired, and the system can be constantly developed to satisfy the changing demands and for improved the performance of its functionaries. It is therefore important for a power system to recognize and control the various possible system failures and to minimize the failure rate to provide uninterrupted power supply to the customers. Time of repair can be shortened by use of standby systems. The measurement of Generation system reliability includes the unavailability, the expected failure frequency and duration, and the expected magnitude of adverse effects

Figure 2. Subdivision of power system reliability

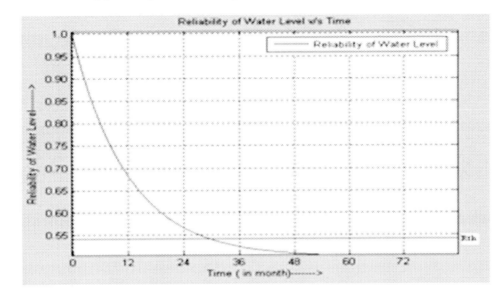

on consumer service. System reliability evaluation is based on the data which make up the system. The collection of subsystem reliability data is therefore a fundamental task in system reliability evaluation. Reliability as applied to a power system is divided into the two general categories viz. system adequacy and system security as shown in Figure 2

System adequacy relates to the existence of sufficient generation, transmission and distribution facilities within the system to satisfy the customer load demand. Adequacy evaluation is therefore associated with system steady state conditions. System security, on the other hand, relates to the ability of the system to cope with disturbances and is consequently associated with transient system conditions. System adequacy is usually associated with system planning for both long and short terms, but is also very important in system operation. System security is concerned with both system planning and operation. Electric power systems are generally categorized into the three segments or functional zones of Generation, Transmission and Distribution. This division is an appropriate one as most utilities are either divided into these zones for the purpose of organization, planning, operation and analysis or are solely responsible for one of these functions. The three functional zones can be combined to give three hierarchical levels, as shown in Figure 3. Hierarchical Level I (HL I) is concerned only with the generation system. Hierarchical Level II (HL II) includes both generation and transmission system and HL III includes all three functional zones. System adequacy analysis is usually conducted in each of the three functional zones or in the three hierarchical levels.

HL II analysis using probability methods is the oldest and most extensively developed area. Considerable effort has been applied over the last two decades to develop acceptable techniques and criteria for HL III analysis. It is also designated as composite power system or bulk power system evaluation. HL I studies are not usually done directly due to the enormity of the problem in a practical system and instead the generation functional zone is analyzed as an independent entity. The present dissertation aims at determining the reliability of Generation System

Figure 3. Hierarchical levels of Power System for Reliability Analysis

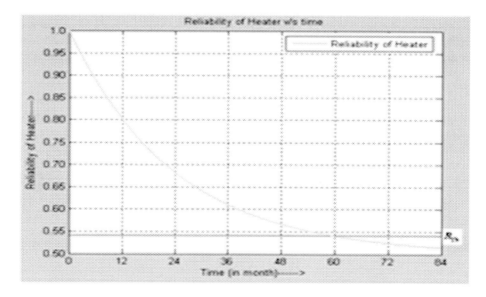

Methods of Reliability Assessments

The methods used in reliability assessments of power systems determine the accuracy of the results. Analytical and simulation approaches are the two types of techniques used in power system reliability analysis. Each approach has its merits and limitations. In this section, the concepts, assumptions, and typical applications of the commonly used methods in both techniques are reviewed. The limitations of analytical approaches are summarized as the reason to select the Artificial Neural Network simulation to perform the reliability analysis in this study and shown in Figure 4.

Analytical Approaches

The analytical analysis methods use mathematical models to provide solutions to a reliability problem. Specific calculation results are obtained for a given set of system topology and input values. Some widely used methods are block diagram, event tree, cut sets, fault tree, state enumeration, and Markov modeling. Using reliability sets in calculation is also proposed in recent years. The common problem is the frequent need to make simplifying assumptions and approximations. The different analytical approaches are discussed below:

Reliability Equivalent Modeling Method

This method utilizes modular concepts to reduce the overall system to a simple failure logic connection diagram, which only contains series and parallel components. The conditional probability approach is then used to calculate this series and parallel arrangements. The application of this method usually needs to oversimplify component reliability parameters in order to transform the system into its functional layout. A typical example is to apply estimated failure rate values according to the number of pieces of equipment falling into each functionality group, such as protection, control, and monitoring groups (R. Billington and J. Oteng-Adjei, 1991).

140

Figure 4. Two Fundamentally different Approaches

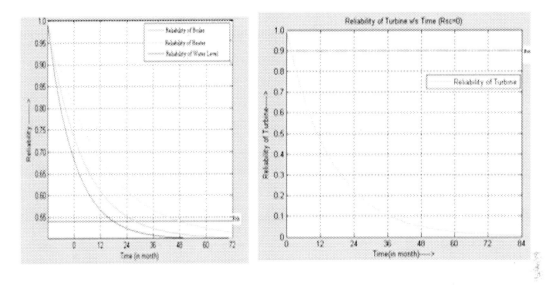

Event Tree Method

An event tree of a system is a visual presentation of all events that may occur in a system. After an initiating event(s) is (are) selected, the possible consequences involving success and failure of the system components are deduced, and fan out as the branches of a tree. Each failure path represents a failure scenario that the system fails if all the components in this path fail. The size of the tree can be staggering for a large system. If each component can reside in either operating or failed status, a complete event tree of a system has $2n$ paths for an n-component system. Therefore, reduced event trees are often used, sacrificing some precision.

Cut Set Method

A cut set contains a set of system components whose failure can lead to the failure of the system. The minimum subset of a cut set is called minimal cut set, which contains the set of components that must fail in order for the system to fail. Each cut in the cut set is in series with other cuts, with the components inside a cut combined using the principle of parallel components. As the failure path of an event tree is equivalent to a cut in the cut set, the minimal cut set can be derived to reduce the size of the tree. However, even the minimal cut set can still be prohibitively large for large systems. As a common case, exhaustive evaluation for precise results is often compromised with fast calculations by using approximations, such as neglecting cut sets greater than a certain value. This is based on an assumption that high order cut sets are much less probable than low order cut sets (G. Haringa; G. Jordan and L. Garver, 1991).

Fault Tree Method

Similar to an event tree, a fault tree is a pictorial representation of the failure logic embedded in the system. The top event of the tree, however, can only be a particular failure condition compared to that of the event tree. The branch events are then constructed as the essential events in order to lead to the top failure event. Therefore, this method is good at mission-oriented evaluation. It is applied particularly for safety assessments and not comprehensive reliability evaluations (G. Oliveira, S. Cunha and M. Pereira, 1987).

State Enumeration Method

The state enumeration method tries to identify the events that have an adverse effect on system reliability, and evaluates the effects. Evaluating all possible contingencies is not practical and not required. The practical enumerative method utilizes contingency screening policies to reduce the number of states evaluated. A common approach is to evaluate the primary contingencies, whose outage frequencies exceed some predetermined value. Another approach is to use the contingency ranking method. It uses the severity level, such as overload condition of the contingency, for screening. The limitation of this method is that the number of contingencies is large if the study system is large, even with the screening approaches (R. Allen; R. Billington and N. Abdel-Gawad, 1986). Moreover, it is not easy to accommodate the stochastic treatment of system loading. Usually, only a few selected load levels based on experience are used to perform the analysis (S. Kumganty, 1994). Planned outages, such as maintenance events, are also not easy to integrate into this approach.

Markov Modeling Method

Markov modeling is a matrix method that is applicable to memory less systems whose components' probability distributions are constant hazard rates (Jorge Munoz-Estrada; Juan Perez-Ruiz and Julian Barquin, 2004). In order to perform the evaluation, the stochastic transitional probability matrix needs to be constructed. It is an $n \times n$ square matrix of an n-state system. As the system topology changes, the system transitional matrix needs to be re-constructed. Therefore, the application of this method is generally limited to simple system configurations.

Reliability Set Calculation Method

In this method, reliability sets are proposed in order to calculate the SAIDI of individual circuits or feeders. The sets are defined based on the segments of the circuits. The system model is simplified to sets, with an average failure rate value of each segment. After that, the optimal placement of a DG for time–varying loads is examined. The major limitation of this method is that it only applies to radial system. Also, it assumes that only one failure takes place at a time, and thus does not handle multiple failures.

Artificial Intelligence Approach

The above review demonstrates that present analytical reliability techniques are not capable of modeling a large number of real system characteristics. For those analytical methods utilizing parallel and series network calculating principles, there are many realistic systems not easily separated into small independent subsystems. All of the analytical methods that model detailed system states and/or enumerate among them have the common problem of system model size for large systems. Stiffness in calculation is also a problem when using traditional matrix methods. Additionally, using the Markov chains implies the events are memory less, which often is not the case. Compared with analytical approaches, the simulation or Artificial Intelligence approach is more universal. A neural network is a powerful data modeling tool that is able to capture and represent complex input/output relationships. Neural Networks are an information processing technique based on the way biological nervous systems, such as the brain, process information. They resemble the human brain in the following two ways:

1. A neural network acquires knowledge through learning.
2. A neural network's knowledge is stored within inter-neuron connection strengths known as synaptic weights.

Neural networks are being applied to an increasing large number of real world problems. Their primary advantage is that they can solve problems that are too complex for conventional technologies; problems that do not have an algorithmic solution or for which an algorithmic solution is too complex to be defined. In general, neural networks are well suited to problems that people are good at solving, but for which computers generally are not. These problems include pattern recognition and forecasting, which requires the recognition of trends in data. The true power and advantage of neural networks lies in their ability to represent both linear and non-linear relationships and in their ability to learn these relationships directly from the data being modeled. Traditional linear models are simply inadequate when it comes to modeling data that contains non-linear characteristics.

Description of Generation System

Generation of electrical energy is nothing but conversion of Mechanical energy into electrical energy. It therefore consists of components viz. Boiler, Turbine, (blade angle and mechanical coupling through shaft), Generator (Generator excitation system, Generator winding), Circuit Breaker (switchgear protection) and Step-Up Transformer as shown in Figure 5.

It is required that the generation system should work without fail and uninterruptedly but such an operation could only be possible when every component of generation system works with acceptable reliability. The Generator's output is fed to rated load. The protection to generator against faults is provided by relay and circuit breaker system. Generation System can be subdivided into Mechanical zone and Electrical Zone. The boiler is used to produce steam of required pressure. The steam impinges upon

Figure 5. Generation System Model

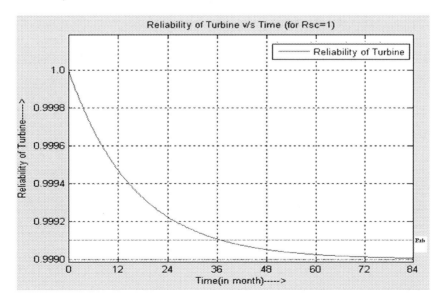

the blades of the turbine. The turbine is therefore put into rotation. This mechanical energy is conveyed to electrical generator through shaft, which in turn gives electrical energy in order to get uninterrupted power supply.

It has been aimed to develop mathematical model for reliability assessment of generation system. The reliability of generation system basically depends upon each factor which is responsible for the flow of energy right from boiler to transmission system. It is therefore a mathematical model has been developed in this dissertation by taking into account the reliability of individual components as a partial reliability & there upon deriving of expression for equivalent reliability of generation system as a whole. Further, the reliability assessment is partially done with the help of Artificial Neural Network. It has been observed that the reliability obtained by both the approaches show close agreement with each other (Vikram Kumar Kamboj et. al, 2012).

Mathematical Approach

The mathematical model for reliability assessment has been developed by way equivalent reliability approach. It represents the overall reliability of generation system. Also, it consists of resultant effect of reliability due to individual components.

Expected Failures of Generation System

In Generation System failures which affects overall system reliability are:

- Boiler
- Turbine
- Generator
- Circuit Breaker

Each of the above is constituted small components and it is therefore the reliability of these parts are functions of partial reliability of internal parts. For example: The Generator is globally reliable only when all of its elements like generator winding, Excitation, Rotor etc. works reliably. Thus, we can say that the reliability of above generation system depends upon so many factors. The failure of generation system can be caused by the individual failure of these elements. The major factors, which causes the failure in generation system are:

1. Heater Failure
2. Water Level Failure
3. Blade Angle Failure
4. Shaft Coupling Failure
5. Generator Winding failure
6. Excitation Failure
7. Circuit Breaker Failure

Expression of mathematical model for reliability of each these parts can be obtained by considering the reliability of the rest of the parts 100%. The Equivalent reliability model has been shown in Figure 6.

Figure 6. Equivalent Reliability Model for Generation System

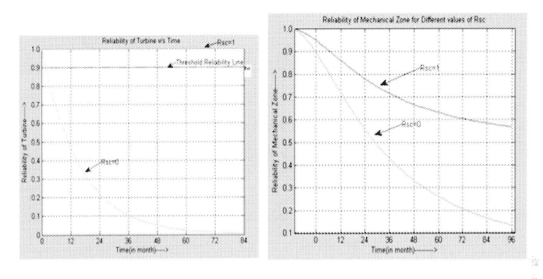

1.   If the failure of an element depends on many factors, AND gate is a logical choice.
2.   If the failure of an element is independent of many factors, OR gate is a logical choice.

Reliability of boiler is assured when Heater and Water Level work within standard limits. Since, the objective is dependent on both the factors, the AND gate has been used. In case of Turbine, our object is that the Turbine should run at the rated speed. This will happen only when both the blade angle and shaft coupling are intact. Thus, logic AND gate has been used to represent the failure of turbine. In case of Generator, e.m.f. production needs simultaneously both the factors i.e. Generator Winding and Excitation. Thus, logic AND gate is required to represent the failure of generator system. In case of Mechanical Zone, our objective is to achieve the rotary motion of turbine with the help of Boiler steam. The rotary motion is possible only when if boiler steam feeds to turbine. Thus, object is dependent on reliable operation of heater as well as Turbine and therefore logic AND gate has been used. In Electrical Zone, The failure of generator is dependent on working of circuit breaker, if fault occurs then circuit breaker operates. Thus, logically AND gate is used to represent the failure of Electrical Zone. Also, in case of overall system, the working of No Load/Full Load of the Generator depends on Mechanical Zone. Thus, logically AND operation is used between Mechanical Zone and Electrical Zone.

Proposed Mathematical Model of Equivalent Reliability

In our discussion in previous section, we have studied that the reliability of overall generation system depends upon so many factors individually and collectively, which is a function of time. After appreciating the equivalent reliability model, it is required to find the mathematical model for individual unit of generation system as well as overall generation system. In this section, we will formulate the mathematical model for individual unit of generation system as well as equivalent reliability of overall generation system. In this process some terms have been used frequently. They have been abbreviated as below:

Mathematical Model for Boiler

It is obvious that the failure of boiler occurs when heater *AND* water level experiences due failure. The failures has been indicated by the bar above the letter e.g $\overline{B}$, the probability of failure of boiler shall be $P(\overline{B})$, then Probability of failure of Boiler is given by $P(\overline{B})$

$$P(\overline{B}) = P(\overline{x}_H \ AND \ \overline{x}_W)$$
$$= P(\overline{x}_H).P(\overline{x}_W) \tag{4.1}$$

If reliability of boiler is indicated by R_B, then the reliability of boiler is given by

$$R_B = 1 - P(\overline{x}_H).P(\overline{x}_W)$$
$$= 1 - (1 - R_H).(1 - R_W) \tag{4.2}$$
$$= R_H + R_W - R_H.R_W$$

Since the reliability of every component decreases with time (H.J. Zimmermann, 1991), in case of Boiler, the of water level decreases continuously with time due to formation of steam. The water is required to be maintained upto the desired level for reliable operation of the boiler. Heater's reliability decreases due to ageing of heating element, because sustainability of heating material is getting degraded with time.

Consider that the reliability of heater, $R_H = \exp(-\lambda_1 t)$ and reliability of water level, $R_W = \exp(-\lambda_2 t)$. Where, λ_1 represents the failure rate of heating element and λ_2 represents the failure rate of water level.

Thus, applying the usual probability laws, we have

$$R_B = \exp(-\lambda_1 t) + \exp(-\lambda_2 t) - \exp(-\lambda_1 t).\exp(-\lambda_2 t) \tag{4.3}$$

Mathematical Model for Turbine

Similarly, for turbine, if probability of failure of turbine is indicated by $P(\overline{T})$, then

$$P(\overline{T}) = P(\overline{x}_{BA} \ AND \ \overline{x}_{SC})$$
$$= P(\overline{x}_{BA}).P(\overline{x}_{SC}) \tag{4.4}$$

where, $P(\overline{x}_{BA})$ represents the probability of failure of blade angel and $P(\overline{x}_{SC})$ represents the probability of failure of shaft coupling.

Here, R_T indicates the reliability of turbine and is given by

$$R_T = 1 - P(\overline{x}_{BA}).P(\overline{x}_{SC})$$
$$= 1 - (1 - R_{BA}).(1 - R_{SC}) \tag{4.5}$$
$$= R_{BA} + R_{SC} - R_{BA}.R_{SC}$$

where, R_{BA} and R_{SC} indicate the reliability of the blade angle and shaft coupling respectively.

Also, the reliability of blade angle decreases with time, consider $R_{BA} = \exp(-\lambda_3 t)$. Where λ_3 is the failure rate of blade angle? But, the Reliability of shaft coupling does not decrease with time. Applying usual probability laws, we have

$$R_T = \exp(-\lambda_3 t) + R_{SC} - \exp(-\lambda_3 t).R_{SC} \qquad (4.6)$$

On the other hand, the reliability of shaft coupling, R_{SC} does not decrease with time, there are two possible discrete states, namely, $s_0 = \overline{x}_1$, the shaft coupling is good ; and $s_1 = x_1$, the shaft coupling is failed.

Case-(i): When shaft coupling is failed (i.e. for state $s_0 = \overline{x}_1$). Reliability of turbine is given by,

$$R_T = \exp(-\lambda_3 t) \qquad (4.7)$$

Case-(ii): When shaft coupling is intact (i.e. for state $s_1 = x_1$). The reliability of turbine is given by

$$R_T = \exp(-\lambda_3 t) + R_{SC} - \exp(-\lambda_3 t).R_{SC} \qquad (4.8)$$

Mathematical Model for Generator

Similarly, for Generator, if probability of failure of Generator is indicated by $P(\overline{G})$, then the failure of Generator is given by

$$\begin{aligned} P(\overline{G}) &= P(\overline{x}_{Wng} \; AND \; \overline{x}_E) \\ &= P(\overline{x}_{Wng}).P(\overline{x}_E) \end{aligned} \qquad (4.9)$$

where, $P(\overline{x}_{Wng})$ and $P(\overline{x}_E)$ represents the probability of failure of Generator winding and excitation respectively.

If R_G represents the reliability of generator unit, then the Reliability of Generator is given by

$$\begin{aligned} R_G &= 1 - P(\overline{x}_{Wng}).P(\overline{x}_E) \\ &= 1 - (1 - R_{Wng}).(1 - R_E) \\ &= R_{Wng} + R_E - R_{Wng}.R_E \end{aligned} \qquad (4.10)$$

Here R_{Whg} and R_E represents the reliability of generator winding and reliability of excitation respectively.

In case of generator the reliability of generator winding, R_{Wng} decay slowly with time due to heating of winding coil and also due to change in insulator property w.r.t. time. Thus the reliability of generator winding, R_{Wng} is also a function of time and can be represented by $\exp(-\lambda_4 t)$. But the reliability of excitation, R_E does not depend on time and has only two possible states 0 and 1.

Therefore, the reliability of generator unit can be expressed as

$$R_G = \exp(-\lambda_4 t) + R_E - \exp(-\lambda_4 t).R_E \tag{4.11}$$

Mathematical Model for Circuit Breaker

If $P(\bar{x}_{CB})$ represents the probability of failure of circuit breaker, then Failure of Circuit Breaker is given by

$$F(CB) = P(\bar{x}_{CB}) \tag{4.12}$$

Reliability of Circuit Breaker is given by

$$
\begin{aligned}
R(CB) &= 1 - P(\bar{x}_{CB}) \\
&= 1 - (1 - R_{CB}) \\
&= R_{CB}
\end{aligned}
\tag{4.13}
$$

where, R_{CB} represents the reliability of the circuit breaker.

Mathematical Model for Mechanical Zone

If combination of Boiler and Turbine is named as Mechanical Zone and if $F(Mech)$ represents the failure of Mechanical Failure of Mechanical Zone is given by

$$
\begin{aligned}
F(Mech) &= P[F(B) \ AND \ F(T)] \\
&= P(\bar{x}_B \ AND \ \bar{x}_T) \\
&= P(\bar{x}_B).P(\bar{x}_T)
\end{aligned}
\tag{4.14}
$$

If R_{Mech} Represents the Reliability of Mechanical Zone, then Reliability of Mechanical Zone is given by

$$
\begin{aligned}
R_{Mech} &= 1 - P(\bar{x}_B).P(\bar{x}_T) \\
&= 1 - [(1 - R_B).(1 - R_T)] \\
&= 1 - [\{1 - \exp(-\lambda_1 t) + \exp(-\lambda_2 t) - \exp(-\lambda_1 t).\exp(-\lambda_2 t)\} \\
&\quad \times \{1 - \exp(-\lambda_3 t) + R_{SC} - \exp(-\lambda_3 t).R_{SC}\}]
\end{aligned}
\tag{4.15}
$$

Mathematical Model for Electrical Zone:

As shown in Fig.4.3, the combination of Generator and Circuit Breaker is named as Electrical Zone. If F (Elec) represents the failure of Electrical Zone, then Failure of Electrical Zone is given by

$$
\begin{aligned}
F(Elec) &= P[F(G) \ AND \ F(CB)] \\
&= P(\bar{x}_G \ AND \ \bar{x}_{CB}) \\
&= P(\bar{x}_G).P(\bar{x}_{CB})
\end{aligned}
\tag{4.16}
$$

If R_{Elec} represents the reliability of Electrical Zone, then Reliability of Electrical Zone is given by:

$$
\begin{aligned}
R_{Elec} &= 1 - P(\overline{x}_G).P(\overline{x}_{CB}) \\
&= 1 - [(1 - R_G).(1 - R_{CB})] \\
&= 1 - [\{1 - (\exp(-\lambda_4 t) + R_E - \exp(-\lambda_4 t).R_E)\} \times (1 - R_{CB})]
\end{aligned}
\tag{4.17}
$$

Mathematical Model of Equivalent Reliability for Overall Generation System

If F(S) represents the failure of overall Generation system, then the failure of overall generation system is given by:

$$
\begin{aligned}
F(S) &= P[F(Mech)\ AND\ F(Elec)] \\
&= P(\overline{x}_{Mech}\ AND\ \overline{x}_{Elec}) \\
&= P(\overline{x}_{Mech}).P(\overline{x}_{Elec})
\end{aligned}
\tag{4.18}
$$

Also, if R_{eq} represents the equivalent reliability of overall Generation system. Then, equivalent reliability of generation system is given by

$$
\begin{aligned}
R_{eq} &= 1 - P(\overline{x}_M).P(\overline{x}_E) \\
&= 1 - [(1 - R_M).(1 - R_E)] \\
&= 1 - [\{1 + \exp(-\lambda_1 t) - \exp(-\lambda_2 t) + \exp(-\lambda_1 t).\exp(-\lambda_2 t)\} \\
&\quad \times \{1 - \exp(-\lambda_3 t) + R_{SC} - \exp(-\lambda_3 t).R_{SC}\}] \\
&\quad \times [\{\exp(-\lambda_4 t) - R_E + \exp(-\lambda_4 t).R_E)\} \times \{1 - R_{CB}\}]
\end{aligned}
\tag{4.19}
$$

Eqn. (4.19) represents the mathematical model of reliability for overall Generation System. It takes into account the reliability of individual model in a collective manner. Further the model is a function of time.

Artificial Neural Network (ANN) Approach

A neural network is a powerful data modeling tool that is able to capture and represent complex input/output relationships. Neural Networks are an information processing technique based on the way biological nervous systems, such as the brain, process information. They resemble the human brain in the following two ways:

1. A neural network acquires knowledge through learning.
2. A neural network's knowledge is stored within inter-neuron connection strengths known as synaptic weights.

Neural networks are being applied to an increasing large number of real world problems. Their primary advantage is that they can solve problems that are too complex for conventional technologies; problems that do not have an algorithmic solution or for which an algorithmic solution is too complex to be defined. In general, neural networks are well suited to problems that people are good at solving, but for which computers generally are not. These problems include pattern recognition and forecasting, which requires the recognition of trends in data. The true power and advantage of neural networks lies in their ability to represent both linear and non-linear relationships and in their ability to learn these relationships directly from the data being modeled. Traditional linear models are simply inadequate when it comes to modeling data that contains non-linear characteristics. The most common neural network model is the multi-layer perceptron (MLP). This type of neural network is known as a supervised network because it requires a desired output in order to learn. The goal of this type of network is to create a model that correctly maps the input to the output using historical data so that the model can then be used to produce the output when the desired output is unknown. A graphical representation of an MLP is shown below in Figure 7.

The MLP and many other neural networks learn using an algorithm called **back-propagation**. With back propagation, the input data is repeatedly presented to the neural network. With each presentation the output of the neural network is compared to the desired output and an error is computed. This error is then fed back (backpropagated) to the neural network and used to adjust the weights such that the error decreases with each iteration and the neural model gets closer and closer to producing the desired output. This process is known as "**training**".

Training of Artificial Neural Networks

Neural networks can be explicitly programmed to perform a task by manually creating the topology and then setting the weights of each link and threshold. However, this by-passes one of the unique strengths of neural nets: the ability to program themselves. The most basic method of training a neural network is trial and error. If the network isn't behaving the way it should, change the weighting of a random link

Figure 7. Graphical representation of Multilayer Perceptron

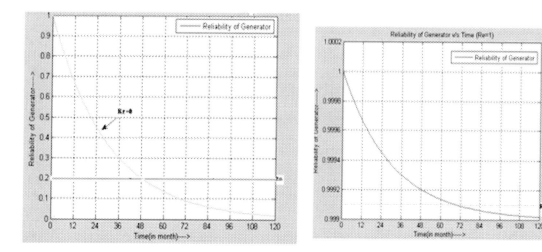

Figure 8. Neural Network

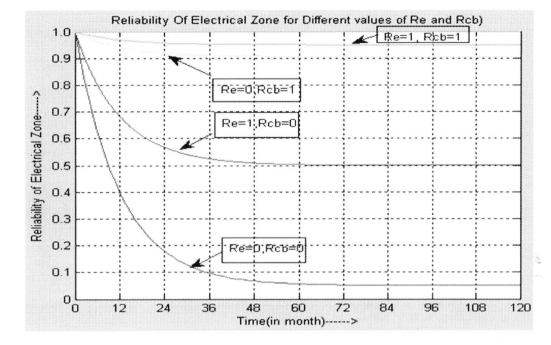

by a random amount. If the accuracy of the network declines, undo the change and make a different one. It takes time, but the trial and error method does produce results. (Figure 8).

The neural network shown in Figure 4.5. Is a simple one that could be constructed using such a trial and error method? The task is to mirror the status of the input row onto the output row. To do this it would have to invent a binary to decimal encoding and decoding scheme with which it could pass the information through the bottle-neck of the two neurons in the centre. Unfortunately, the number of possible weightings rises exponentially as one adds new neurons, making large general-purpose neural nets impossible to construct using trial and error methods. In the early 1980s two researchers, David Rumelhart and David Parker, independently rediscovered an old calculus-based learning algorithm. The back-propagation algorithm compares the result that was obtained with the result that was expected. It then uses this information to systematically modify the weights throughout the neural network. This training takes only a fraction of the time that trial and error method take. It can also be reliably used to train networks on only a portion of the data, since it makes inferences. The resulting networks are often correctly configured to answer problems that they have never been specifically trained on. As useful as back-propagation is, there are often easier ways to train a network. For specific-purpose networks, such as pattern recognition, the operation of the network and the training method are relatively easy to observe even though the networks might have hundreds of neurons.

Testing of Artificial Neural Network

One common approach used to test learned regularities is to divide the data set into two parts and use one part for training and another part for validating the discovered patterns. This process is repeated several times and if results are similar to each other than a discovered regularity can be called reliable for data. Three major methods for selecting subsets of training data are known as:

Figure 9. Flow Chart for Training of Artificial Neural Network

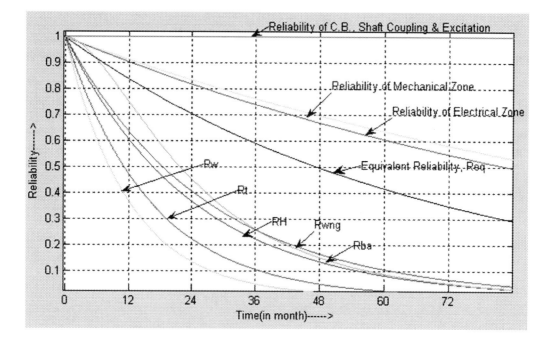

1. Random selection of subsets,
2. Selection of disjoint subsets,
3. Selection of subsets according the probability distribution.

These methods are sometimes called, respectively, bootstrap aggregation (bagging), cross validated committees, and boosting.

Testing of Mathematical Model

The mathematical model of equivalent reliability (R_{eq}) has been developed for the generation system by taking into account the factors which affect the equivalent reliability such as

1. Heater
2. Water Level
3. Blade Angle
4. Shaft Coupling
5. Generator Winding
6. Excitation
7. Circuit Breaker

The mathematical model of equivalent reliability takes into account the reliability for each of above parameters. The equivalent reliability of any Generation System is said to be acceptable if the effects of failure of individual factors does not allow equivalent reliability to fall below the threshold value (R_{Th}).

Thus, to have an overall idea about the reliability, it is necessary to develop and test the model for failure against singular parameter with rest of the parameters considered intact. The model is therefore tested on singular basis to determine its effects on the equivalent reliability.

Reliability of Boiler

The mathematical model for determining the reliability of Boiler, R_B has been developed and it is given by eqn.(4.3) and reproduced as below:

$$R_B = \exp(-\lambda_1 t) + \exp(-\lambda_2 t) - \exp(-\lambda_1 t).\exp(-\lambda_2 t)$$ (ref. eqn. (4.3))

The reliability is basically a function of time and is dependent on water level as well as heater quality.

Case-1: Reliability of Water Level vs. Time

The mathematical model for determining the reliability of water level, R_W has been developed in chapter 4 and it is given in article 4.3.2.1 (page no.-33) and reproduced as below:

$$R_W = \exp(-\lambda_2 t)$$

The reliability is basically a function of time and depends on rate of fall of water level λ_2. When this model is tested, the results are as follows:

Figure 10. Reliability of Water Level vs. Time

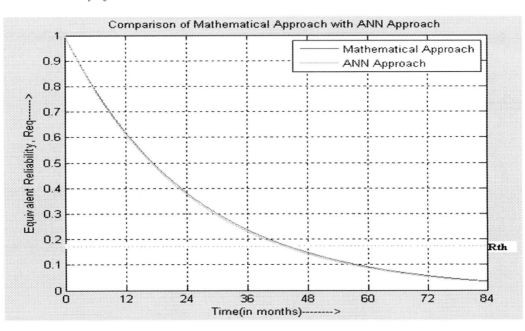

Figure 11. Reliability of Heater vs. Time

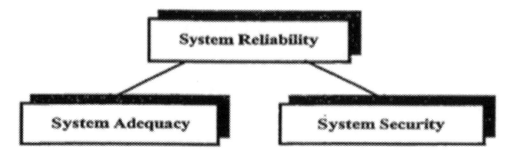

Case-2: Reliability of Heater vs. Time

The mathematical model for determining the reliability of Heater, R_H has been developed in chapter 4 and it is given in article 4.3.2.1 and reproduced as below:

$$R_H = \exp(-\lambda_1 t)$$

The reliability is basically a function of time and depends on failure of heating element, λ_1. When this model is tested, the results are as follows:

Case-3: Integrated Boiler Reliability

The mathematical model for determining the reliability of Boiler, R_B has been developed in chapter 4 and it is given by eqn.4.3 (page no.-33) and reproduced as below:

$$R_B = \exp(-\lambda_1 t) + \exp(-\lambda_2 t) - \exp(-\lambda_1 t) . \exp(-\lambda_2 t) \qquad \text{(ref. eqn.4.3)}$$

The reliability is basically a function of time and is dependent on rate of fall of water level, λ_2 as well as rate of failure of heating element, λ_1. When this model is tested, the results are as follows:

Reliability of Turbine

The mathematical model for determining the reliability of Turbine, R_T has been developed in chapter 4 and it is given by eqn. (4.6) and reproduced as below:

$$R_T = \exp(-\lambda_3 t) + R_{SC} - \exp(-\lambda_3 t) . R_{SC}$$

The reliability is basically a function of time and is dependent on Blade Angle as well as Shaft Coupling.

Figure 12. Integrated Reliability of Heater and Water Level and Boiler

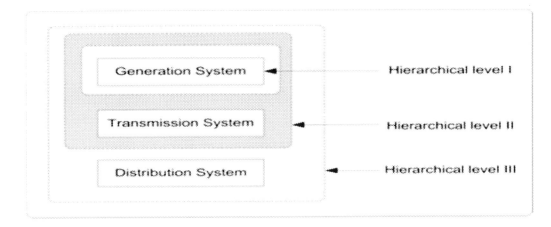

Figure 13. Reliability of Turbine vs. Time

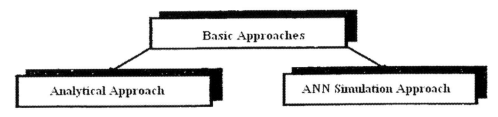

Case-1: For Rsc=0 i.e. Shaft Coupling: Failure State

The mathematical model for determining the reliability of Turbine, R_T has been developed in chapter 4 and it is given by eqn. 4.6 (page no.-34) and reproduced as below:

$$R_T = \exp(-\lambda_3 t) + R_{SC} - \exp(-\lambda_3 t).R_{SC} \qquad \text{(ref. eqn.4.6)}$$

The reliability is basically a function of time and is dependent on rate of failure of blade angle as well as Shaft Coupling. When this mathematical model is tested for Rsc=0, the results are as follows

Case-2: For Rsc=1 i.e. Shaft Coupling: Perfect State

The mathematical model for determining the reliability of Turbine, R_T has been developed in chapter 4 and it is given in article 4.3.2.2 by eqn.4.6, (page no.-34) and reproduced as below:

$$R_T = \exp(-\lambda_3 t) + R_{SC} - \exp(-\lambda_3 t).R_{SC}$$

The reliability is basically a function of time and is dependent on rate of failure of blade angle as well as Shaft Coupling. When this mathematical model is tested for Rsc=1, the results are as follows:

Figure 14. Reliability of Turbine vs. Time (Shaft coupling: perfect state)

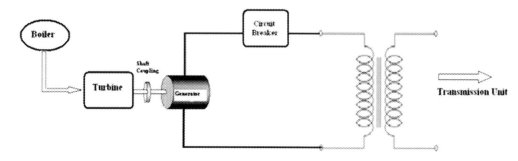

Case-3: Integrated Study

The mathematical model for determining the reliability of Turbine, R_T has been developed in chapter 4 and it is given by eqn. 4.6 and reproduced as below:

$$R_T = \exp(-\lambda_3 t) + R_{SC} - \exp(-\lambda_3 t).R_{SC} \qquad \text{(ref. eqn.(4.6))}$$

The reliability is basically a function of time and is dependent on rate of failure of blade angle as well as Shaft Coupling. When this mathematical model is tested for different values of Rsc (i.e. 0 or 1) the results are as follows:

Figure 15. Reliability of Turbine v/s Time (Integrated Study)

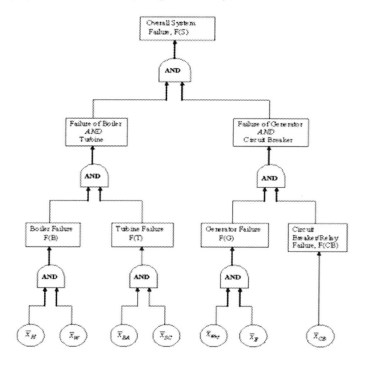

Reliability of Mechanical Zone

The mathematical model for determining the reliability of Mechanical Zone, R_{Mech} has been developed in chapter 4 and it is given in article 4.3.2.5 by eqn.4.15, and reproduced as below:

$$R_{Mech} = 1 - [\{1 - \exp(-\lambda_1 t) + \exp(-\lambda_2 t) - \exp(-\lambda_1 t).\exp(-\lambda_2 t)\}$$
$$\times \{1 - \exp(-\lambda_3 t) + R_{SC} - \exp(-\lambda_3 t).R_{SC}\}]$$

(ref. eqn.4.15)

The reliability is basically a function of time and is dependent on rate of failure of heating element, rate of fall of water level, rate of failure of blade angle as well as Shaft Coupling. When this mathematical model is tested for different values of Rsc (i.e. 0 or 1) the results are as follows:

Reliability of Generator

The mathematical model for determining the reliability of Generator, R_G has been developed in chapter 4 and it is given in article 4.3.2.3 by eqn.4.10, and reproduced as below:

$$R_G = R_{Wng} + R_E - R_{Wng}.R_E$$

The reliability is basically a function of time and is dependent on Generator Winding as well as Excitation.

Case-1: For RE =0 i.e. Exciter: Failure State

The mathematical model for determining the reliability of Generator, R_G has been developed in chapter 4 and it is given by eqn.4.11, and reproduced as below:

$$R_G = \exp(-\lambda_4 t) + R_E - \exp(-\lambda_4 t).R_E$$

(ref. eqn.4.11)

The reliability is basically a function of time and is dependent on rate of failure of generator winding as well as excitation. When this mathematical model is tested for RE =0, the results are as follows:

Case-2: For RE =1 i.e. Exciter: Perfect State

The mathematical model for determining the reliability of Generator, R_G has been developed and it is given by eqn.4.11, and reproduced as below:

$$R_G = \exp(-\lambda_4 t) + R_E - \exp(-\lambda_4 t).R_E$$

(ref. eqn.4.11)

The reliability is basically a function of time and is dependent on rate of failure of generator winding as well as excitation. When this mathematical model is tested for RE =1, the results are as follows:

Figure 16. Reliability of Exciter: failure state

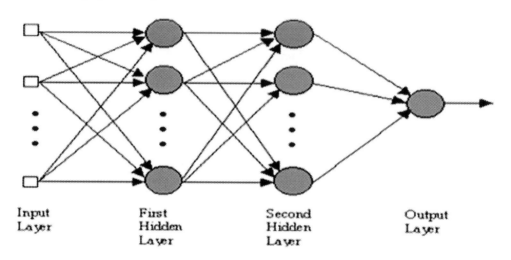

Input Layer   First Hidden Layer   Second Hidden Layer   Output Layer

Case-3: Integrated Study

The mathematical model for determining the reliability of Generator, R_G has been developed in chapter 4 and it is given in eqn.4.11, and reproduced as below:

$$R_G = \exp(-\lambda_4 t) + R_E - \exp(-\lambda_4 t).R_E$$

The reliability is basically a function of time and is dependent on Generator winding as well as Excitation. When this mathematical model is tested for different states of RE (i.e.0 & 1).

Reliability of Circuit Breaker

The mathematical model for determining the reliability of circuit breaker has been developed in chapter 4 and it is given by eqn.4.13, and reproduced as below:

$$R(CB) = R_{CB}$$ (ref. eqn.4.13)

Case-1: Circuit Breaker: Failure State

The mathematical model for determining the reliability of circuit breaker has been developed in chapter 4 and it is given in eqn.(4.13) and reproduced as below:

$$R(CB) = R_{CB}$$

The reliability of circuit breaker depends upon two states (i.e. 0 & 1). When this mathematical model is tested for RCB=0, the results are as follows:

Case-2: Circuit Breaker: Perfect state

The mathematical model for determining the reliability of circuit breaker has been developed in chapter 4 and it is given in eqn.4.13, and reproduced as below:

$$R(CB) = R_{CB}$$

The reliability of circuit breaker depends upon two states (i.e. 0 & 1). When this mathematical model is tested for RCB=1.

Case-3: Integrated Study

The mathematical model for determining the reliability of circuit breaker has been developed in chapter 4 and it is given in eqn.4.13 and reproduced as below:

$$R(CB) = R_{CB}$$

(ref eqn.4.13)

The reliability of circuit breaker depends upon two states (i.e. 0 & 1). When this mathematical model is tested for RCB=0 and RCB=1, the results are as follows:

Reliability of Electrical Zone

The mathematical model for determining the reliability of Electrical Zone, R_{Elec} has been developed in chapter 4 and it is given in eqn.4.17, and reproduced as below:

$$R_{Elec} = 1 - [\{1 - (\exp(-\lambda_4 t) + R_E - \exp(-\lambda_4 t).R_E)\} \times (1 - R_{CB})]$$

(ref. eqn.4.17)

The reliability is basically a function of time and is dependent on rate of failure of generator winding as well as on different states of Exciter and Circuit Breaker. When this mathematical model is tested for different values of RE and R_{CB}, the results are as follows:

Overall Reliability Study

The mathematical model for determining the equivalent reliability, R_{eq} of Generation System, has been developed and it is given in article 4.3.2.7 by eqn. 4.19,and reproduced as below:

$$R_{eq} = 1 - [\{1 + \exp(-\lambda_1 t) - \exp(-\lambda_2 t) + \exp(-\lambda_1 t).\exp(-\lambda_2 t)\}$$
$$\times \{1 - \exp(-\lambda_3 t) + R_{SC} - \exp(-\lambda_3 t).R_{SC}\}]$$
$$\times [\{\exp(-\lambda_4 t) - R_E + \exp(-\lambda_4 t).R_E)\} \times \{1 - R_{CB}\}]$$

(Ref. eqn. 4.19)

Figure 17. Reliability of Electrical Zone vs. Time for Different RE and R_{CB}

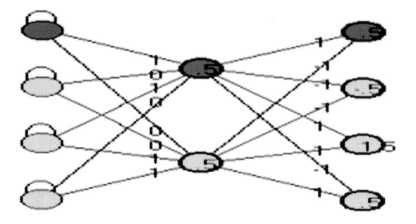

The reliability is basically a function of time and is dependent on rate of fall of water level, rate of failure of heating element, rate of failure of blade angle, rate of failure of generator winding as well as on different states of Exciter, Shaft Coupling and Circuit Breaker. When this mathematical model is tested for the reliability of individual parameters as well equivalent reliability R_{eq}, the results are as follows:

Testing Method of Artificial Neural Network

In order to test the reliability of Generation System on Artificial Neural Network (ANN) a proposed model has been developed, which has been trained as well as tested on MATLAB. The ANN developed model has been trained for seven inputs (i.e. Rw, RH, RSC, RBA, RWng, RE, RCB) and one output (i.e.Req).

Figure 18. Reliability of individual units as well as equivalent reliability, R_{eq} of Generation System

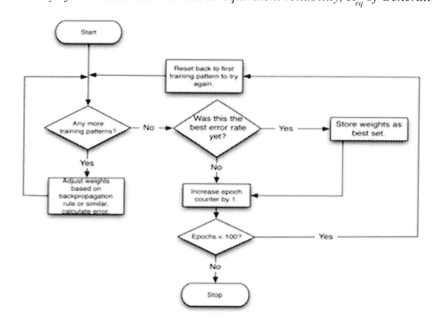

In order to obtain ANN model to test the reliability of Generation System, which is mainly affected by seven parameters. A proposed ANN model is required to be simulated. Thus, the ANN model which represents the dynamics of generation system its failure has been derived by trial and error method on the platform of MATLAB.

In order to test the reliability of generation system on ANN toolbox, the following pprocedure has to adopt to train and test the neural network model:

Step 1: firstly, Open the MATLAB Toolbox and look for the Neural Network Toolbox.

Step 2: Now, select Network/Data Manager tool from the toolbox of Artificial Neural Network.

Step 3: After Selecting the Network/Data Manager we have specified the inputs data, Output data and target in this tool box.

Step 4: To train the proposed ANN model, we have selected the following parameters:

Network Type = Feedforword

Backprop Input Ranges = [0 1; 0 1; 0 1;0 1; 0 1; 0 1; 0 1]

Train Function = TRAINLM

Adaption Learning Function = LEARNGDM

Performance Function = MSE

Numbers of Layers = 2

Step 5: Select Layer 1, type in 2 for the number of neurons, & select TANSIG as Transfer Function.

Select Layer 2, type in 1 for the number of neurons, & select TANSIG as Transfer Function.

Step 6: Then, confirm by hitting the Create button, which concludes the ANN network implementation phrase.

Step 7: Now, highlight ANN Model of generation system with ONE click, then click on Train network button.

Step 8: On Training Info, select INPUTS as Inputs, TARGET as Targets.

On Training Parameters, specify:

epochs = 1000 (since we would like to train the network for a longer duration).

goal = 0.000000000000001 (since we would like to see if the ANN MODEL OF GENERATION SYSTEM that we implemented earlier is capable of producing precise results).

max_fail = 50

After, confirming all the parameters have been specified as indented, hit Train Network. This will give you a training and performance plot. The 350*7 data are used for training and testing the proposed ANN model.

In the training process, some weights are also adjusted to provide the accuracy in training process. After the successful training of 300*7 data, the proposed ANN model is tested for 50*7 data. Which has also shown the similar results.

Illustration Testing and Results of

1. Mathematical Model
2. ANN Model

The data for equivalent reliability are given in Appendix A. Also, the data for ANN Model are given in Appendix B. Both the Mathematical and ANN models were tested for reliability and the results are shown in Fig.5.11.

The reliability study was carried and it was found that it depends upon so many factors, individually and collectively. It is a function of time. The factors were also tested on Artificial Neural Network. After a through training about 300*7data, the Artificial Neural Network also gave similar and Agreeable results for 50*7 testing data. The results of both the studies provided a good validation of reliability model.

CONCLUSION

1. Reliability of operation of any system is the prime attribute of that system which leads to its selection and rejection.
2. Generation System is the most vital part of any power system having many links where the reliability can be question marked. It therefore needed fine study of all the constituent parts, the reliability of each of which ensures the total or equivalent reliability of the Generation System.
3. A Generation System is said to possess the reliability provided its Mechanical unit and Electrical units each possesses the reliability. Also, the reliability of mechanical unit further depends on the reliability of boiler and turbine & reliability of electrical unit further depends on the reliability of Generator and Circuit Breaker.
4. Reliability v/s time study has been carried out for each part that affect the reliability of generation system.
5. Mathematical model for equivalent reliability for both mechanical and electrical units have been found as a function as a function of time and is shown in Fig.5.8 and 5.15 in chapter- 5.
6. The ANN model has been developed by trial and error method for assessing the reliability of generation system. The developed model finds its base in Neural Network Toolbox in MATLAB. The training data set consists of (7:1)*300 data & testing data set consists of (7:1)*50 data. After a successful training of the ANN model by adopting the network type as feed forward and train function as TRAINLM, testing data follows the same pattern, which proves that Artificial Intelligence system also support the mathematical model.
7. Both models gave agreeable results and the reliability of generation system could be assessed along with the fine outcome of the knowledge of threshold value of reliability R_{Th} beyond which the generation system would fail and vice-versa.

This chapter presents the mathematical model of reliability of individual components and equivalent reliability of entire generation system. It suggests the approach to determine the critical reliability of both individual and equivalent reliability of the generation system. The results have been equally tested with Artificial Neural Network (ANN) approach. It has been found that both the methodologies show good and acceptable agreement and therefore the results obtained by mathematical approach stands validated.

162

REFERENCES

Allen, R., Billington, R., & Abdel-Gawad, N. (1986). The IEEE Reliability Test System -Extensions to and Evaluation of the Generating System. *IEEE Trans. on PWRS, 1*(4), 1–7.

Billington, R., & Oteng-Adjei, J. (1991). Utilization Of Interrupted Energy Assessment Rates In Generation And Transmission System Planning. *IEEE Trans. on PWRS, 6*(3), 1245–1253.

Haringa, G., Jordan, G., & Garver, L. (1991). Application of Monte Carlo Simulation to Multi-Area Reliability Evaluations. *IEEE Computer Applications in Power, 4*(1), 21–25. doi:10.1109/67.65031

Kamboj, Bhardwaj, Bhullar, Arora, & Kaur. (2012). Mathematical model of reliability assessment for generation system. *2012 IEEE International Power Engineering and Optimization Conference*, 258-262. 10.1109/PEOCO.2012.6231118

Kumganty, S. (1994). Effect of HVDC Component Enhancement on the Overall Reliability Performance. *IEEE Transactions on Power Delivery, 9*(1).

Munoz-Estrada. Perez-Ruiz, & Barquin. (2004). *Including Combined- Cycle Power Plants in Generation System Reliability Studies*. In The 8th International Conference on Probabilistic Methods Applied to Power Systems, Iowa State University, Ames, IA. doi:10.1007/978-94-015-7949-0

Oliveira, G., Cunha, S., & Pereira, M. (1987). A Direct Method for Multi-Area Reliability Evaluation. *IEEE Trans. on PWRS, 2*(4), 934–942.

Zimmermann, H. J. (1991). *Fuzzy Set Theory and its Applications* (2nd ed.). Kluwer Academic Publishers.

Chapter 9
High Level Transformation Techniques for Designing Reliable and Secure DSP Architectures

Jyotirmoy Pathak

https://orcid.org/0000-0002-9927-7231
Lovely Professional University, India

Abhishek Kumar
Lovely Professional University, India

Suman Lata Tripathi
Lovely Professional University, India

ABSTRACT

Reverse engineering (RE) has become a serious threat to the silicon industry. To overcome this threat, the ICs need to be made secure and non-obvious in order to find their functionality with their architecture. Real-time signal processing algorithms need to be faster and more reliable. Adding up additional circuits for increasing the security of the IC is not permittable due to increase in overhead of the IC. In this chapter, the authors introduce a few high-level transformations (HLT) that are used to make the circuit more reliable and secure against the reverse engineering without having overhead on the IC.

INTRODUCTION

Reverse Engineering (RE) has become a serious threat to the silicon industry. To overcome this threat the IC's needed to be made secure and non-obvious in order to find their functionality with their architecture. Real-time signal processing algorithms need to be faster and more reliable. Adding up additional circuit

DOI: 10.4018/978-1-7998-1464-1.ch009

for increasing the security of the IC is not permit-able due to an increase in overhead of the IC. In this chapter, we introduce a few High-Level Transformations (HLT) which are used to make the circuit more reliable and secure against reverse engineering without having overhead on the IC.

With the sudden blast in smart devices in our home, work, military services it has come into notice that software is not the only thing that needs protection (Anderson, 2006). The gigantic use of IC chips in all devices and the optimum use of the functionality of devices has to lead to increasing in the number of transistors in the chip parabolic-ally as predicted in called as Moore's Law. The increase in chip density results in decreases in the size of the chip (Iqbal, Potkonjak, Dey, Parker, 1993). This decrease in the range of 6nm-7nm has to lead to the design and fabrication of IC chips as a tedious and time-consuming task. This has resulted in the development of various outsourcing companies for Fabrication, Masking, and IP generation task. The increase in outsourcing (Lao & Parhi, 2015) of the chips has made the designer lose his/her control in his design. The security of the IC chips can be breached at many levels. There has been a tremendous increase in cases of Hardware theft. This demand for a method to provide protection against Hardware Theft and Reverse Engineering attacks. This paper deals with a technique called Obfuscation. Obfuscation is a technique of making the design functionally and structurally tedious to reverse engineer. A VLSI circuit designed by an engineer, re-sed by several other people. The circuit is available in various format like RTL/SOC/ASIC (J. Zhang, 2013). This circuit can be protected in order to avoid malpractice. This architecture can be protected by means of intestate property (IP). VLSI industry pays millions of dollars to keep their IP safe. There are two types of IP exits (Moore, 1998) (1) Hard IP and (ii) Soft IP. Soft IP is synthesizable register transfer level (RTL) while Hard IP contains (GDSII format) IC which can be used for direct implementation. Vendors provide these IP on purchase with an agreement for authenticates use. (Gu & Zhou, 2019) A different method of IP protection is (a) legal means such as patent, copyright and trade (b) license agreement and encryption technique (c) determining unauthorized means of usage like hardware watermarking. Addition of IP protection technique enhances the circuit cost but increase the security level; but their severe threat in VLSI industry to protect their IP. Misuse of the IP leads to heavy loss of revenue, it is reported that industry going in loss more than 10000$/per year. There are following mechanism through which data is stealing from hardware of the circuit. Only software mechanism is not sufficient to protect the IC. A complex combination of encryption technology and few hardware based features incorporated.

1. **IP Piracy:** A person or firm purchase IP from a vendor and create another copy by cloning. After making a few changes the sell to another firm ar higher cost (Zhou, 2017).
2. The untrusted firm makes an illegal copy of GDSII (Hard IP) and sells to another firm,
3. Untrusted foundry manufacturer sells IP at different brand
4. Adversary performs post-silicon reverse engineering on IC to create a clone.
5. The designer may misuse the IP during design
6. While an IP is a communication with the other IPs on SoC, those may extract valuable information from that IP.
7. Adversary may misused the IP while SoC communicates with remotely hardware
8. Hardware trojan horse (HTH) (Li & Lach, 2018), (Potkonjak, Nahapeti, Nelson & Massey, 2009) based attack leaks side-channel information while ASCI/SOC perform computation in term of power, delay, radiation, etc.

Complete set of VLSI architecture contains IP from various vendors; all parties are involved d in the IC design flow are vulnerable to different forms of IP (Cocchi, Baukus, Chow & Wang, 2014) infringement which can result in loss of revenue and market share. There is a need for piracy proof VLSI design flow (Lao & Parhi, 2014). Existing design flow do not ensure hardware protection to all parties o IP vendors, chip designers, system designers, and end-users. To protect IP right at all stages require anti-piracy (Alkabani, Koushanfar, and Potkonjak, 2007) VLSI design flow in which IP vendors, designers, and manufacturers take an active part in securing their own rights and ensure that the customer gets an authentic and trustworthy product. Reverse engineering like Obfuscation is a technique that changes a design into other that is functionally equivalent to the original one. The increase in time and effort of Reverse engineering design is directly proportional to the secure the device is. Major two classification of Obfuscation are

Structural Obfuscation- We makes the design structurally complicated so that the adversary cannot understand the working by looking at its RTL.

Functional Obfuscation- We make the output incorrect till a correct key is entered making it impossible for an attacker to understand the functionality till he/she is aware of the key (Koushanfar, 2012). With the nature of activity obfuscation classifieds passive and active obfuscation. A passive obfuscation does not directly affect the functionality of the system. In passive obfuscation soft IP like RTL code, HDL description, inserting scripting in the program and active obfuscation directly affect the functionality. Active obfuscation is also known as key-based obfuscation technique, the obfuscated design works normally at pre-determined secret key otherwise circuit's works incorrectly.

This paper deals with Structural Obfuscation by using High-Level Transformations like Pipelining and Folding. Section 2 describes the methodology of obfuscation based IP protection. The details have been explained in Section 3 of this paper. High-Level Transformation results in a complex DSP design of the circuits with switches timing control which is difficult to understand by looking; in addition to that, the same structure is used for different functions. It can be achieved by controlling the switches present in (Yu, Dofe, Zhang & Frey, 2017) the circuit to behave in a different matter. The switches determine the mode in which the circuit operates. If the switches are correctly synched then the output obtained is correct else it results in a wrong output.

OBSFUCATION BASED IP PROTECTION

Observation based IP protection is a method to keep soft IP protected (Chakraborty, Bhunia, 2009) by making the HDL program difficult to understand for a normal person it relies on cryptographic technique to encrypt the program. Methods of Obsfucation based IP protection are (Parhi & Messerschmit, 1989)

1. Removing the comment from program
2. Changes the internal nodes name
3. Encrypt the source code by a secret key provided by valid vendors and decrypt from the same key by valid customer i.e. symmetrical encryption
4. Decompilation-based RTL obfuscation approach, which requires converting the RTL design into a gate-level netlist
5. Physical unclonable function (PUF) authenticate IC instance based on circuit inheritance property like power or delay. PUF is widely used in hardware fingerprinting.

None of these techniques prevent possible reverse-engineering effort at later stages of the design and manufacturing flow (Kroft, 1974). An alternative method of IP protection is

Authentication based IP protection: Approaches proposed are directed toward embedding a digital watermark in the design. During design modification, it generates challenge-response pair which alert the concern person but no response pair generated during normal functioning. A different way to achieve the same functionality by changing the internal architecture is (Inoue, Henmi, Yoshikawa, Ichihara, 2011)

Watermarking: Watermakrin in IP is achieved through adding unique binary code in the source code. This technique is based on encryption technology mostly used for copyright protection and data hiding

Fingerprinting: Process of including IP buyer identity into the hardware is known as fingerprinting

User identification technique- This technique is also known as a hardware signature, it includes both watermarking and fingerprinting. Watermarking requires a secret key which is a sequence of binary code fuzzified with fingerprint ID of the buyer through a function to generate hardware signature that can be embedded into the hardware IP (Meguerdichian & Potkonjak, 2010).

HIGH-LEVEL TRANSFORM

High-Level Transformations was known since a long back as a technique to reduce area, power, and delay of a DSP circuit. It includes (Sengupta & Roy, 2017) pipelining, folding, unfolding, retiming, etc. Folding is a technique of obfuscation to obtain a circuit with similar structure but different functionality. It makes it difficult for the adversary to understand the structure of the instances in which switches are turned on is not known to him. The design can be opted back to the original structure using the unfolding technique. Folding tremendously decreases the power and delay constraint in the design. There have been researches in the field of Hardware Obfuscation through HLT techniques (Parhi, 1989). This shows the use of folding techniques in case of a filter and how with the use of this the same structure for the filter can work as a 5- tap filter and 3-tap filter on the basis of switch instance. This HLT (Nie & Toyonaga, 2007) technique was a bit modified to provide a bit better obfuscation for security and to make it more difficult to reverse engineer.

RETIMING

It is a transformation technique where the delay element present in the circuit is being re-lactated within the circuit without affecting the overall functionality of the circuit.

This can be well described with a filter expression as

$$w(n) = ay(n-1) + by(n-2)$$

$$y(n) = w(n-1) + x(n)$$
$$= ay(n-2) + by(n-3) + x(n)$$

Figure 1. Retiming approach of DSP architecture

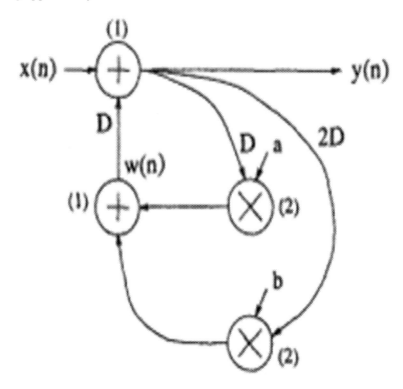

Now changing the locations of delay elements in the filter we get the new expression as

$$w1(n) = ay(n-1)$$

$$w2(n) = by(n-2)$$

$$y(n) = w1(n-1) + w2(n-1) + x(n)$$
$$= ay(n-2) + by(n-3) + x(n)$$

In the above example, we changed the location of the delay element but the overall character of the filter remains the same (Parhi & Messerschmitt, 1991).

This technique is a well-known technique which was used for making the circuit as a pipelined circuit for increasing the iteration bound. Here the same technique is being used with other HLT techniques for increasing the complexity of the circuit so that the reverse engineering (Parhi, Wang & Brown, 1992) could become difficult.

FOLDING

In this technique, different kind of algorithms is being implemented on the same hardware functional unit with the systemematic way of operation through time-multiplexed manner. For making the circuit pipelined with a reduced number of the functional unit we use the folding technique (Sengupta, Roy, Mohanty, Corcoran, 2017).

The technique can be well explained with the following expression

$$y(n) = a(n) + b(n) + c(n)$$

Now applying the folding technique to the above architecture we can reduce the functional unit from 2 adders to 1 single adder with one stage of pipelining.

By applying the folding technique (Sengupta, Roy, Mohanty & Corcoran, 2017) on the above circuit we notice that input sample must remain the same for the 2 clock cycle. In general, in any folded architecture the input values have to remain the same for N clock cycle, where N is the number of different algorithms applied on the same hardware functional unit.

Figure 2. Folding in DSP architecture

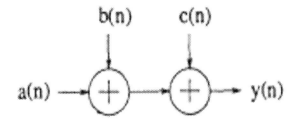

Figure 3. Folding technique to reduce functional unit

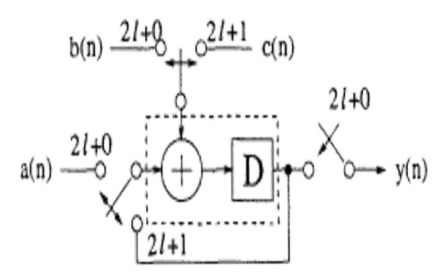

WALLACE TREE MULTIPLIER

Wallace Tree multiplier is a parallel DSP multiplier (Rostami, Koushanfar, Rajendran & Karri, 2013) which is used to multiply 2 unsigned numbers. Many research has been done on this architecture but none have been used to provide a Folded secure version of architecture which will be done in this paper. The Wallace tree multiplier the number in the steps given below:-

In the conventional architecture of Wallace tree multiplier total of 66 full adder blocks are required as the basic functional unit. Now applying (Potkonjak, 2010) Folding and retiming technique on the conventional Wallance tree architecture the architecture is made pipelined and the total number of adders is reduced to 8 in numbers.

Figure 4. Wallace tree multiplier algorithm

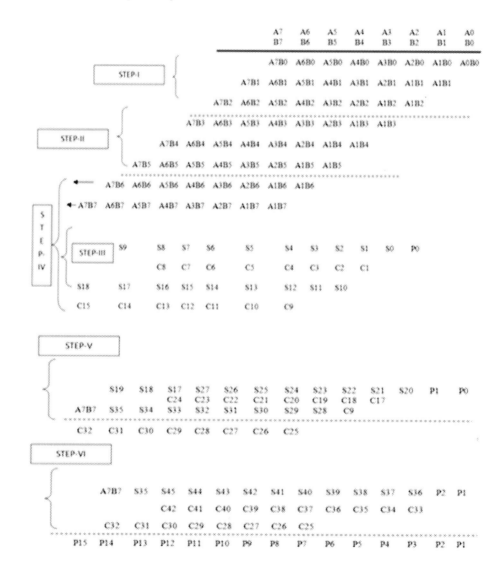

Figure 5. Modified Wallace tee multiplier

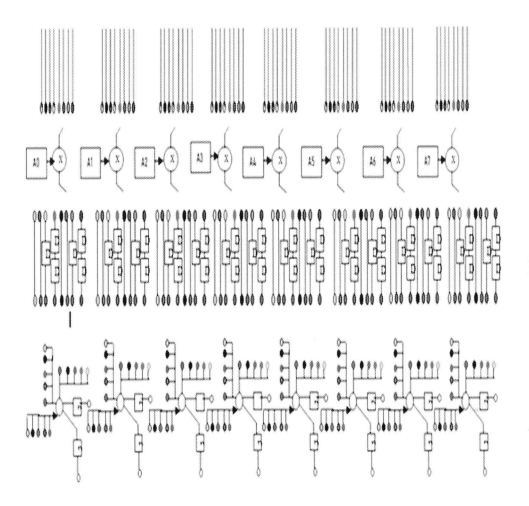

The new modified pipelined Wallance tree architecture have reduced the number of functional units without changing the overall functionality of the circuit with proper time multiplexing the signals.

RELIABILITY AND SECURITY

A system is considered to be secure and more reliable if reverse engineering is not possible or is very difficult to perform. A folded architecture of digital signal processing algorithm have time-multiplexed systematically arranged signals for using the same functional block which not only reduces the number of hardware units required (Dutt & Li, 2009) but also provide the security as without the proper information of the switching clock the functionality which is being generated through the system is not being possible to find out. This way of providing security to the system by changing the architecture of the system is known as structural obfuscation.

Table 1. Comparative study of conventional and modified Wallace tree more multiple

Architecture Constraints	Wallace Tree (Conventional)	Folded Wallace Tree
Area	909	1710
Power (mW)	338	267
Delay(ns)	26.988	3.324

SIMULATION RESULT

The conventional Wallance tree multiplier architecture and the folded Wallance tree multiplier architecture are being implemented on Cadence with 180nm technology and the power delay area is being calculated and compared. The achieved results are being shown in the table.

On comparing the data obtained the folded architecture of Wallace Tree multiplier is speed have increased and the power dissipation is being reduced with an increase in the area which is due to the increasing the number of switching elements into the architecture. This folded architecture is much faster and secure against reverse engineering which increases the reliability of the system.

REFERENCES

Alkabani, Y., Koushanfar, F., & Potkonjak, M. (2007). Remote activation of ICs for piracy prevention and digital right management. *Proc. Int. Conf. Comput.-Aided Design*, 674-677. 10.1109/ICCAD.2007.4397343

Anderson, D. R. (2006). US patent infringement statute extends to the international market for copies of US software. *Journal of Intellectual Property Law & Practice*, 1(4), 234–235.

Chakraborty, R. S., & Bhunia, S. (2009). HARPOON: An obfuscation-based SoC design methodology for hardware protection. *IEEE Transactions on Computer-Aided Design of Integrated Circuits and Systems*, 28(10), 1493–1502. doi:10.1109/TCAD.2009.2028166

Cocchi, R. P., Baukus, J. P., Chow, L. W., & Wang, B. J. (2014, June). Circuit camouflage integration for hardware IP protection. *Proceedings of the 51st Annual Design Automation Conference*, 1-5 10.1145/2593069.2602554

Dutt, S., & Li, L. (2009). Trust-based design and check of FPGA circuits using two-level randomize ECC structure. *ACM Transactions on Reconfigurable Technology and Systems*, 2(1), 1–36. doi:10.1145/1502781.1508209

Gu, J., Qu, G., & Zhou, Q. (2019), Information hiding for trusted system design. *Proceedings of the 46th Design Automation Conference (DAC '09)*, 698–701.

Inoue, T., Henmi, H., Yoshikawa, Y., & Ichihara, H. (2011). High-Level Synthesis for Multi-Cycle transient fault Tolerant Datapaths. *Proc. 17 the IEEE International On-Line Testing Symposium*, 13-18. 10.1109/IOLTS.2011.5993804

Iqbal, Z., Potkonjak, M., Dey, S., & Parker, A. C. (1993). Critical Path Minimization Using Retiming and Algebraic Speed-Up. *30th ACM/IEEE Design Automation Conference*, 573-577. 10.1145/157485.165046

Koushanfar, F. (2012). *Hardware metering: A survey. In Introduction to Hardware Security and Trust* (pp. 103–122). New York, NY: Springer. doi:10.1007/978-1-4419-8080-9_5

Kroft, D. (1974). Comments on A Twos Complement Parallel Array Multiplication Algorithm. *IEEE Transactions on Computers*, *C-23*(12), 1327–1328. doi:10.1109/T-C.1974.223863

Lao, Y., & Parhi, K. K. (2014). Protecting DSP circuits through obfuscation. *IEEE International Symposium on Circuits and Systems (ISCAS)*, 798-801. 10.1109/ISCAS.2014.6865256

Lao, Y., & Parhi, K. K. (2015). Obfuscating DSP Circuits via High-Level Transformations. *IEEE Transactions on Very Large Scale Integration (VLSI) Systems*, *23*(5), 819–830.

Li, J., & Lach, J. (2018). At-speed delay characterization for IC authentication and Trojan horse detection. *Proceedings of the IEEE International Workshop on Hardware-Oriented Security and Trust (HOST '08)*, 8–14.

Moore, G. (1998). Cramming More Components Onto Integrated Circuits. *Proceedings of the IEEE*, *86*(1), 82–85. doi:10.1109/JPROC.1998.658762

Nie, T., & Toyonaga, M. (2007). An efficient and reliable watermarking system for IP Protection. *IEICE Transactions on Fundamentals of Electronics, Communications and Computer Science*, *E90-A*(9), 1932–1939. doi:10.1093/ietfec/e90-a.9.1932

Parhi, K. (1989). Algorithm transformation techniques for concurrent processors. *Proceedings of the IEEE*, *77*(12), 1879–1895. doi:10.1109/5.48830

Parhi, K., & Messerschmitt, D. (1989). Pipeline interleaving and parallelism in recursive digital filters. I. Pipelining using scattered look-ahead and decomposition. *IEEE Transactions on Acoustics, Speech, and Signal Processing*, *37*(7), 1099–1117. doi:10.1109/29.32286

Parhi, K., & Messerschmitt, D. (1991). Static rate-optimal scheduling of iterative data-flow programs via optimum unfolding. *IEEE Transactions on Computers*, *40*(2), 178–195. doi:10.1109/12.73588

Parhi, K., Wang, C.-Y., & Brown, A. (1992). Synthesis of control circuits in folded pipelined DSP architectures. *IEEE Journal of Solid-State Circuits*, *27*(1), 29–43. doi:10.1109/4.109555

Potkonjak, M., Nahapetian, A., Nelson, M., & Massey, T. (2009). Hardware trojan horse detection using gate-level characterization. *Proceedings of the 46th Design Automation Conference (DAC '09)*, 688–693. 10.1145/1629911.1630091

Potkonjak. (2010). Synthesis of trustable ICs using untrusted CAD tools. *Proceedings of the Design Automation Conference*, 633–634.

Rostami, M., Koushanfar, F., Rajendran, J., & Karri, R. (2013, November). Hardware security: Threat models and metrics. In *Proceedings of the International Conference on Computer-Aided Design*, (pp. 819-823). IEEE Press.

Sengupta, A., & Roy, D. (2017). Protecting an intellectual property core during architectural synthesis using high-level transformation based obfuscation. *Electronics Letters*. doi:10.1049/el.2017.1329

Sengupta, A., Roy, D., Mohanty, S., & Corcoran, P. (2017). DSP Design Protection in CE through Algorithmic Transformation Based Structural Obfuscation. *IEEE Transactions on Consumer Electronics*, *63*(4), 467–476. doi:10.1109/TCE.2017.015072

Sengupta, A., Roy, D., Mohanty, S. P., & Corcoran, P. (2017). DSP design protection in CE through algorithmic transformation based structural obfuscation. *IEEE Transactions on Consumer Electronics*, *63*(4), 467–476. doi:10.1109/TCE.2017.015072

Wei, S., Meguerdichian, S., & Potkonjak, M. (2010). Gate-level characterization: foundations and hardware security applications. *Proceedings of the Design Automation Conference (DAC '10)*, 222–227. 10.1145/1837274.1837332

Yu, Q., Dofe, J., Zhang, Y., & Frey, J. (2017). Hardware Hardening Approaches Using Camouflaging, Encryption, and Obfuscation. *Hardware IP Security and Trust*, 135–163.

Zhang, J. (2013). FPGA IP protection by binding finite state machine to physical unclonable function. *Proc. 23rd Int. Conf. Field Program. Logic Appl. (FPL)*, 1-4. 10.1109/FPL.2013.6645555

Zhou, H. (2017). Structural Transformation-Based Obfuscation. Hardware Protection through Obfuscation, 221–239.

Chapter 10
Design for Testability of High–Speed Advance Multipliers:
Design for Testability

Suman Lata Tripathi
ⓘD https://orcid.org/0000-0002-1684-8204
School of Electronics and Electrical Engineering, Lovely Professional University, India

ABSTRACT

An efficient design for testability (DFT) has been a major thrust of area for today's VLSI engineers. A poorly designed DFT would result in losses for manufacturers with a considerable rework for the designers. BIST (built-in self-test), one of the promising DFT techniques, is rapidly modifying with the advances in technology as the device shrinks. The increasing complexities of the hardware have shifted the trend to include BISTs in high performance circuitry for offline as well as online testing. Work done here involves testing a circuit under test (CUT) with built in response analyser and vector generator with a monitor to control all the activities.

INTRODUCTION

This chapter covers a basic introduction to testing, problems incorporated, Built-in-Self Test (BIST) (Stroud, 2002 & Hussain, 2013) including the major advantages and disadvantages of BIST along with its hardware implementation. The aim here is to launch the basic ethics of BIST and how it fits into the overall testing process. Therefore, a general description of testing is illustrated to explain the imperative aspects of testing that are needed for implementation of BIST module on FPGA kit. Secondly, a fault model is proposed for the 4-bit Wallace tree multiplier circuit and test pattern generated. The verification of different fault models has been carried out using FPGA board for reliable operation of the multiplier. In high performance systems such as a microprocessor, DSP, etc. the arithmetic addition and multiplication of two binary numbers are fundamental and most often used operations. This work presents the process of design implementation for a complete BIST working on both normal operation mode as well as test mode for a 4 -bits Wallace tree multiplier circuitry.

DOI: 10.4018/978-1-7998-1464-1.ch010

MULTIPLIER

Multipliers are now becoming an integral part of today's digital circuit such as a microprocessor, DSP, etc. The addition and multiplication are fundamental arithmetic operations for two binary numbers. It was found that 70% DSP and microprocessor instructions are based on addition and multiplication operations and dominates the overall execution time. So, the high-speed processing has become a demand with expanding computer and signal processing applications. The DSP based multiplier must be designed for low power dissipation. An optimized number of operation leads to a reduction in dynamic power consumption and hence reducing total power consumption. So, the designer must concentrate on low power-efficient and high-speed circuit design. The objective should be designing a physically packed multiplier with high speed and low power dissipation. A two operand addition circuit with radix-4 partial-product generation and an addition were used to implement 2's-complement array multipliers. A modern Xilinx FPGAs was used that is based on a 6-input LUT architecture for the implementation of high performance array multiplier (Walter, 2016). A fast Montgomery multiplier was proposed using Vertex-7 FPGA that works iteratively where a digit of an operand is multiplied by another digit of an operand, the accumulated result is reduced by Montgomery method (Erdem, 2018). Here, in the present chapter, a 4-bit Wallace tree multiplier is designed on FPGA Artix-7 board for the delay, power and fault analysis. The fault analysis is described with stuck-at and bridging fault models for 4-bit Wallace tree multiplier.

WALLACE TREE

A Wallace tree multiplier is an efficient digital hardware implementation that multiplies two numbers devised by scientist Chris Wallace. The Wallace tree uses carry select adders for the addition of optimum generated partial products. The Wallace tree has three steps:

- Each bit of the arguments is multiplied by each bit of the other, resulting n^2 results. Resulting wires carry different weights depending on the position of the multiplied bits.
- Reduced the number of partials products to two by layers of full and half adders.
- Two-wire groups are there that are added with a conventional adder.

FIELD PROGRAMMABLE GATE ARRAY (FPGA)

The Nexys4 board is an user-friendly digital circuit development board based on the latest Artix-7™ Field Programmable Gate Array (FPGA) from Xilinx. The large and high- capacity FPGA (Xilinx part number XC7A100T-1CSG324C) has generous external memories, Ethernet, collection of USB and other ports. Designers can implement introductory combinational circuits to powerful embedded processors on Nexys4 board (Figure 2). Nexys4 board has several built-in peripherals such as accelerometer, MEMs digital microphone, temperature sensor, speaker amplifier and I/O devices that allow a wide range of designs implementation without needing any other components.

The Artix-7 FPGA offers more capacity, higher performance, and more resources than earlier designs. Artix-7 100T description include:

Figure 1.Wallace tree multiplier block diagram

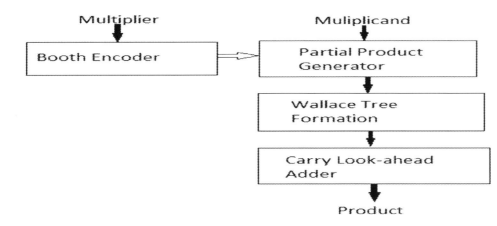

- 15,850 logic slices, each slice have four 6-input Look-Up-Tables(LUT) with 8 flip-flops
- A fast block RAM with 4,860 Kbits
- Six clock management tiles including phase-locked loop (PLL) in it
- 240 DSP slices
- Internal clock speeds more than 450MHz
- On-chip analog-to-digital converter (XADC)

TESTING FUNDAMENTALS

The life-cycle of a product has three stages design, fabrication and testing. The testing phase is of extreme importance and useful in reliable and assured circuit performance. Firstly, testing performed at design phase in the form of design authentication that seeks to detect errors and recognize in order to ensure

Figure 2. Nexys4 FPGA board

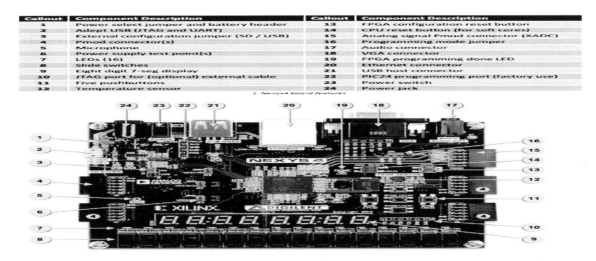

the fabricated product performance according to its planned functionality. In the fabrication phase, testing seeks to detect any fabricating defects that would prevent a given part of the printed circuit board, integrated circuit etc. and ensure the performance in the final design for which it is intended. Lastly, testing seeks to recognize different faults persist during operation that would produce the inconsistent operation of that system. In some systems, testing also identifies the faulty components and replaces it by spare one in order to resume the fault-free operation of the system. The quality and quantity of in-situ testing during the life-cycle of the product depends on the assets of the final product. For instance, a low-cost, high-volume simple calculator when compared to a high-reliability, high-availability, complex telephone system. A calculator may never undergo testing beyond that of manufacturing tests, and if the fault occurs during operation, the product is thrown away. As per necessities and perception of the quality of the product, the designers, product engineers and test engineers must think over the possible product life-cycle for their components.

In every testing problem, the design and development of a product must be provided with superior testing process in a gainful manner. This goal has become more problematic to achieve as Very Large Scale Integrated (VLSI) circuit and printed circuit board (PCB) component concentrations increase to a large extent. Several companies reported that testing costs are sometimes more than 55% of the total product cost. This includes product expansion cost, where test progress accounts for half the total progress cost, as well as manufacture costs where the testing cost is half of the total manufacturing cost.

DESIGN VERIFICATION AND PRODUCT DEVELOPMENT

Design confirmation is an important testing feature of product progress. The intent is to verify that the design meets the system necessities and conditions. Typically, a number of different simulation techniques are used including high-level simulation through a combination of behavioral modeling and test benches (Figure 3). Register Transfer Level (RTL) models of the detailed design are then established and confirmed with the same test benches that were used for confirmation of the architectural design. These latter levels of design notions provide the capability to perform supplementary design confirmation through logic, switch-level, and timing simulations. These three levels of design notions also provide the basis for fault models that will be used to evaluate the efficiency of manufacturing tests.

Changes in system necessities or conditions later in the design cycle endanger the plan as well as the value of the design. Late changes to a design represent one of the two most significant risks to the overall mission, the other risk being insufficient design confirmation. The excellence of the design verification process is dependent on the ability of the test benches, functional vectors, and the designers who analyze the simulated responses to detect design errors.

TESTING AT MANUFACTURING LEVEL

Once a design has been confirmed and is ready for fabricating, testing at the various points throughout the process tries to find faults that may have resulted during fabrication. For VLSI devices, the first test is wafer-level testing (Figure 4). Each chip (or die) on the wafer is tested and when a defective chip is found, the wafer-level test machine jets a dot of ink onto the chip to indicate it is defective. The wafer is then cut into chips the pass chips are packaged. The packaged chips are then again tested; this is known as

Figure 3. Design verification and Product development

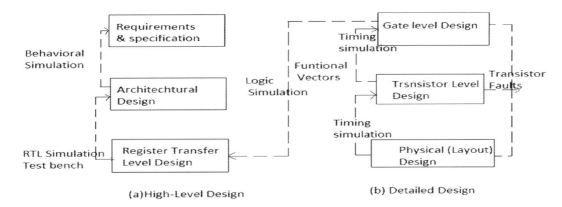

(a)High-Level Design   (b) Detailed Design

device-level testing or package testing. Wafer and device-level tests find any defects that could have been continued during the manufacturing or packaging processes. Here, test vectors are used are functional Vectors produced during the designing stage with addition to target vectors that aim to locate potential faulty sites. These faulty sites are produced from the gate, transistor, and physical designs. The efficiency of the complete set of test vectors used by the ATE can be assessed through fault simulation modeling where fault models are matched to determine if the set of test vectors used will certainly detect these possible faults. Functional vectors depend on system-level complete design with the conditions for which it was designed.

When too expensive and complex PCBs and systems cannot afford to be discarded when faults are detected during these tests, a repair/finding procedure (or some similar testing procedure) is required to recognize the faulty part for replacement in order to make the system or PCB working again. There are situations wherein the system produces fault after the system is sold, referred to as field faults. In this case, this system is called back to the manufacturing unit for repair. This process includes troubleshooting the faults, identification, replacement, retest to check for the system operation functionality. This process is cumbersome and yields in economic, logistics, manpower and time loss.

Figure 4. Wafer-level testing

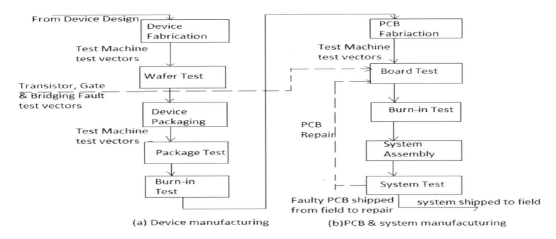

(a) Device manufacturing   (b)PCB & system manufacuturing

THE TESTING PROBLEM

Testing has converted into a tougher problem over the years. As the number of input-output pins for most chips increases by an order of exponential growth, the number of transistors contained in many chips also has increased by four times that of this growth. Manufacturing defects have increased due to device sizes becoming larger, creating a larger area to sustain them. Transistor and wire sizes have become smaller, increasing the number and types of defects that occur during the manufacturing process. For example, thinner wires lead to more open spaces while wires that are near together lead to more shorts. Higher operating frequencies related to the smaller device sizes also enhance the testing problem generating defects and ultimately faulty operation of the system.

HDLs like VHDL and Verilog descriptions, along with computerized synthesis tools have detached the designer from the gate and transistor-level knowledge with their design. It is not shocking that test procedure costs have grown from about 25% to around 50% of the total product manufacturing costs in just over a decade. In addition, more expensive ATE test machines are required to handle the greater number of I/O pins, higher operating frequencies, and the larger sets of test vectors typically related with the more complex VLSI devices. As a result, test machines in excess of one million dollars are now generic. Forecasts for the future indicate that the testing problem will only get poorer by the year 2020.

Finally, due to the increasing difficulties of VLSI devices and PCBs, the ability to provide some level of fault analysis during fabricating, testing is needed to assist failure mode analysis (FMA) for yield improvement and repair measures. BIST is considered to be one of the chief solutions to these forecasted and rising testing difficulties.

BIST AND ITS OPERATION

The basic idea of BIST, in its most modest form, is to strategize a circuit so that the circuit can test itself and conclude whether it is 'good' or 'bad' or 'fault-free' or 'faulty', respectively. This typically requires that extra circuitry and functionality be combined into the design of the circuit to help the self-testing feature (Stroud, 2002). This extra functionality must be capable of producing test vectors as well as providing a mechanism to check if the output responses of the Wallace tree to the test vectors corresponding to that of a fault-free circuit.

BASIC BIST ARCHITECTURE

An illustrative architecture of the BIST circuitry as it might be combined into the CUT is demonstrated in the block diagram (Figure 5). This BIST architecture includes two crucial functions as well as two additional functions that are required to simplify the execution of the self-testing feature while in the circuit. The two vital purposes comprise the test pattern generator (TPG) and output response analyzer (ORA). While the TPG produces a sequence of vectors for testing the Wallace tree multiplier, the ORA compacts the output replies of the Wallace tree multiplier into Pass/Fail signal. BIST works with two different modes, normal and test. Normal mode accepts test vectors from primary inputs and feeds it to Wallace tree multiplier and produces respective output, while in the test mode inputs vectors are gener-

Figure 5. Basic BIST architecture

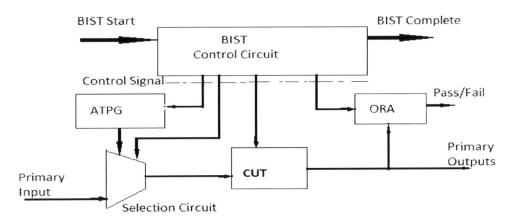

ated from CFSR and given to Wallace tree multiplier. The Output Response Analyzer is a comparator which compares Wallace tree Multiplier output with ROM values stored (Hussain, 2013).

ROM is very crucial at this stage for correct response comparison. The response is either pass or fail, pass meaning BIST is able to sensitize the errors in the Multiplier, fail meaning otherwise. Aside from the usual system I/O pins, the integration of BIST may also require extra I/O pins for activating the BIST sequence (the BIST Start control signal), reporting the results of the BI.

DESIGNING AND IMPLEMENTATION

Multiplication is another common operation that is found in the ALUs of most modern processors. As for all electronic logic circuits, an arithmetic multiplier is used based on two binary operands, referred to as the multiplier and the multiplicand. A set of partial products is obtained from each multiplier bit similar to that a human performs long multiplication by hand but in binary. Each partial product results in a zero value if its corresponding multiplier bit has zero value, and equal to the multiplicand shifted left by the suitable number of positions if the multiplier bit is one. All partial products are then added with multiple adders to obtain the final product.

WALLACE -TREE MULTIPLIER

The Wallace-tree multiplier includes the partial products of multiplication and different groups of constituent's bits according to their weight (Figure 6). The weight of a particular bit depends on its position – for instance, the least significant bit has weight $2^0 = 1$ while bit 3 bit has weight $2^3 = 8$. These bits are then reduced by layers of half adders and full adders in a tree structure to compute the final product from the partial products. The Wallace structure operates with multiple layers that reduce the number of bits with the same weight at each stage. The operation performed depends on the number of bits in the layer

One: Pass the bit down to the next layer.

Two: Add the bits together with a half adder, passing the sum to the same weight in the next layer and the carry-out to the next weight in the next layer.

Three or more: Add any three bits together with a full adder. Pass the sum and any remaining bits to the same weight next layer. Pass the carry –out to the next weight in the next layer.

The primary advantage of the Wallace tree is that it has a significantly smaller number of logic gates as compared to an array multiplier. The tree structure requires log(n) reduction layers with each layer containing at most 2n adder, so on more than 2nlog(n) adders are required as opposed to the n^2 adders required for an array multiplier. The use of an MBE halves the number of partials products, so that the number of adders required is further reduced to 2nlog (n/2) in this instance. The MBE itself requires n/2 partial product logic blocks, with each logic block requiring approx. twelve gates to evaluate the partial product.

The disadvantage of the Wallace-tree is that in contrast to an array multiplier, it has an irregular layout and wiring structure. This is because the higher weights have more wires and therefore require more adders than the lower weight wires. These extra adders also require additional internal wiring to connect them all up correctly.

LINEAR FEEDBACK SHIFT REGISTERS

The LFSR is one of the most commonly used TPG executions in BIST applications. One motive for this is that an LFSR is more area effective than a counter, requiring less combinational logic per flop. There are two basic kinds of LFSR implementations (Figure 7), the internal *LFSR* feedback and external feedback LFSRs (Kasunde, 2013 & Praveen, 2018 & Rao, 2013). The external feedback LFSR in Figure 7b best demonstrates the source of the name of the circuit: a shift registers with feedback paths that are linearly joint via the exclusive-OR gates. Both LFSRs are duals of each other. Both employments require the same quantity of logic in terms of Ex-OR gates and flops. In the external LFSR, there are two Ex-OR gates in the worst-case path from the output of the last flop in the shift register to the input

Figure 6. Algorithm of Wallace-Tree Multiplier

| | | | | a3 | a2 | a1 | a0 |
				b3	b2	b1	b0
				a3b0	a2b0	a1b0	a0b0
			a3b1	a2b1	a1b1	a0b1	
		a3b2	a2b2	a1b2	a0b2		
	a3b3	a2b3	a1b3	a0b3			
		r5	r4	r3	r2	r1	r0
		c4	c3	c2	c1		
	a3b3	a2b3	a1b3	a0b3			
	p6	p5	p4	p3	p2	p1	p0
	c_5	c_4	c_3	c_2			
c6 s6		s5	s4	s3	s2	s1	s0

Figure 7. LFSR implementations

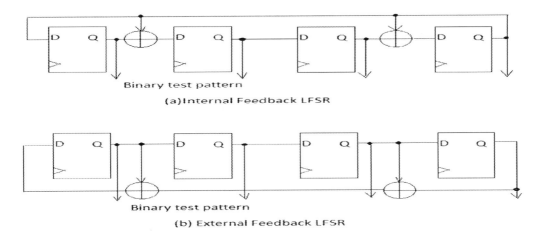

(a)Internal Feedback LFSR

(b) External Feedback LFSR

of the first flop in the shift register. On the other hand, the internal LFSR (Figure 7a) has one Ex-OR gate on any path between flip-flops (Walter, 2016). Therefore, the internal feedback LFSR provides employment with the highest extreme operating frequency for use in high-performance applications. The main advantage of external feedback LFSRs is the consistency of the shift register; hence, in some applications, external feedback LFSRs is favored.

The order of test vectors produced at the outputs of the flip-flops in an LFSR is a function of the exclusive-OR gates placed in the feedback path. This is illustrated by the state diagrams associated with the two LFSR employments. Both LFSRs are internal feedback type employments but when set to any state other than all 0s state the LFSR in Figure 8b goes through all possible states except the all 0s state repeats the order. The longest possible arrangement of vectors for any n-bit LFSR (Figure 8) will be 2^n-1 since any LFSR initialized to the all 0s state will remain in that same state itself. An LFSR that generates a sequence of 2^n-1 unique vectors before repeating is referred to as a maximum length sequence LFSR.

Figure 8. LFSR state diagrams

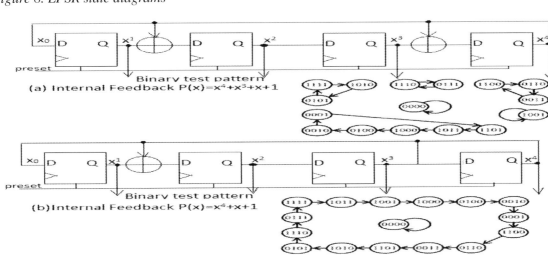

READ ONLY MEMORY (ROM)

Read-only memory can read in normal system operation but not changed. It is a non-volatile, that is, the stored information is retained even if the power is removed. It is classified as follows:

- Mask Programmable ROM (MROMs)
- UV Erasable Programmable ROM(EPROM)
- One Time Programmable EPROM
- Electrically Erasable and Programmable ROM
- Flash Memory

COMPARATOR

Comparator is used here to compares two binary words that indicate whether they are equal or not. Some comparators also tell about signed or unsigned numbers and the arithmetic relationship between the words. These are also called as magnitude comparators that indicate Pass/Fail showing a voluntary indication to BIST completion.

IMPLEMENTATION OF BIST MODULE AND FAULT MODELING

Meanwhile, the devices become more portable and handy; power reduction in today's VLSI circuits is a key parameter step for designing and testing. Interestingly, the power consumed in the testing process is higher than that in normal operation. Four major reasons assumed for power rise when the circuit being tested are

- Rapid transition of the test bits,
- Simultaneous activation of the whole circuit during the test,
- Extra test circuitry, and
- Low correlation among test vectors

Built-In Self-Test (BIST) is a DFT methodology that tests the aimed circuit by checking for faults and produces a response saying it has passed the test or not. This self-test feature comes with a tradeoff – More area requirement and higher power consumption due to extra circuit added to the functional circuit. Automatic Test Vector Generator (ATPG) and Output Response Analyzer (ORA) are the most essential part of this additional circuit. Most widely accepted vector generator is a linear feedback shift register (LFSR) which outputs pseudorandom test vectors. Modifications are performed on the classical model to improve any of the above-mentioned problems. One of them is used in proposed BIST implementation. Counters, compression-based, accumulators, signature analysis, concentration, etc. are some of ORAs which lessens the output response and makes it easier for the controller to give results. The so-called golden vectors (fault-free responses) are stored in ROM memory of the circuit.

Figure 9. Implemented BIST Blocks

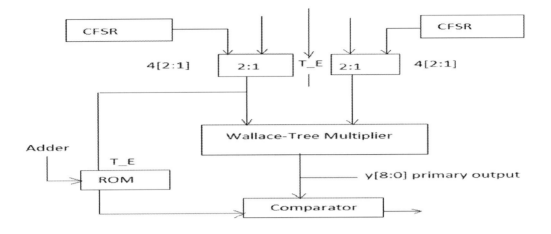

IMPLEMENTED BIST ARCHITECTURE

The architecture implemented uses bit-swapping CFSR (Issa, 2008 & Mohd, 2017) as shown in Figure 9, instead of classic LFSR to reduce the dynamic power consumption in the whole circuit with a penalty of very little area overhead. A 4-bit BS-LFSR was designed with an active area of 1241.1588um² and power consumption of 53.8844nW (power savings=19.43%) (Mohd, 2017). The CUT is predetermined to be of 4-bit which here is an RC (Ripple Carry) adder and CL (Carry Look ahead) adder. The BIST controller generates a control signal scan enable 'SE' to various blocks in the BIST to activate the test mode. A selection circuit which is a 2:1 multiplexer isolates primary inputs during test modes from those inputs coming from test vector generator.

BIST works with two different modes, normal and test. Normal mode accepts test vectors from primary inputs and feeds it to CUT and produces respective output, while in the test mode inputs vectors are generated from BS-CFSR and given to CUT. The Output Response Analyzer is a comparator which compares the CUT output with ROM values stored. The synchronization between TPG and ROM is very crucial at this stage for correct response comparison. The response is either pass or fail, pass meaning BIST is able to sensitize the errors in the CUT, fail meaning otherwise. Figure 9 shows the implemented BIST architecture using BS-CFSR as vector generator.

FAULT MODELING

Structural testing with fault models involves verifying each unit (gate and flip flop) is free from faults of the fault model (Dislis, 1991 & Bushnell, 2000).

WHAT IS FAULT MODELS?

A model is an abstract representation of a system. The abstraction is such that modeling reduces the complexity in representation but captures all the properties of the original system required for the application in question.

WIDELY ACCEPTED FAULT MODELS

Many fault models were proposed, but the ones widely accepted are as follows

a) **Stuck-at fault model**: In this model, faults are fixed (0 or 1) value to a net which is an input or an output of a logic gate or a flip-flop in the circuit. If the net is stuck to 0, it is called stuck-at-0 (s-a-0) fault and if the net is stuck to 1, it is called stuck-at-1 (s-a-1) fault. If it is assumed that only one net of the circuit can have a fault at a time, it is called single stuck-at fault model. Without this assumption, it is called multiple stuck-at fault model. This has been observed through fabrication and testing history, that if a chip is verified to be free of single stuck-at faults, then it can be stated with more than 99.9% accuracy that there is no defect in the silicon or the chip is functionally (logical) normal. Further, single-stuck at fault is extremely simple to handle in terms of DFT required, test time, etc. this will be detailed in consequent lectures. So, a single stuck-at fault model is the most widely accepted model.

b) **Bridging Fault**: A bridging fault represents a short between groups of nets. In the most widely accepted bridging fault models, short is assumed between two nets in the circuit. The logic value of the shorted net may be modeled as 1-dominant (OR bridge), 0-dominant (AND) bridge. It has been observed that if a circuit is verified to be free of stuck-at faults, then with high probability it can be stated that there is no bridging fault also.

Next, we elaborate more on the stuck-at fault model which is most widely accepted because of its simplicity and quality of test solution. Here A and B are 4-bit input vectors. P represents 8-bit product value generated through Wallace tree multiplier.

DESIGN AND SIMULATION RESULTS

The analysis of multiplier design carried out with implementation on FPGA Artix-7 board using Verilog programs on Xilinx ISE simulator. Xilinx ISE (Integrated Synthesis Environment) is a software tool developed by Xilinx for analysis and synthesis of HDL designs, allowing the developer to synthesize ("compile") their designs, examine RTL diagrams, perform timing analysis, simulate a design's reaction to different stimuli, and configure the target device with the programmer. RTLs helps one to determine the blocks made by the design, they usually are made with help of multiplexers and LUTs. This helps in determining the number of cells used for that particular implementation and finally the area analysis. Another aspect is timing waveform check, where functionality is checked and inferred to the correctness of the implemented system. They also provide dynamic or switching power to be analyzed in brief. Wherein, one must know that switching or the dynamic power consumption has the majority share in the total power dissipation of a system. Another possible analysis is a delay and the worst path calculation. Here, the designer makes sure that there are no timing violations found in the implementation. It has been noted that asynchronous systems produce most violations than any other designs, and hence it becomes crucial to check for such designs.

RTL ANALYSIS

Figure 10 represent RTL analysis of multiplier generated through data flow model using Verilog on Xilinx tool. RTL analysis is an easier way to design and verify any complex circuit to obtain an optimum gate-level netlist. Figure 11 describes the RTL level generated for Wallace tree multiplier. There are four blocks of multiplier testing module that are logic block, test pattern generator(CFSR), output response analyzer(ORA) and comparator. The logic block is developed with a suitable connection of 2x1 multiplexer blocks as shown in Figure 12. The ORA, CFSR is depicted by Figure 13 and 14 respectively. A suitable test pattern is generated through the CFSR circuit that is further applied to logic block to generate response for stuck –at-fault and bridging fault. The responses generated through logic blocks are compared with ORA and comparator circuit with standard data set to test multiplier circuit for stuck-at and bridging fault. The overall analysis is done through built –in-self test module to check the pass or failure of design for a particular set of fault models.

DELAY ANALYSES

The delay analysis has been done for different technology node with consideration of worst path delays. The critical paths are extremely important when delays are considered for any VLSI circuit. The timing reports were generated to check any kind of violations in the design. The worst path delay was measured 609 psec and 1982 psec for 45nm technology fast and slow libraries respectively.

Figure 10. RTL for Multiplier

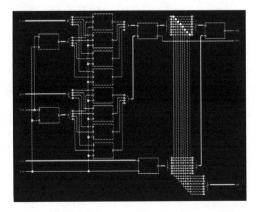

Figure 11. RTL for Wallace Tree Multiplier

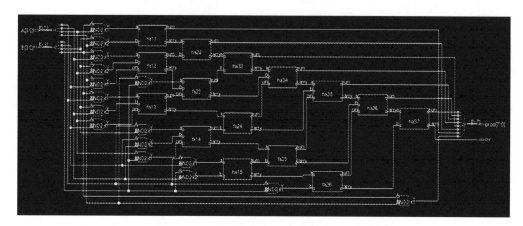

Figure 12. RTL for 2*1 MUX .

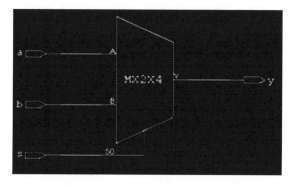

Figure 13. RTL for 4-bit comparator (ORA).

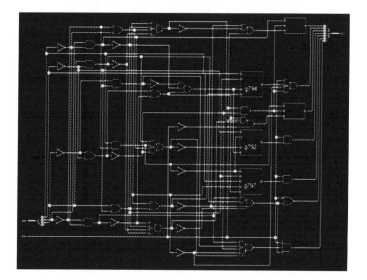

Figure 14. RTL for 8-bit CFSR.

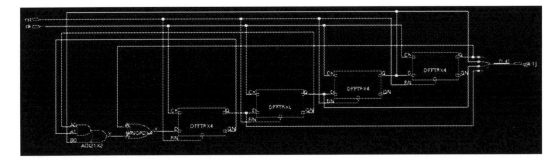

FREQUENCY ANALYSIS

The frequency analysis is carried out for slow and fast 45nm, 90nm and 180nm technology node on Cadence circuit simulator. The measured results are depicted here for a comparison of different technology.

Slow 45→1/536=1865.67Ghz
Slow 90→1/848.60=1178.411Ghz
Slow 180→1/1203.30=831.047Ghz
Fast 45→1/22.60=4492.36Ghz
Fast 90→1/278.40=3591.95Ghz
Fast 180→1/501.60=1993.62Ghz

CONCLUSION

A 4 bit Wallace tree multiplier circuit is designed for testing of stuck-at and bridging faults using BIST module. In a conventional BIST module, LFSR is generally used for test pattern generator. In this work, a low power BIST using CFSR for 4-bit Wallace tree Multiplier was implemented with a CUT that gave more power reduction compared to Dadda Multiplier. CFSR reduces the transition count in test vector without affecting the pseudo-random nature, observing a reduction of 6.5% overall power in the BIST when CL adder was implemented. Cadence RTL compiler was used to implement BIST architecture with the 45nm technology node with fast and slow library and did not show any critical path violations.

REFERENCES

Binti, Choong, Bin, Kamal, & Badal. (2017). Bit swapping linear feedback shift register for low power application using 130nm complementary metal oxide semiconductor technology. *IJE Transactions B. Applications*, *30*, 1126–1133.

Bushnell, M., & Agrawal, V. (2000). *Essentials of Electronic Testing for Digital, Memory and Mixed-Signal VLSI Circuits*. Boston: Kluwer Academic Publishers.

Dislis,, C., Ambler,, A.,, & Dick,, J. (1991). Economic Effects in Design and Test. *IEEE Design & Test of Computers, 8,* 64-77. doi:10.1109/54.107206

Erdem, O., & Serdar, S. (2018). A fast digit based Montgomery multiplier designed for FPGAs with DSP resources. *Michroprocessor and Microsystem, Elsevier, 62,* 12–19. doi:10.1016/j.micpro.2018.06.015

George, W. E. (2016). Array Multipliers for High Throughput in Xilinx FPGAs with 6-Input LUTs. *Computers MDPI, 5,* 1–25.

Hussain & Padma. (2013). Test Vector Generator (TPG) for Low Power Logic Built-In Self-Test (BIST). International Journal of Advanced Research in Electrical. *Electronics and Instrumentation Engineering, 2,* 1634–1640.

Issa, S. (2008). Bit-swapping LFSR for low-power BIST. *Electronics Letters, 44*(6), 401–402. doi:10.1049/el:20083481

Praveen, & Shiva, & Kurian. (2013). Improved Design of Low Power TPG Using LPLFSR. *International Journal of Computer & Organization Trends, 3,* 101–106.

Praveen, & Swamy, & Shanmukha. (2018). Design of BIST with Low Power Test Vector Generator. *Journal of Circuits, Systems, and Computers, 27,* 1–18. doi:10.1142/S0218126618500780

Stroud Charles, E. (2002). *A Designer's Guide to Built-in Self-Test.* Springer.

Vara Prasada, R. R., Anjaneya, V. N., Sudhakar, B. G., & Murali, M. C. (2013). Power Optimization of Linear Feedback Shift Register (LFSR) for Low Power BIST implemented in HDL. *International Journal of Modern Engineering Research, 3,* 1523–1528.

Chapter 11
A Novel Moth–Flame Algorithm for PID–Controlled Processes With Time Delay

Shamik Chatterjee
Lovely Professional University, India

Vivekananda Mukherjee
Indian Institute of Technology, Dhanbad, India

ABSTRACT

This chapter proposes a classical controller to control the industrial processes with time delay. A new population-based metaheuristic technique, called moth flame optimization (MFO) algorithm, is employed to tune the parameters of the classical proportional-integral-derivative (PID) controller for achieving the desired set point and load disturbance response. MFO-based PID controller may deal with wide ranges of processes, which includes integrating and inverse response as well as it may control the processes of any order with time delay. The transient step response profile obtained from the proposed MFO-based PID controller is juxtaposed with those obtained from other methods for optimizing the gains of the PID controller to control the processes with time delay. The proposed controller is analyzed by implementing step disturbance in the process at a specific simulation time. For few time delay processes, reference models are employed for better transient response as well as to analyze the controller for controlling the overall system with reference model.

INTRODUCTION

In the process industries, proportional integral derivative (PID) controllers are still being broadly used even after significant development of control theory. Since the genesis of the automation systems, for the process industry, the PID controller has been the cardinal technique of feed forward and feedback control. Process industries have accepted the PID controller for its cost-to-benefit ratio.

DOI: 10.4018/978-1-7998-1464-1.ch011

BACKGROUND

In the process industries, proportional-integral-derivative (PID) controllers are still being broadly used even after significant development of control theory. Since the genesis of the automation systems, for the process industries, the PID controller has been the cardinal technique of feed forward and feedback control. Process industries have accepted the PID controller for its cost-to-benefit ratio (McMillan, 2011).

The controller has proved to be simple in control, easily perceived, cheap in cost and can be maintained easily. In the last few decades, various methods have been proposed for its proper tuning. Numerous aspects of control performance requirements are considered in these tuning methods, such as, controller output, set-point response, load disturbance rejection and robust operation. The parameters of the PID controller sway the performance of the control system. The PID controller has been designed by many researchers using various methods (Astrom & Hagglund, 1995), such as, Ziegler-Nichols method (Ziegler & Nichols, 1942), cohen-coon method (Cohen and Coon, 1953) and so on.

The IMC structure is a well-accepted technique with ameliorate robustness. The desired closed loop time constant can be used by the user to specify the performance in terms of a single parameter. The IMC structure is used by Rivera *et al.* in (1986) to design the controller whose main objective is to show that IMC leads the PID controller in a natural way. Generally, the procedure of the IMC design is felicitous despite of the system considered. There is no requirement of any allocation to bestow with a system of very single type. The IMC configuration can be employed in the classical feedback configuration to enumerate a PID controller. The IMC design method is exploited mainly for low-order processes but, by employing model reduction technique, the high-order process can be reduced to low-order one to apply the technique for high-order processes. Lee *et al.* (1998) and Shamsuzzoha and Lee (2007) have employed the IMC configuration based Maclaunin series expansion of the equivalent classical feedback controller to obtain the gains of the PID controller. Panda in (Panda, 2008; Panda, 2009) has used the Laurent series expansion for obtaining the desired controller performance. In 2012, for stabilizing the unstable process, Vijayan and Panda (2012) used the double feedback loops with the inner loop to propose the PID controller based on the IMC scheme. Wang *et al.* proposed another IMC approach based on minimization of frequency response error (refer (Wang *et al.*, 2001)).

For obtaining the parameters of the PID controller, model matching technique can be applied in the frequency domain, without model reduction in high-order processes. In the year 1995, Wang *et al.* (1995) proposed new frequency domain design method for the PID controller. Later on, optimization technique and multiple frequency points for matching purpose have been employed by Wang *et al.* (1997). For stable and unstable process, Rao *et al.* (2009), Vanavil *et al.* (2015) as well as Chen and Seborg (2002) have employed the direct synthesis (DS) approach for the design of the PID controller, where the design is based on a desired closed loop transfer function (Seborg *et al.*, 1989). The DS approach also requires model reduction technique like IMC approach, as it is also based on low-order model. Skogested (2003) presented analytic rules for tuning the PID controller model reduction which is simple. For a first order or second order time delay model, Skogested (2003) has employed a single tuning rule and also used 'half rule' for achieving time delay. Jeng and Lin (2012) have designed a robust PID controller for inverse response and time delay based stable/integrating processes. Chen *et al.* (2006) have designed an analytical PID controller based on IMC structure to transform inverse processes with single adjustable control parameters. Ho *et al.* (1995) have derived a tuning formula for PI and PID controllers to obtain the user-specified gain and phase margins. The integral squared error criterion is minimized in (Ali & Majhi, 2010) to propose the tuning formulas for integrating processes. The researchers have proposed it with the constraint that there is a user specified value at gain crossover frequency in the slope of the Nyquist curve. For an integrating plus dead time transfer function model, a single method has been proposed by Chidambaram and Sree (2003), which is based on complementing the co-efficient of respective power of s in the denominator and that in the numerator for proportional-integral (PI), proportional-derivative (PD), PID controller settings.

Yang *et al.* (2011) have proposed a method to tune the PID controller parameters which is achieved from process data directly collected from an off-line experiment. A sub optimal PID controller is designed in (Zhang *et al.*, 2002) to obtain the required time domain for the nominal system or uncertain systems with time delay. A H_∞ PID controller based on minimum control theory has been proposed in (Zhang *et al.*, 2002). Shamsuzzoha (2015) has proposed an analytical design method for PI/PID controller tuning for different types of time delay processes. There is a single tuning formula for this IMC based approach to adjust the performance of the controller. Anwar *et al.* have proposed a design method for the PID controller, which is based on DS approach for obtaining the desired set point or load disturbance response (Anwar *et al.*, 2015). Umamaheshwari *et al.* (2016) have designed the PID controller by employing the concept of Mikhalevich technique and the crossover frequency is changed on the basis of desired phase margin.

The overall response may be sluggish or even unstable when the method of dominant pole placement is used for the PID controller (Persson & Astrom, 1992) as the dominancy of poles placed is not assured. Wang *et al.* have removed this drawback in (Wang *et al.*, 2009) by employing dominant pole placement method for PID controllers. Fruehauf *et al.* have introduced simplified IMC methods for PID tuning (Fruehauf *et al.*, 1994). Wang *et al.* (1995) and Karimi *et al.* (2003) have used various properties of frequency response in frequency domain design method. Zhuang and Atherton (1993) have presented few simple tuning methods for tuning the PID controller. They have formulated the technique by using the integral of error performance criteria, which is a time weighted method, for controlling the first order plus dead time (FOPDT) plant. Pan and Anwar (2013) have designed a technique for PID controller, based on frequency response matching method, for industrial processes with time delay. At the specified frequency points, the authors (Pan & Anwar, 2013) have matched the frequency response of the desired system to that of the designed control system.

Wang *et al.* (1999) have proposed a simple PID controller design method which is based on a second order plus dead time modeling technique and a closed-loop pole allocation strategy through root locus approach. It offers better performance for a big range of linear self regulating processes. In (Chien *et al.*, 2003), Chien *et al.* have derived rules for tuning the PID controller based on DS method to control the inverse and large overshoot response with dead time systems. Anwar and Pan have proposed a design method for PID controller based on the IMC principle by imprecise frequency response matching at low frequency points (refer (Anwar & Pan, 2013)).

MAIN FOCUS OF THE CHAPTER

Literature survey unfurls that the precursory researchers have implemented different techniques to optimize the gains of the PID controller for controlling different processes. Rise time (T_R), settling time (T_s), overshoot (M_P) and steady state error (E_{SS}) have to be optimum to yield better response than the previous research works based on different optimization techniques for the optimization of PID controller to control different processes. It would be very much pertinent to adopt modern advanced optimization technique to tune the PID controller's parameters and these techniques may help to achieve the optimum transient response of the processes.

As per the literature, the optimization techniques which are studied have various problems and limitations of their own such as in the case of GA, the main limitations are (a) elucidation of early convergence, (b) in each iteration cycle, implication of oodles of crossover and mutation performance and (c) necessity of additional time for complete execution. In the case of PSO, it deals with the concept of simulation of bird musters in search spaces which are multi-dimensional. PSO has undergone many hardheaded studies, which reveals that infinity may occur, *i.e.* there is a chance of divergence of the particle yet on presently interpreting the maximum velocity and acceleration constants (Clerc &Kennedy, 2002). Ziegler-Nichols (Z-N) method yields large value of M_p and T_s, which leads to an unacceptable performance, for this drawback the experienced operator refines the values of the PID controller (Visioli, 2001).

The moth flame optimization (MFO) technique has been considered in the present work (Mirjalili, 2015). The parameters of the controller are optimized to its optimal value by this population based metaheuristic search technique. It is unfolded from (Mirjalili, 2015) that, in comparison to other prominent metaheuristic algorithms, the MFO furnishes very accurate and relative responses while considering mathematical benchmark test functions. There are two main advantages of MFO algorithm (Hafez *et al.*, 2014; Mirjalili, 2015) *viz.* (a) the local mimima problem is avoided by MFO, while it is faced by many popular optimization techniques and (b) the process of exploitation and exploration is high in MFO which may expedite to surpass other algorithms. In this research work, the gains of the PID controllers are optimized by the MFO algorithm and collated with other notable state-of the-art techniques.

The paramount contributions of the present work are to:

1. Optimize the PID controller's parameters by MFO technique to control the studied processes.
2. Collate the step response rendered by MFO and those proffered by other optimization techniques as reported in the recent state-of-the-art literatures.
3. Analyze the ability of the proposed controller to control a process having a step disturbance of specific value being injected at certain time.
4. Study the overall system response by implementing the reference models in few selected processes.

STRUCTURE OF THE PID CONTROLLER

The PID controller has lingered the most extensively used controller in process control after its inception in the market in 1939. After 50 years, it has been inspected that more than 80% controllers, successfully used in process industry houses, are PID controllers or amended PID controllers. It is easily comprehensible and as well as vigorous in nature that can proffer outstanding control performance in spite of the altered dynamic characteristics of the process plant.

Generally, PID controllers are acclimatized on-site and the assimilation is done by proposing different genre of tuning rules which succors intricate and top-notch tuning of the controllers. Proportional, integral and derivative are the three gains which together forms the PID controller. The E_{SS} cannot be eliminated by the proportional controller but it can minimize the value of T_R. If the value of the proportional gain is made too high, then the system may become unstable whereas on decreasing its value, the system may yield smaller output in reference to larger input error and the controller may lead to become less sensitive and responsive. The value of E_{SS} can be terminated by adjusting the value of the integral controller gain but on exceeding its range, it may lead to poor transient response. On increasing the value of the integral gain, higher will be the value of M_p whereas the system will become sluggish on decreasing the value of M_p. The stability of the system may increase, if the value of the derivative gain is made high which will minimize the value of M_p and it will result in more improved response of the system whereas the system may become unstable on increasing the gain of the derivative controller (Astrom & Hagglund, 1995; Cominos & Munro, 2002). The PID controller is designed on comprising of its three parameters *viz.* proportional gain (K_p), integral gain (K_I) and derivative gain (K_D). The parameters are tuned by using trial and error method based on the designer's experience and the nature of the plant. In Figure 1, block diagram of basic structure of conventional feedback control system is shown. The main objective, as per this figure, is to control the process so that the output signal achieves the value of reference signal.

The output of the PID controller, as shown in Figure 1, is presented in (1)

$$u(t) = K_P e(t) + K_I \int_0^t e(x)\,dx + K_D \frac{de(t)}{dt} \tag{1}$$

where $e(t)$ = Reference signal – Output signal.

In Laplace domain, the transfer function of PID controller may be expressed by (2).

$$T_F = K_P + \frac{K_I}{s} + sK_D \tag{2}$$

Figure 1. Block diagram of conventional PID controller

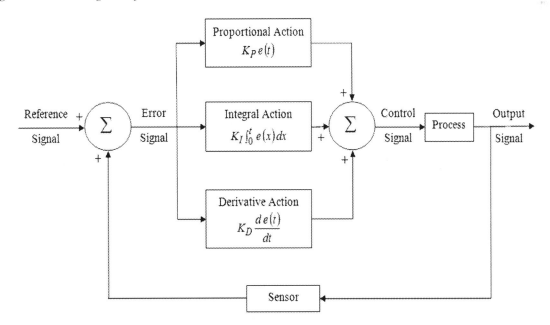

MATHEMATICAL PROBLEM FORMULATION

Mathematical problem formulation is accomplished to obtain the exposition of an optimization problem. The appropriate definition of the constraints of the task and the outline of the objective function are two foremost strands of an optimization errand. In the succeeding two sub-fragments, these two strands are embellished, in succession, in the current accord.

Design of Objective Function

The performance of the studied processes is analyzed by adopting a time domain objective function. To ally with this analysis of performance regarding time domain, involvement of the performance parameters is required. The performance index of the step response profile (*i.e.* T_R, T_S, M_P and E_{SS}) has to be optimum and congruous with the step input perturbations for obtaining better performance. A time domain objective function, named as figure of demerit (FOD), has to be minimized which is expressed in (3) (Banerjee *et al.*, 2013).

$$FOD = \left(1 - e^{-\beta}\right)\left(M_P + E_{SS}\right) + e^{-\beta}\left(T_S - T_R\right) \tag{3}$$

To reduce M_P and E_{SS}, the value of β in (3) is designated to be greater than 0.7 but the values of T_S and T_R have to be reduced for that it will be better to employ the value of β less than 0.7. In the current task, 1.0 is taken as the value of β. The values of T_S, T_R, M_P and E_{SS} are delivered by the step response of the process. Thus, the main aim of the current optimization task is to obtain the optimum value of FOD, is defined in (3).

Constraints of Present Work

The parameters of the controller should be bounded within certain pre-itemized limits. The curtailment of the parameters, which are to be optimized, may be given by (4)

$$\left.\begin{array}{l} K_P^{\min} \leq K_P \leq K_P^{\max} \\ K_I^{\min} \leq K_I \leq K_I^{\max} \\ K_D^{\min} \leq K_D \leq K_D^{\max} \end{array}\right\} \tag{4}$$

where the minimum and the maximum values, of the respective variables, are denoted, respectively, by the superscripts min and max. For the current work, the range of variable K_P is set between 0.01 and 30, the value of K_I lies in the range of 0.0001 and 10 and within the range of 0.01 and 20, the value of K_D lies.

MOTH FLAME OPTIMIZATION TECHNIQUE

The MFO technique is studied in this chapter for optimizing the controller parameters to control the studied industrial processes. MFO algorithm is based on the flying characteristics of moths. In characteristics, moths are almost similar to butterflies.

Moths use a special mechanism, called transverse orientation, to navigate in night, which is a very interesting competency. Moths use this special mechanism for traversing in nights by retaining a constant angle with respect to the moon using moon light. Moths cruise long distances in a straight route by applying this mechanism (Gaston *et al.*, 2013). Convergence of moth towards light has been described by Frank *et al.* (2006). Mirjalili used this phenomenon and proposed an optimization technique, called as MFO algorithm.

In MFO, moths are figured to be the candidate solution and the problem's variables are being the position of moths in the space. With change in their position vectors, the moths can traverse in 1-*D*, 2-*D*, 3-*D* or hyper dimensional space. MFO is a population based algorithm and the set of moths is displayed in a matrix, as presented in (5) (Mirjalili, 2015),

$$
M = \begin{bmatrix}
m_{1,1} & m_{1,2} & \cdots & \cdots & m_{1,d} \\
m_{2,1} & m_{2,2} & \cdots & \cdots & m_{2,d} \\
\cdot & \cdot & \cdot & \cdot & \cdot \\
\cdot & \cdot & \cdot & \cdot & \cdot \\
\cdot & \cdot & \cdot & \cdot & \cdot \\
m_{n,1} & m_{n,2} & \cdots & \cdots & m_{n,d}
\end{bmatrix}
\tag{5}
$$

where *d* is the number of variables whereas number of moths is *n*. An array has been assumed for all moths to keep the respective fitness values, as presented in (6) (Mirjalili, 2015).

$$
OM = \begin{bmatrix} OM_1 & OM_2 & \cdot & \cdot & OM_n \end{bmatrix}^T
\tag{6}
$$

Fitness value is the return value of the fitness (objective) function for each moth. Towards the fitness function, the position vector (first row in the matrix *M*, for instance) is promoted. By considering the respective moth as its fitness value (OM_1 in the matrix, defined in (6), for instance), the output of the fitness function is allocated.

Flames are another key component of MFO algorithm, like moth matrix, named flames which are considered as matrix and it is presented in (7).

$$
F = \begin{bmatrix}
F_{1,1} & F_{1,2} & \cdots & F_{1,d} \\
F_{2,1} & F_{2,2} & \cdots & F_{2,d} \\
\cdot & \cdot & \cdot & \cdot \\
\cdot & \cdot & \cdot & \cdot \\
\cdot & \cdot & \cdot & \cdot \\
F_{n,1} & F_{n,2} & \cdots & F_{n,d}
\end{bmatrix} \tag{7}
$$

From (7), it may be stated that there is an equality in dimensions of M and F. The corresponding fitness value is captured in an array as per consideration for the flames defined in (8).

$$
OF = \begin{bmatrix} OF_1 & OF_2 & \cdot & \cdot & OF_n \end{bmatrix}^T \tag{8}
$$

On the basis of modification of the values of moths and flames (both are solutions) in each iteration, it may be reported that both have different values. The veridical searching agents are the moths (which roam in search space) while the flames are the foremost position as obtained by the moths.

The three-tuple MFO algorithm reckons the global optimal issue of the optimization issues and it is presented in (9).

$$
MFO = \left(I,\, P,\, T \right) \tag{9}
$$

I is a function in (9) that unfurls a random population of moths and corresponding fitness values. The methodical model of this function is stated in (10).

$$
I : \varphi \rightarrow \{ M, OM \} \tag{10}
$$

The moths are moved by the function P in the search space which is the main function. The updated version of matrix M is returned by this function, which is received eventually, as stated in (11).

$$
P : M \rightarrow M \tag{11}
$$

The function T, as stated in (9), returns true otherwise false if the termination criterion is satisfied, as presented in (12).

$$
T : M \rightarrow \{ true,\, false \} \tag{12}
$$

In Algorithm 1 (Mirjalili, 2015), the general body of the MFO algorithm is described. The initial conditions are generated after which the objective function is calculated by using the function I. The utilization of any random distribution is allowed in this function. In Algorithm 2 (Mirjalili, 2015), the shown method is used by default.

Algorithm 1: General structure of the MFO algorithm
  M=I();
  while T(M) is equal to false
  M=P(M);
  end

There are two arrays (called as *ub* and *lb*) in Algorithm 2, which states the upper and lower bounds of the variables, respectively. The two arrays are shown in (13) and (14), in order,

$$ub = \left[ub_1, \, ub_2, \, ub_3, \, ..., \, ub_{n-1}, \, ub_n \right] \qquad (13)$$

where the upper bound of the ith variable is ub_i.

Algorithm 2: Calculation of the objective function
    for i = 1: *n*
    for j = 1: *d*
    M(i,j)=(ub(i)-lb(i))*rand()+lb(i);
    end
    end
    OM=Fitness Function (M)

$$lb = \left[lb_1, \, lb_2, \, lb_3, \, ..., \, lb_{n-1}, \, lb_n \right] \qquad (14)$$

where the lower bound of the ith variable is lb_i.

Modification of each moth's position is done with respect to a flame (15)

$$M_i = S\left(M_i, F_j \right) \qquad (15)$$

where the ith moth is indicated by M_i whereas j^{th} flame is indicated by F_j and the spiral function is denoted by *S*.

In (16) a logarithmic spiral is presented, as used in the MFO algorithm

$$S\left(M_i, F_j \right) = D_i . \, e^{bt} . \, \cos\left(2\pi t \right) + F_j \qquad (16)$$

where D_i, for the jth flame, is the distance of the ith moth and a constant, *b* is used for stating the logarithmic spiral shape whereas a random number within [-1, 1] is denoted by *t*.

As per (17), the calculation of the variable *D* can be done (Mirjalili, 2015)

$$D_i = \left| F_j - M_i \right| \qquad (17)$$

where the jth flame is denoted by F_j and the ith moth is denoted by M_i.

In (16), the simulation of spiral flying way of moths is carried out. In respect to flame, the forthcoming position of a moth is defined. The moths become the important part of the proposed method by being dictated to update their positions around the flames from the spiral method. According to the spiral equation, a moth is permitted to fly around a flame. Thus, exploration and exploitation of moths are guaranteed through the search space.

The best solutions (*i.e.,* flames) guarantee the probability for finding better solutions. Hence, in the matrix *F*, recently achieved *n* best solutions are included. With respect to this matrix, the position of the moths has to be updated during the process of optimization. In order to heighten further exploitation, the value of *t* is conjectured to be a random number lying in the range [*r*,1], whereas throughout the course of iteration, *r* is linearly declined from -1 to -2. According to the number of iterations, in this process, the moths are mastered to utilize their respective flames more exquisitely. The moths are succored by the corresponding flames to update their positions. The position of the first moth gets updated based on the best flame whereas according to the worst flame, the position of the last moth gets updated. Depending upon *n* different locations in the search space, the process of updating the position of moths degrades the exploitation of the best solution. To obtain the solution for the number of flames, an attuned mechanism is used.

The presented formula of (18) may be utilized in this respect (Mirjalili, 2015)

$$flame\ number\ =\ round\left(N-l*\frac{N-1}{T}\right) \tag{18}$$

where the current number of iteration is denoted by *l*, *N* denotes the maximum number of flames and the maximum number of iterations is denoted by *T*. The decrement in the number of flames balances the exploitation and exploration of the search space. In Algorithm 3 (Mirjalili, 2015), the steps of the *P* function are displayed. The *P* function goes on executing till a true value will be returned by the *T* function. The best optimized approximated value of the best moth is returned after the *P* function is terminated.

Algorithm 3: General steps of function *P*
    Update flame number using (18)
    OM=Fitness Function (M)
    If iteration==1
    F=sort(M);
    OF=sort(OM);
    else
    F=sort(M_{l-1},M_l);
    OF=sort(M_{l-1},M_l);
    end
    for i = 1: *n*
    for j = 1: *d*
    Update *r* and *t*
    Calculate *D* using (17) with respect to the corresponding moth

Update M(i,j) using (15) and (16) with respect to the corresponding moth
end
end

SOLUTIONS AND RECOMMENDATIONS

The processes, considered in the present work as per the state-of-the-art literatures, are controlled by the proposed PID controller tuned by MFO. Analyzation of the step response is the main purpose of this segment of presentation. In MATLAB/SIMULINK (version 7.10), the simulation has been carried out and the computer used for this research work is of 2.77 GHz, Intel Core™$i7$. For the proposed algorithm, maximum population size is chosen as 50 and 100 is set for the maximum number of iterations.

The overall model, which is presented as block diagrams, is different for different examples depending upon the analysis of the proposed controller and the performance of the studied process (without controller). The block diagram, portrayed in Figure 2, is used in examples 1-8. Similarly, for example 9 and example 16, the block diagram, shown in Figure 3, is utilized. Block diagram, presented in Figure 4, is used for examples 10-13.

Figure 2. Block diagram of PID controller for examples 1-8.

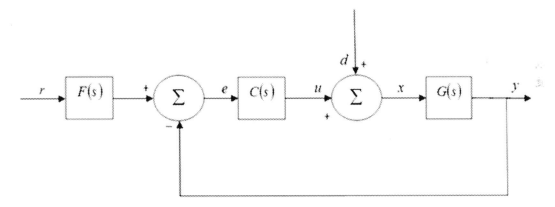

Figure 3. Block diagram of PID controller for examples 9 and 16.

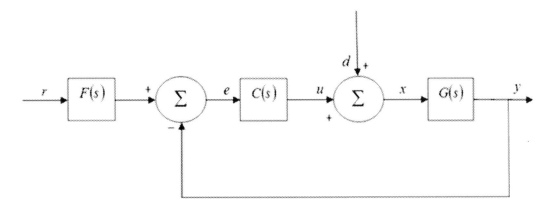

Figure 4. Block diagram of PID controller for examples 10-13.

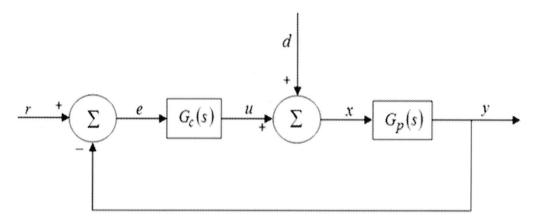

Example 1

A FOPDT model is considered for this example and the transfer function of the process is represented in (19) (Zhang *et al.*, 2002).

$$G(s) = \frac{e^{-0.5s}}{s+1} \tag{19}$$

A step disturbance of magnitude 1.0 p.u. is implemented in the system at *t*=8 s. The proposed MFO algorithm gives the tuned PID controller for the control of the process, whose transfer function is presented in (20).

$$C(s) = 1.00 + \frac{1.00}{s} + 0.15s \tag{20}$$

In Figure 5 (a), a collocation of the step response profile of the proposed MFO based PID controlled FOPDT process and step response profiles achieved by other optimization techniques *viz.* Shamsuzzoha and Lee (Anwar *et al.*, 2015), Anwar *et al.* (Anwar *et al.*, 2015) and Zhang *et al.* (Zhang *et al.*, 2002) is depicted.

The convergence profile of the proposed MFO algorithm is achieved by proclaiming the optimum FOD value for the proposed PID controlled process. In Figure 5 (b), the MFO based convergence profile of FOD for the present system is portrayed and it is perceived that the MFO based FOD value converges in 42 iterations at a value of 5.2135.

Example 2

For this example, an inverse response process is studied whose transfer function is represented in (21) (Jeng & Lin, 2012).

Figure 5. (a) Comparative step response analysis of example 1 and **(b)** *MFO based convergence profile of FOD value for example 1.*

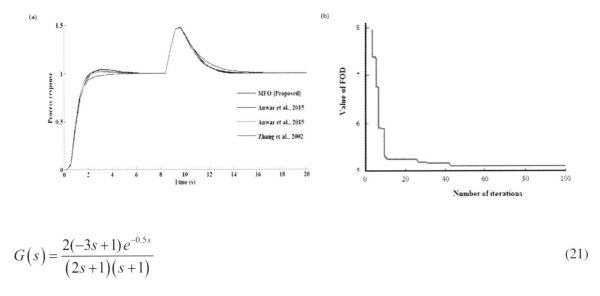

$$G(s) = \frac{2(-3s+1)e^{-0.5s}}{(2s+1)(s+1)} \tag{21}$$

In the system, at $t=50$ s, a step disturbance of magnitude 1.0 p.u. is implemented. The optimized transfer function of the proposed MFO algorithm based PID controller, which is used to control the process, is presented in (22).

$$C(s) = 0.2 + \frac{0.068}{s} + 0.13s \tag{22}$$

The comparison of the unit step response profile of the proposed MFO based PID controlled inverse response process and unit step response profiles obtained from other optimization techniques (*viz.* Jeng and Lin (2012), Chen *et al.* (Anwar *et al.*, 2015) and Anwar *et al.* (2015)) is portrayed in Figure. 6 (a).

For the proposed PID controlled inverse response process, the demonstration of the optimum FOD value led to accomplishment of the convergence profile of the proposed MFO algorithm. The convergence profile is depicted in Figure 6 (b). From this figure, it is observed that in 35 iterations, the MFO based PID controller converges at a FOD value of 26.0923.

Example 3

An oscillatory process's transfer function, which is contemplated for this example, is presented in (23) (Wang *et al.*, 1997).

$$G(s) = \frac{e^{-0.4s}}{(s^2+s+1)(s+3)} \tag{23}$$

Figure 6. (a) Comparative step response analysis of example 2 and (b) MFO based convergence profile of FOD value for example 2.

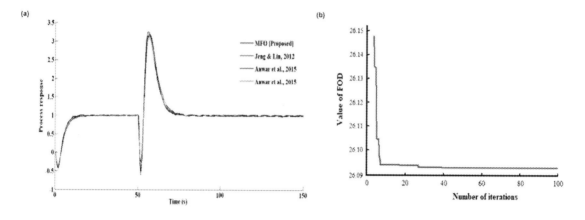

Implementation of a step disturbance, of magnitude 1.0 p.u., is carried out at t=60 s. To control the studied oscillatory process, the parameters of PID controllers are tuned by MFO algorithm, whose transfer function is given by (24).

$$C(s) = 1.12 + \frac{1.19}{s} + 1.4s \tag{24}$$

The unit step response profile of the proposed MFO based PID controlled oscillatory process is compared with the unit step response profiles achieved from other optimization techniques *viz.* Wang *et al.* (1997), Ho *et al.* (Anwar *et al.*, 2015), Anwar *et al.* (2015), which is rendered in Figure 7 (a).

The convergence profile is depicted in Figure 7 (b), which is portrayed on the basis of minimum FOD value obtained, for tuning the PID controller's gains to control the studied oscillatory process. It is delineated from the figure that the MFO based PID controller converges to optimum FOD value in 27 iterations at a value of 24.3606.

Example 4

An integrating plus first order plus time delay is studied in this example, whose transfer function is presented in (25) (Ali & Majhi, 2010).

$$G(s) = \frac{e^{-4s}}{s(s+1)} \tag{25}$$

The proposed MFO tuned PID controller for controlling the integrating plus first order plus time delay process is represented by the transfer function, as in (26).

Figure 7. (a) Comparative step response analysis of example 3 and (b) MFO based convergence profile of FOD value for example 3.

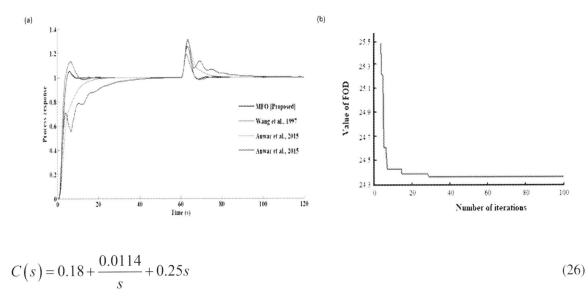

$$C(s) = 0.18 + \frac{0.0114}{s} + 0.25s \tag{26}$$

To improve servo response, a set point filter is considered whose transfer function is presented in (27) (Anwar *et al.*, 2015).

$$F(s) = \frac{1}{12s+1} \tag{27}$$

Anwar *et al.* (2015) have only used the set point filter, presented in (27), to improve the response. The same is considered in the present work. The proposed MFO based PID controller has been compared to other optimization techniques *viz.* Ali and Majhi (2010), Anwar *et al.* (2015) and Skogestad (2003), on the basis of unit step response portrayed in Figure 8 (a). At *t*=100 s, magnitude of 1.0 p.u step disturbance is implemented.

For this integrating plus first order plus time delay process, on the basis of minimum FOD value, the convergence profile is portrayed in Figure 8 (b). As noted in this figure, in 25 iterations with a FOD value of 49.3446, the MFO based PID controller converges.

Example 5

An integrating plus time delay, represented by the transfer function presented in (28), is studied (Chidambaram & Sree, 2003) as example 5.

$$G(s) = \frac{0.0506}{s} e^{-6s} \tag{28}$$

Figure 8. (a) Comparative step response analysis of example 4 and (b) MFO based convergence profile of FOD value for example 4.

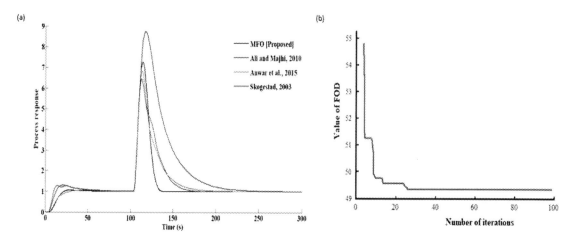

The tuned parameters of the MFO based PID controller, represented in the form of transfer function, is given in (29).

$$C(s) = 3.6 + \frac{0.169}{s} + 12.2s \qquad (29)$$

The set point filter is considered, to improve servo response, whose transfer function is presented in (30) (Anwar *et al.*, 2015).

$$F(s) = \frac{1}{15s+1} \qquad (30)$$

In the present work, Anwar *et al.* (2015) have used the set point filter which is expressed in (29). For the studied integrating plus time delay process, the PID controller, tuned by MFO, is used to control the process and the step response obtained by the proposed controller is compared to those obtained by other techniques surfaced in the literature *viz.* Anwar *et al.* (2015), Ali and Majhi (2010) and Chidambaram and Sree (2003) (see Figure 9 (a)). A disturbance of magnitude 1.0 p.u. is implemented at *t*=100 s.

A convergence profile has been portrayed, in Figure 9 (b), on the basis of FOD value obtained from the proposed MFO based PID controller for controlling the integrating plus time delay process. The MFO based PID controller for the studied process converges in 51 iterations yielding a FOD value of 45.6370.

Example 6

In this example, a high-order process is studied whose transfer function is represented in (31) (Yang *et al.*, 2011).

Figure 9. (a) Comparative step response analysis of example 5 and (b) MFO based convergence profile of FOD value for example 5.

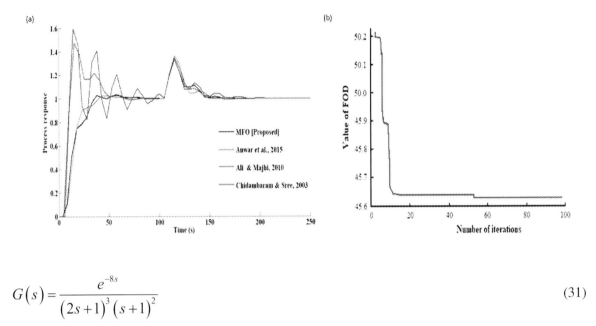

$$G(s) = \frac{e^{-8s}}{(2s+1)^3 (s+1)^2} \tag{31}$$

The transfer function of the MFO based PID controller for above specified high-order process is presented in (32).

$$C(s) = 0.46 + \frac{0.056}{s} + 0.99s \tag{32}$$

The optimization techniques, as per the state-of-the-art of literatures employed by the researchers *viz.* Anwar *et al.* (2015), Umamaheshwari *et al.* (2016) and Yang *et al.* (2011), are considered for a comparison with the proposed MFO based PID controller for controlling the studied process.

The step responses displayed in Figure 10 (a), are used for detailed comparative analysis of all optimization techniques. At *t*=150 s, a disturbance of magnitude 1.0 p.u. is applied in this process.

On the support of minimum FOD value and number of iterations, a convergence profile has been depicted in Figure 10 (b) for the MFO based PID controlled studied process. In 34 iterations, the MFO based PID controller converges to a FOD value of 70.8210.

Example 7

A process has been considered for the analysis of MFO based PID controller, which is represented in (33) (Yang *et al.*, 2011).

$$G(s) = \frac{e^{-2.2s}}{(4s^2 + 2.8s + 1)(s+1)^2} \tag{33}$$

Figure 10. (a) Comparative step response analysis of example 6 and (b) MFO based convergence profile of FOD value for example 6.

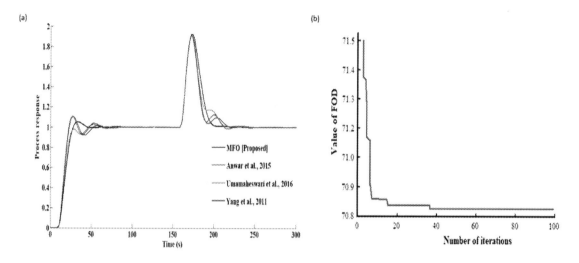

For the above given process, using MFO, the tuned parameters of PID controller is represented in the form of transfer function presented in (34).

$$C(s) = 0.57 + \frac{0.153}{s} + 1.03s \tag{34}$$

The step response of the proposed MFO based PID controller, for studied process including disturbance of magnitude 1.0 p.u at $t=200$ s, is compared to other optimization techniques *viz.* Anwar *et al.* (2015), Anwar *et al.* (2015), Yang *et al.* (2011), and the comparison is depicted in Figure 11 (a).

The convergence profile of the proposed MFO based PID controller for controlling the studied process is portrayed in Figure 11 (b) on the basis of the minimum FOD value and number of iterations. The proposed technique offers a FOD value of 79.8718 in 41 iterations.

Example 8

A process of 20th order has been considered in this example and its transfer function is presented in (35) (Yang *et al.*, 2011).

$$G(s) = \frac{1}{(s+1)^{20}} \tag{35}$$

At $t=100$ s, a step disturbance of magnitude 1.0 p.u. is implemented in the system to encounter the performance of the proposed controller. The MFO based optimal PID controller parameters, for the above process, is represented in (36) in the form of transfer function.

Figure 11. (a) Comparative step response analysis of example 7 and (b) MFO based convergence profile of FOD value for example 7.

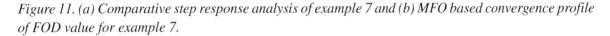

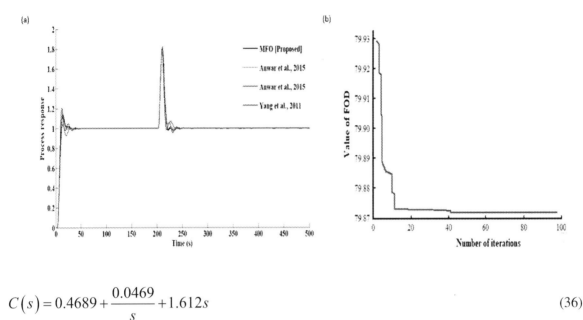

$$C(s) = 0.4689 + \frac{0.0469}{s} + 1.612s \tag{36}$$

The step response of the proposed MFO based PID controller for controlling the process has been compared with the other optimization techniques *viz.* Anwar *et al.* (2015), Yang *et al.* (2011) and Umamaheshwari *et al.* (2016), (see Figure 12 (a)).

In Figure 12 (b), the convergence profile of the proposed MFO based FOD value is portrayed. In 47 iterations and at a FOD value of 54.5015, the proposed controller converges to its optimal value.

Figure 12. (a) Comparative step response analysis of example 8 and (b) MFO based convergence profile of FOD value for example 8.

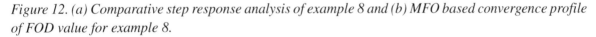

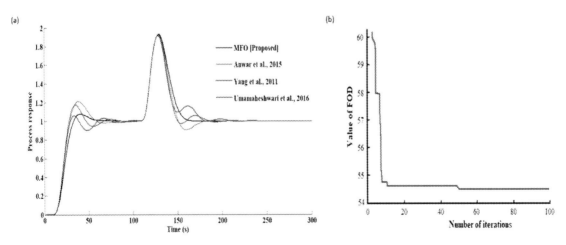

Example 9

A FOPDT industrial process has been considered in example 9, whose transfer function is given in (37) (Wang *et al.*, 1995).

$$G(s) = \frac{e^{-5s}}{10s+1} \tag{37}$$

A reference model has been chosen for the requirement for designing the PID controller, whose transfer function is presented in (38).

$$M(s) = \frac{e^{-5s}}{5s+1} \tag{38}$$

The MFO is used to optimize the parameters of PID controller whose transfer function is represented in (39).

$$G_{PID}(s) = 1.7 + \frac{0.2}{s} + 2.3s \tag{39}$$

Figure 13 (a) portrays the juxtaposition of the step response obtained from MFO based PID controller for the FOPDT industrial process and the unit step response obtained from other optimization techniques *viz.* Rivera *et al.* (1986), Wang *et al.*(1995) and Pan and Anwar (2013).The proposed controller is analyzed by introducing a step disturbance of magnitude 1.0 p.u. at time 80 s.

The portrayal of the convergence profile, based on the minimum FOD value and number of iterations (*i.e.*, MFO) may be observed in Figure 13 (b). In 45 iterations, the proposed MFO based PID controller converges with a FOD value of 36.8797.

Example 10

The transfer function of a FOPDT process, which has been considered in this case, is represented in (40) (Panda, 2008).

$$G_P(s) = \frac{1}{s+1} e^{-0.25s} \tag{40}$$

The transfer function of the tuned parameters of MFO based PID controller is represented in (41).

$$G_C(s) = 1.4 + \frac{1.7}{s} + 0.1s \tag{41}$$

Figure 13. (a) Comparative step response analysis of example 9 and (b) MFO based convergence profile of FOD value for example 9.

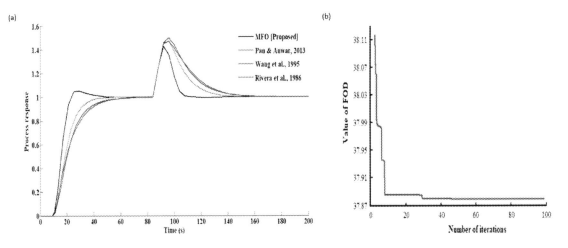

The unit step responses obtained from other optimization techniques (*viz.* Panda (2008), Lee *et al.* (1998) and Anwar and Pan (2013)) are considered for the comparison with the step response obtained from the proposed MFO based PID controller (see Figure 14 (a)). For this example, the step load disturbance of magnitude 1.0 p.u. is injected at *t*=15 s.

From the Figure 14 (b), it is observed that the proposed MFO based PID controller converges in 60 iterations at an optimal value FOD value of 7.2647.

Example 11

In this example, a second order plus dead time (SOPDT) process is considered, whose transfer function is presented in (42) (Panda, 2008).

Figure 14. (a) Comparative step response analysis of example 10 and (b) MFO based convergence profile of FOD value for example 10.

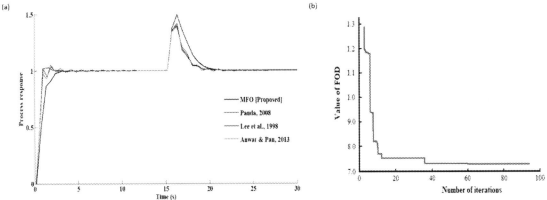

$$G_P(s) = \frac{2}{(10s+1)(5s+1)} e^{-1s} \tag{42}$$

The parameters of the PID controllers are optimized by MFO and the transfer function of the tuned parameters is represented in (43).

$$G_C(s) = 4.5 + \frac{0.35}{s} + 13s \tag{43}$$

The proposed MFO based PID controller's step response is compared with the unit step response achieved from other optimization techniques as per the state-of-the-art literatures *viz.* Panda (2008), Lee *et al.*(1998) and Anwar and Pan (2013) (refer Figure 15 (a)). At t=90 s, the step disturbance of value 1.0 p.u is injected to analyze of the proposed controller for controlling SOPDT process.

At a FOD value of 42.9986, the proposed MFO based PID controller converges in 33 iterations (as shown in Figure 15 (b)) for this process.

Example 12

The transfer function of a third order oscillatory system with time delay, as considered for this example, is presented in (44) (Wang *et al.*, 1999).

$$G_P(s) = \frac{1}{(s^2+2s+3)(s+3)} e^{-0.3s} \tag{44}$$

The transfer function of tuned parameters of PID controller is presented in (45), which is optimized by MFO algorithm.

Figure 15. (a) Comparative step response analysis of example 11 and (b) MFO based convergence profile of FOD value for example 11.

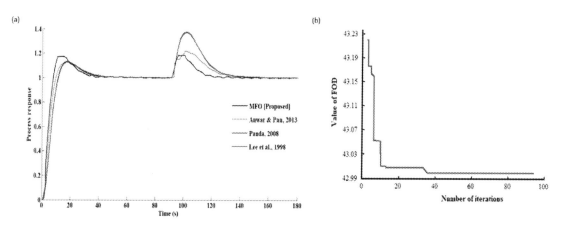

$$G_C(s) = 5.1 + \frac{5.75}{s} + 2.3s \tag{45}$$

The figure having comparison of the step response obtained from the proposed MFO based PID controller and the step response achieved from other optimization techniques *viz.* Wang *et al.* (1999), Ho*et al.* (1995)and Anwar and Pan (2013), with the step disturbance injected at *t*=15 s of value 1.0 p.u., is portrayed in Figure 16 (a).

The convergence profile of FOD value offered by the proposed MFO, while tuning the PID controller for this process, is displayed in Figure 16 (b). As per this figure, it is observed that the proposed controller converges in 25 iterations at a FOD value of 7.0459.

Example 13

In (46), the transfer function of the second order time delay system with inverse response, as considered for this example, is presented (Jeng & Lin, 2012).

$$G_P(s) = \frac{2(-3s+1)}{(2s+1)(s+1)} e^{-0.5s} \tag{46}$$

In (47), the transfer function of the tuned parameters of MFO based PID controller for this process is expressed.

$$G_C(s) = 0.178 + \frac{0.07}{s} + 0.128s \tag{47}$$

Figure 16. (a) Comparative step response analysis of example 12 and (b) MFO based convergence profile of FOD value for example 12.

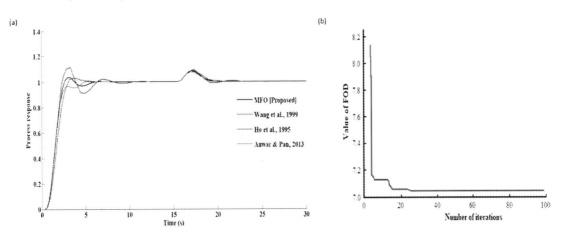

For controlling the second order time delay system with inverse response, the step response obtained from the MFO based PID controlled system is compared to the step response achieved from other optimization techniques, such as Jeng and Lin (2012), Chien *et al.* (2003) and Anwar and Pan (2013), for the analysis of the proposed controller. For further analysis, a step disturbance of value 1.0 p.u. is implemented in the process at *t*=40 s, as portrayed in Figure 17 (a).

For controlling the studied process, the convergence profile, on the basis of minimum number of iterations and FOD value, of the proposed MFO based PID controller is depicted in Figure 17 (b). From this figure, it may be observed that the proposed controller converges in 46 iterations at a FOD value of 21.6081.

Example 14

From the literature, a FOPDT industrial process is considered in this example. Its transfer function is presented in (52) (Wang *et al.*, 1995).

$$G(s) = \frac{e^{-50s}}{10s+1} \tag{52}$$

For the detailed analysis, to design the PID controller, reference model is chosen followed by the closed loop process. In comparison to the process's time constant, the process's time delay is very large. Designing task of a controller is very difficult in such cases. To describe this, four reference models are being considered for which different tuned parameters of MFO based PID controller are obtained.

Reference Model 1

The transfer function of the first reference model is given in (53) (Pan & Anwar, 2013).

Figure 17. (a) Comparative step response analysis of example 13 and (b) MFO based convergence profile of FOD value for example 13.

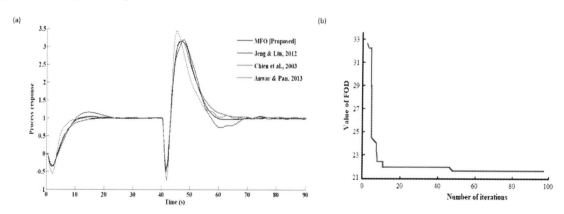

$$M(s) = \frac{e^{-50s}}{30s+1} \tag{53}$$

The optimized parameters of MFO based PID controller for the above stated process along with the reference model $M(S)$ is presented in the form of transfer function (see (54)).

$$G_{PID}(s) = 0.5 + \frac{0.017}{s} + 4s \tag{54}$$

Reference Model 2

The second reference model is considered for the analysis of the example 16, whose transfer function is presented in (55) (Pan & Anwar, 2013).

$$M(s) = \frac{e^{-50s}}{20s+1} \tag{55}$$

The transfer function of the tuned parameters, for the specified process along with the reference model $M(S)$ of the MFO based PID controller is presented in (56).

$$G_{PID}(s) = 0.45 + \frac{0.0156}{s} + 4s \tag{56}$$

Reference Model 3

For further analysis, the third reference model is considered whose transfer function is presented in (57) (Pan & Anwar, 2013).

$$M(s) = \frac{e^{-50s}}{15s+1} \tag{57}$$

For controlling the studied inverse response process, $G(S)$, along with the reference model $M(S)$, the tuned parameters of the MFO based PID controller is presented in the form of transfer function in (58).

$$G_{PID}(s) = 0.44 + \frac{0.0154}{s} + 3.989s \tag{58}$$

Reference Model 4

The fourth reference model is considered, whose transfer function is presented in (59) (Pan & Anwar, 2013).

$$M(s) = \frac{e^{-50s}}{10s + 1} \tag{59}$$

The transfer function of the tuned parameters, for the process and reference model of the MFO based PID controller is given in (60).

$$G_{PID}(s) = 0.49 + \frac{0.0168}{s} + 5.8s \tag{60}$$

The step response obtained from the MFO tuned PID controlled FOPDT industrial process, with the reference models, are being compared with the step response obtained from the technique adopted by Pan and Anwar (2013). The procured numerical data of the time domain simulation responses encapsulating the FOD value, time domain performance characteristics, (such as M_p, T_R and T_S) and the controller gains are drafted in Table 1. The time constant is in decreasing order from reference model 1 to 4 and it is observed from the table that the FOD value is not decreasing after reference model 1 (*i.e.*, from reference model 2 to 4). Thus, the reference model 1 is considered for further collation, by comparing the step response obtained from the MFO based PID controller to control the studied process (along with reference model 1) with step response obtained from other optimization technique, such as Wang *et al.* (1995), Rivera *et al.* (1986), Cohen and Coon (1953) and Zhuang and Atherton (1993), to control the same process without reference model. In Figure 18 (a), the comparison of the step responses is displayed with the step load disturbance of value 0.5 p.u. at *t*=600 s.

From this figure, it may be reported that the PID controller, tuned by MFO, manifests a stupendous unit step response for the FOPDT industrial process, with reference model 1, in juxtaposition with the step responses obtained from other optimization techniques. It may be divulged from the Table 1 that in comparison with other compared optimization techniques, the best performance index values and the FOD values are gratified by the proposed MFO based PID controller for the FOPDT industrial process, with reference model 1.

The convergence profile of the MFO based PID controller for controlling the FOPDT industrial process, with all reference models, are merged and displayed in Figure 18 (b). It is observed from Figure 18 (b) that the proposed controller for controlling the studied process, with reference model 1, converges at a value of 257.2222 in 42 iterations while it converges at a FOD value of 257.6797 in 46 iterations for the process with reference model 2 and for controlling the considered model with reference model 3, the proposed controller converges in 34 iterations at a FOD value of 260.3049 and finally at a FOD value of 261.4582, the proposed controller converges in 50 iterations for controlling the process with reference model 4. Hence, it may be particularized that the proposed MFO algorithm has unveiled promising convergence mobility on optimizing the PID controller for the studied inverse response process.

Figure 18. (a) Comparative step response analysis of example 14 and (b) MFO based convergence profile of FOD value for example 14.

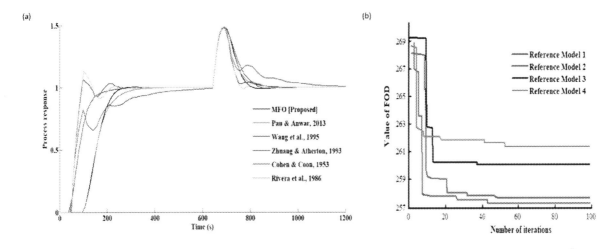

5.1 Discussion of the Result obtained from Examples 1-13

The step responses obtained from the MFO based PID controller for controlling the processes presented in examples 1-13 are compared to different optimization techniques as surfaced in the recent state-of-the-art literatures. The comparative analysis of the optimization techniques, based on the step response obtained from the process output on being controlled by the PID controller, are displayed in aforementioned different figures. The unit step response purveyed by MFO based PID controller intimates a phenomenal consequence, as portrayed in those figures, in comparison to other optimization techniques based on unit step response. Time domain performance characteristics, which confines M_P, T_R, T_S and E_{SS} along with other numerical data of the time domain simulation responses(*viz.* FOD value and controller gains), are outlined in Table 2. From this table, it may be reported that the best performance index and the FOD values are indulged by the proposed MFO based PID controller in comparison to other optimization techniques for the processes stated in examples 1-13.

The convergence profiles of the proposed MFO based PID controller are shown in respective examples. The minimum number of iterations and value of FOD shows the ability of the controller to control the processes with ease and robustness. Thus, from the displayed figures, it may be divulged that for the studied process (as considered in examples 1-13), the proposed MFO algorithm has exhibited propitious convergence mobility on optimizing the PID controller installed in the respective processes with delay.

CONCLUSION

The MFO algorithm is considered for tuning the parameters of the classical PID controller for controlling the industrial processes with time delay. The control of liquid level of three tanks and speed control of DC motor is performed using the proposed controller for achieving the best value. The dynamic response profile of the process controlled by the proposed MFO based PID controller manifests a stupendous outcome compared to other tuning methods based PID controller for the same process. The

Table 1. MFO optimized PID controller gains and transient response parameters for controlling example 14.

Reference Models	Technique	K_P	K_I	K_D	M_P	$T_S(s)$	$T_R(s)$	E_{SS}	FOD
Reference	MFO [Proposed]	**0.3**	**0.0129**	**1.8**	**1.4832**	**794.1740**	**97.3701**	**00**	**257.2222**
Model 1	Pan and Anwar [Pan & Anwar, 2013]	0.335	0.012	1.878	1.4825	836.5290	115.6357	00	266.0818
Reference	MFO [Proposed]	0.47	0.0156	3.5	1.4891	756.0119	57.9740	00	257.6797
Model 2	Pan and Anwar [Pan & Anwar, 2013]	0.412	0.014	3.72	1.4905	791.5071	78.3266	00	263.2501
Reference	MFO [Proposed]	0.44	0.0154	3.989	1.4926	763.3956	58.2264	00	260.3049
Model 3	Pan and Anwar [Pan & Anwar, 2013]	0.51	0.016	4.99	1.4899	816.8025	50.3245	00	282.8525
Reference	MFO [Proposed]	0.49	0.0168	5.8	1.4884	751.7580	43.4458	00	261.4582
Model 4	Pan and Anwar [Pan & Anwar, 2013]	0.53	0.018	6.67	1.4879	784.5031	38.7763	00	275.2189
	Wang et al. [Wang et al., 1995]	0.275	0.011	1.023	1.4874	845.4363	85.4269	00	280.4718
Without	Zhuang and Atherton [Zhuang & Atherton, 1993]	0.508	0.015	6.74	1.4866	824.4431	37.1763	00	290.4966
Reference	Cohen and Coon [Cohen & Goon, 1953]	0.517	0.008	4.92	1.4827	1032.7	243.0180	0.009	291.3881
	Rivera et al. [Rivera et al., 1986]	0.538	0.015	3.84	1.4924	852.0137	34.2327	00	301.7233

Table 2. MFO optimized PID controller gains and transient response parameters for controlling examples 1-13.

Process	Technique	K_P	K_I	K_D	M_P	$T_S(s)$	$T_R(s)$	E_{SS}	FOD
Example 1	MFO [Proposed]	**1.00**	**1.00**	**0.15**	**1.4786**	**12.7841**	**1.1508**	**00**	**5.2135**
	Shamsuzzoha and Lee [Anwar et al., 2015]	1.08	1.02	0.11	1.4784	12.9339	1.0382	00	5.3099
	Anwar et al. [Anwar et al., 2015]	1.12	1	0.12	1.4722	13.2334	1.0400	00	5.4155
	Zhang et al. [Zhang et al., 2002]	1.11	0.88	0.11	1.4789	14.2297	1.2360	00	5.7140
Example 2	MFO [Proposed]	**0.2**	**0.068**	**0.13**	**3.1980**	**71.1912**	**5.7757**	**0.017**	**26.0923**
	Jeng and Lin [Jeng & Lin, 2012]	0.191	0.063	0.127	3.1508	76.3712	7.1857	0.0009	27.4389
	Chen et al. [Anwar et al., 2015]	0.21	0.067	0.134	3.2492	76.7864	5.5215	0.0119	28.2729
	Anwar et al. [Anwar et al., 2015]	0.212	0.066	0.176	3.1466	83.4119	6.9041	0.0166	30.1393
Example 3	MFO [Proposed]	**1.12**	**1.19**	**1.14**	**1.2567**	**66.6138**	**2.5405**	**00**	**24.3606**
	Wang et al. [Wang et al., 1997]	1.96	1.75	3.74	1.1912	66.6773	2.0274	00	24.5313

continues on following page

Table 2. Continued

Process	Technique	K_P	K_I	K_D	M_P	$T_s(s)$	$T_R(s)$	E_{ss}	FOD
	Anwar *et al.* [Anwar *et al.*, 2015]	0.74	0.612	0.827	1.3027	74.1634	8.5238	0.0004	24.9660
	Ho *et al.* [Anwar *et al.*, 2015]	1.60	0.41	0.0	1.3165	85.0362	18.4380	0.001	25.3277
Example 4	MFO [Proposed]	**0.18**	**0.0114**	**0.25**	**7.2844**	**133.1568**	**11.5164**	**0.0001**	**49.3446**
	Ali and Majhi [Ali & Majhi, 2010]	0.19	0.0084	0.53	6.4659	163.3290	4.2186	00	62.6085
	Anwar *et al.* [Anwar *et al.*, 2015]	0.20	0.0083	0.357	6.8295	174.3766	15.1517	00	62.8805
	Skogestad [Skogestad, 2003]	0.13	0.0039	0.126	8.7558	208.2765	6.3263	0.003	79.8146
Example 5	MFO [Proposed]	**3.6**	**0.169**	**12.2**	**1.3318**	**144.4250**	**22.6333**	**0.0001**	**45.6370**
	Anwar *et al.* [Anwar *et al.*, 2015]	3.5	0.16	6.62	1.3624	149.6734	14.7974	0	50.4687
	Ali and Majhi [Ali & Majhi, 2010]	3.39	0.17	9.96	1.4744	144.9437	4.2750	00	52.6701
	Chidambaram and Sree [Chidambaram & Sree, 2003]	4.06	0.15	10.97	1.5952	163.4942	4.6995	0.0001	59.4132
Example 6	MFO [Proposed]	**0.46**	**0.056**	**0.99**	**1.9232**	**199.3760**	**12.0546**	**0.001**	**70.8210**
	Anwar *et al.* [Anwar *et al.*, 2015]	0.543	0.055	2.30	1.9124	211.6736	11.0591	0.001	74.9957
	Umamaheshwari *et al.* [Umamaheshwari *et al.*, 2016]	0.6454	0.0621	2.3447	1.9088	210.8644	8.7930	0.001	75.5292
	Yang *et al.* [Yang *et al.*, 2011]	0.63	0.060	1.745	1.7796	234.1880	4.4906	0.001	85.6071
Example 7	MFO [Proposed]	**0.57**	**0.153**	**1.03**	**1.8082**	**219.2181**	**5.1670**	**0.001**	**79.8718**
	Anwar *et al.* [Anwar *et al.*, 2015]	0.558	0.143	0.857	1.8166	219.7801	5.4965	00	79.9619
	Anwar *et al.* [Anwar *et al.*, 2015]	0.58	0.124	0.650	1.8240	233.1747	5.6933	0.0022	84.8222
	Yang *et al.* [Yang *et al.*, 2011]	0.73	0.170	1.467	1.7796	234.1880	4.4906	0.001	85.6071
Example 8	MFO [Proposed]	**0.4689**	**0.0469**	**1.612**	**1.9309**	**159.0578**	**14.1946**	**0.0001**	**54.5015**
	Anwar *et al.* [Anwar *et al.*, 2015]	0.525	0.055	1.66	1.9286	174.8583	11.9199	00	61.1480
	Yang *et al.* [Yang *et al.*, 2011]	0.62	0.052	2.21	1.9213	180.3996	11.1229	00	63.4746
	Umamaheshwari *et al.* [Umamaheshwari *et al.*, 2016]	0.6206	0.04585	3.057	1.9141	200.5961	11.9340	00	70.6000
Example 9	MFO [Proposed]	**1.7**	**0.2**	**2.3**	**1.4292**	**106.9054**	**9.0909**	**00**	**36.8792**
	Pan and Anwar [Pan & Anwar, 2013]	1.12	0.1	1.2	1.4722	132.3341	17.7192	00	43.0860

continues on following page

Table 2. Continued

Process	Technique	K_P	K_I	K_D	M_P	$T_s(s)$	$T_R(s)$	E_{ss}	FOD
	Wang *et al.* [Wang *et al*, 1995]	1	0.08	2	1.4730	144.5310	25.3707	00	44.7584
	Rivera *et al.* [Rivera *et al.*, 1986]	0.966	0.08	0.88	1.5022	143.8507	22.5360	00	45.5692
Example 10	MFO [Proposed]	**1.4**	**1.7**	**0.1**	**1.4221**	**18.0158**	**0.7084**	**00**	**7.2647**
	Panda [Panda, 2008]	1.58	1.48	0.103	1.3970	19.0906	0.7013	00	7.6468
	Lee *et al.* [Lee *at el.*, 1998]	1.55	1.48	0.065	1.4197	19.1458	0.7062	00	7.6796
	Anwar and Pan [Anwar & Pan, 2013]	1.55	1.481	0.1467	1.3889	19.8547	0.7189	0.0015	7.9172
Example 11	MFO [Proposed]	**4.5**	**0.35**	**13**	**1.1874**	**119.3223**	**4.4589**	**0.0019**	**42.9986**
	Anwar and Pan [Anwar & Pan, 2013]	3.149	0.207	11.12	1.2206	130.4446	5.7139	0.002	46.6489
	Panda [Panda, 2008]	1.93	0.130	6.123	1.3700	134.4880	7.0625	0.001	47.7338
	Lee *et al.* [Lee *at el.*, 1998]	1.91	0.130	5.777	1.3747	134.4801	6.9266	0.001	47.7839
Example 12	MFO [Proposed]	**5.1**	**5.75**	**2.3**	**1.0853**	**18.6739**	**1.3825**	**00**	**7.0459**
	Wang *et al.* [Wang *et al.*, 1999]	3.88	5.38	2.15	1.0924	18.9627	1.6267	00	7.06679
	Ho *et al.* [Ho *et al.*, 1995]	5.06	5.92	1.08	1.1166	18.5562	1.2542	00	7.0695
	Anwar and Pan [Anwar & Pan, 2013]	5.78	5.62	3.66	1.0787	18.9029	1.4824	00	7.0979
Example 13	MFO [Proposed]	**0.178**	**0.07**	**0.128**	**3.1452**	**58.5496**	**5.2113**	**0.003**	**21.6081**
	Jeng and Lin [Jeng & Lin, 2012]	0.19	0.063	0.128	3.1671	63.8392	7.8420	0.013	22.6062
	Chien *et al.* [Chien *et al.*, 2003]	0.15	0.078	0.1563	3.2011	81.0127	4.6492	0.0148	30.1196
	Anwar and Pan [Anwar & Pan, 2013]	0.25	0.076	0.213	3.4398	80.2755	3.9943	0.0155	30.2407

FOD values obtained by the proposed controller exhibits the value which is close to nominal, collated to other methods. Additionally, the proposed controller has delivered good performance even after the injection of step load disturbance of certain value at specific time in the system compared to the other established methods. The process output, controlled by the proposed controller, has delivered optimal simulation results closer to the desired output on being compared with the reference models which are designed specifically for some processes for better analysis of the controller and achieving more desired response. The transient response of the industrial processes of various orders with time delay controlled by the proposed MFO based PID controller may be employed in this area as it has yielded effective performance in the field of process control applications.

REFERENCES

Ali, A., & Majhi, S. (2010). PID controller tuning for integrating processes. *ISA Transactions, 49*(1), 70–78. doi:10.1016/j.isatra.2009.09.001 PMID:19782358

Anwar, M. N., & Pan, S. (2013). Synthesis of the PID controller using desired closed-loop response. *IFAC Proc., 46*(32), 385-390.

Anwar, M., Shamsuzzoha, M., & Pan, S. (2015). A frequency domain PID controller design method using direct synthesis approach. *Arabian Journal for Science and Engineering, 40*(4), 995–1004. doi:10.100713369-015-1582-4

Astrom, K. J., & Hagglund, T. (1995). *PID Controllers Theory Design and Tuning* (2nd ed.). Instrument Society of America.

Banerjee, A., Mukherjee, V., & Ghoshal, S. P. (2013). Modeling and seeker optimization based simulation for intelligent reactive power control of an isolated hybrid power system. *Swarm and Evolutionary Computation, 13*, 85–100. doi:10.1016/j.swevo.2013.05.003

Chen, D., & Seborg, D. E. (2002). PI/PID controller design based on direct synthesis and disturbance rejection. *Industrial & Engineering Chemistry Research, 41*(19), 4807–4822. doi:10.1021/ie010756m

Chen, P., Zhang, W., & Zhu, L. (2006) Design and tuning method of PID controller for a class of inverse response processes. *Proceedings of the 2006 American Control Conference.* 10.1109/ACC.2006.1655367

Chidambaram, M., & Sree, R. P. (2003). A simple method of tuning of PID controller for integrating/dead time processes. *Computers & Chemical Engineering, 27*(2), 211–215. doi:10.1016/S0098-1354(02)00178-3

Chien, I. L., Chung, Y. C., Chen, B. S., & Chuang, C. Y. (2003). Simple PID controller tuning method for processes with inverse response plus dead time or large overshoot response plus dead time. *Industrial & Engineering Chemistry Research, 42*(20), 4461–4477. doi:10.1021/ie020726z

Cohen, G., & Coon, G. (1953). Theoretical consideration of retarded control. *Transaction of ASME, 75*(1), 827-834.

Cominos, P., & Munro, N. (2002). PID controllers: recent tuning methods and design to specification. *Proc. IEE Control Theory and Appl., 149*(1), 46-53. 10.1049/ip-cta:20020103

Frank, K. D., Rich, C., & Longcore, T. (2006). Effects of artificial night lighting on moths. Ecological Consequences of Artificial Night Lighting, 305-344.

Fruehauf, P. S., Chien, I. L., & Lauritsen, M. D. (1994). Simplified IMC-PID tuning rules. *ISA Transactions, 33*(1), 43–59. doi:10.1016/0019-0578(94)90035-3

Gaston, K. J., Bennie, J., Davies, T. W., & Hopkins, J. (2013). The ecological impacts of nighttime light pollution: A mechanistic appraisal. *Biological Reviews of the Cambridge Philosophical Society, 88*(4), 912–927. doi:10.1111/brv.12036 PMID:23565807

Hafez, A. L., Zawbaa, H. M., Hassanien, A. E., & Fahmy, A. A. (2014). Networks community detection using artificial bee colony swarm optimization. In *Proc. of the Fifth Int. Conf. on Innov. in Bio-Inspired Comp. and Appl. (IBICA)*. Ostrava, Czech Republic: Springer.

Ho, W. K., Hang, C. C., & Cao, L. S. (1995). Tuning of PID controllers based on gain and phase margin specification. *Automatica, 31*(3), 497–502. doi:10.1016/0005-1098(94)00130-B

Jeng, J. C., & Lin, S. W. (2012). Robust proportional–integral–derivative controller design for stable/integrating processes with inverse response and time delay. *Industrial & Engineering Chemistry Research, 51*(6), 2652–2665. doi:10.1021/ie201449m

Juang, C.F. (2004). A hybrid of genetic algorithm and particle swarm optimization for recurrent network design. *IEEE Trans. Syst. Man. Cybern., B, 34*(2), 997-1006.

Karimi, A., Garcia, D., & Longchamp, R. (2003). PID controller tuning using Bode's integrals, IEEE Transaction. *Control System Technol, 11*(6), 812–821. doi:10.1109/TCST.2003.815541

Lee, Y., Park, S., Lee, M., & Brosilow, C. (1998). PID controller tuning for desired closed-loop responses for SI/SO systems. *AIChE Journal. American Institute of Chemical Engineers, 44*(1), 106–115. doi:10.1002/aic.690440112

McMillan, G. K. (2011). *Industrial applications of PID control, PID control in the third Millennium: Lessons learned and new approaches.* Springer.

Mirjalili, S. (2015). Moth-flame optimization algorithm: A novel nature-inspired heuristic paradigm. *Knowledge Syst., 89,* 228–249. doi:10.1016/j.knosys.2015.07.006

Pan, S., & Anwar, N. M. (2013). A frequency response matching method for PID controller design for industrial processes with time delay, ICAC3 2013. *CCIS, 361,* 636–646.

Panda, R. C. (2008). Synthesis of PID tuning rule using the desired closed loop response. *Industrial & Engineering Chemistry Research, 47*(22), 8684–8692. doi:10.1021/ie800258c

Panda, R. C. (2009). Synthesis of PID controller for unstable and integrating processes. *Chemical Engineering Science, 64*(12), 2807–2816. doi:10.1016/j.ces.2009.02.051

Persson, P., & Astrom, K. J. (1992). Dominant pole design – A unified view of PID controller tuning. *IFAC Proc., 25*(14), 377-382.

Rao, A. S., Rao, V. S. R., & Chidambaram, M. (2009). Direct synthesis-based controller design for integrating processes with time delay. *Journal of the Franklin Institute, 346*(1), 38–56. doi:10.1016/j.jfranklin.2008.06.004

Rivera, D. E., Morari, M., & Skogestad, S. (1986). Internal model control 4. PID controller design. *Industrial & Engineering Chemistry Process Design and Development, 25*(1), 252–265. doi:10.1021/i200032a041

Seborg, D. E., Edgar, T. F., & Mellichamp, D. A. (1989). *Process dynamics and control.* New York: Willey.

Shamsuzzoha, M. (2015). A unified approach for proportional–integral– derivative controller design for time delay processes. *Korean Journal of Chemical Engineering*, *32*(4), 583–596. doi:10.100711814-014-0237-6

Shamsuzzoha, M., & Lee, M. (2007). IMC–PID controller design for improved disturbance rejection of time-delayed processes. *Industrial & Engineering Chemistry Research*, *46*(7), 2077–2091. doi:10.1021/ie0612360

Skogestad, S. (2003). Simple analytic rules for model reduction and PID controller tuning. *Journal of Process Control*, *13*(4), 291–309. doi:10.1016/S0959-1524(02)00062-8

Umamaheshwari, G., Nivedha, M., & Prisci Dorritt, J. (2016). Design of tunable method for PID controller for higher order system. *Int. J. Engg. and Comp Sc*, *5*(7), 17239–17242.

Vanavil, B., Chaitanya, K. K., & Rao, A. S. (2015). Improved PID controller design for unstable time delay processes based on direct synthesis method and maximum sensitivity. *Int. J. Syst. Sci.*, *46*(8), 1349–1366.

Vijayan, V., & Panda, R. C. (2012). Design of PID controllers in double feedback loops for SISO systems with set-point filters. *ISA Transactions*, *51*(4), 514–521. doi:10.1016/j.isatra.2012.03.003 PMID:22494496

Wang, L., Barnes, T. J. D., & Cluett, W. R. (1995) New frequency domain design method for PID controllers. *IEE Proc. Control Theory Appl.*, *142*(4), 265–271.

Wang, Q. G., Hang, C. C., & Bi, Q. (1997). A frequency domain controller design method. *Transactions of the Institution of Chemical Engineers*, *75*(1), 64–72. doi:10.1205/026387697523228

Wang, Q. G., Hang, C. C., & Yang, X. P. (2001). Single-loop controller design via IMC principles. *Automatica*, *37*(12), 2041–2048. doi:10.1016/S0005-1098(01)00170-4

Wang, Q. G., Lee, T. H., Ho, W. H., Bi, Q., & Zhang, Y. (1999). PID tuning for improved performance. *IEEE Transactions on Control Systems Technology*, *7*(4), 457–465. doi:10.1109/87.772161

Wang, Q. G., Zhang, Z., Astrom, K. J., & Chek, L. S. (2009). Guaranteed dominant pole placement with PID controllers. *Journal of Process Control*, *19*(2), 349–352. doi:10.1016/j.jprocont.2008.04.012

Yang, X., Xu, B., & Chiu, M. S. (2011). PID controller design directly from plant data. *Industrial & Engineering Chemistry Research*, *50*(3), 1352–1359. doi:10.1021/ie100784k

Zhang, W., Xi, Y., Yang, G., & Xu, X. (2002). Design PID controllers for desired time-domain or frequency-domain response. *ISA Transactions*, *41*(4), 511–520. doi:10.1016/S0019-0578(07)60106-2 PMID:12398281

Zhuang, M., & Atherton, D. P. (1993). Automatic tuning of optimum PID controllers. *IEE Proc. D-Control Theory Appl.*, *140*(3), 216–224.

Ziegler, J. G., & Nichols, N. B. (1942). Optimum Settings for Automatic Controllers. *Trans. of ASME*, *64*, 759–768.

Chapter 12
Artificial Intelligence for Interface Management in Wireless Heterogeneous Networks

Monika Rani
IKG PTU, Jalandhar, India

Kiran Ahuja
ⓘ https://orcid.org/0000-0002-1213-8010
DAVIET, Jalandhar, India

ABSTRACT

Wireless communication/networks are developing into very complex systems because of different requirements and applications of consumers. Today, mobile terminals are equipped with multi-channel and multiple access interfaces for different kinds of applications (or services). The combination of these access technologies needs an intelligent control to interface the best channel, interface/access or link for best services. In interface management, an arrangement is used to assign channels to interfaces in the multi-channel multi-interface environment. Artificial intelligence is one of the upcoming areas with different techniques which is used now a days to meet user's requirements. Quality of service (QoS) and quality of experience (QoE) are the performance parameters on which the success of any technique depends upon. Reliability of any system plays an important role in user satisfaction. This chapter shows some of the artificial techniques that can be used to make a reliable system.

INTRODUCTION

Artificial Intelligence (AI) is related with two terms "artificial" and "intelligence". Artificial related to anything which is manmade. Things prepared / generated by human with the help of machines are synthetic or artificial. Intelligence shows the capability of an individual to realize, learn or assume. Artificial

DOI: 10.4018/978-1-7998-1464-1.ch012

intelligence is an arrangement. It is not a system although implemented in the system. AI is defined in many ways .One researcher termed AI as "It is the study of how to train the computers so that computers can do things which at present human can do better". Thus, it is an arrangement of intelligence where human can include the entire the qualities in a device that he has itself.

Thus, AI has the ability of a computer system or a device to imagine, assume and learn. It is related with area of research which tries to make a system "smart and intelligent".

In other language, AI is an area of computer science that emphasizes the design of intelligent machines that acts like human brain. Some of the actions of system designed with artificial intelligence are: thinking, voice recognition, capturing and learning etc. AI systems typically express some of the following nature linked with individual intelligence: scheduling, learning, interpretation, trouble solving, information designing, examination, movement managing, social intelligence and imagination in somewhere.

AI is not limited to just computer or technology industry. Instead, it is being broadly used in other areas such as medical, business, education, law, manufacturing and wireless communication. Some of the examples of AI which used now a days are: Siri (which uses machine-learning technology to interacts with the user on a daily routine), Netflix (information-on-demand service), Drone (translate the environment into a 3D model through sensors and video cameras.), Alexa (friendly female voice-activated assistant) and many more.

In simple way, AI is the implementation of individual intelligence processes by machines, mostly computer systems. The process include following steps to execute any task.

Learning: Gathering of information and set rules or protocols for implementation this information.
Reasoning: Implement the rules to acquire estimated or definite conclusions.
Manipulation: Making adjustments without human intervention.

The somewhat same process is followed by one of the branch of AI which is termed as machine learning although it is different.

Machine Learning (ML)

ML is a sub-branch of artificial intelligence. In ML devices understand, execute and get better their operations by exploiting the process knowledge and experience obtained in the form of outcome data. It is also considered as a function of AI that provides the ability to self learns and improves from experience. The simple definitions is *"Machine Learning is said to learn from experience E w.r.t some class of task T and a performance measure P if learner's performance at the task in the class as measured by P improves with experiences"*. ML focuses on the development of algorithms that can analyze data and make predictions. ML also differs from normal computer programming. In computer programming the set of instructions are given by individual to solve the problem whereas ML made the program itself. It is named as an algorithm, a model and sometimes an agent learning from the data it is given. Computer programming will not learn or get any better with experience whereas ML has ability to solve the problem gets better with experience.

ML also defined as automating and improving the learning process of computers depends on their past experiences with no help of programming i.e. without any human help. The operation begins with giving a high-quality quality data and then guiding our machines (computers) by creating machine learn-

Table 1. Dissimilarity between AI and ML

Artificial intelligence (AI)	Machine Learning(ML)
AI has the ability to get and implement the knowledge.	ML can only acquire knowledge not implement it.
The main concern is to raise the chance of success not accuracy.	The main concern is to increase accuracy. Success rate doesn't matter.
AI has decision making ability.	ML has learning ability.
It works as an algorithm having set of instructions that does smart work.	It is a simple concept machine takes data and learn from data.
The aim is to reproduce natural intelligence to solve difficult problem.	The aim is to understand from data on particular operation to improve the performance of machine at its max.
AI set out for searching the best possible solution.	ML set out for only solution whether it is best possible or not.
AI show the way of intelligence or wisdom.	ML show the way of knowledge.

ing models using the data and different programs. The selection of program depends on what kind of data do we have and what kind of job we are demanding to computerize.

Difference Between AI and ML

There is a misreading that artificial intelligence is similar to machine learning but it is not true. Some of the points are listed below which differentiate both terms.

Artificial intelligence has taken an amazing significance in the last couple of years. A numerous types of AI are available to help many systems or organizations to operate intelligently.

Different Types of Artificial Intelligence

The present intelligent system has the ability to handle large amounts of information and solving problematical calculations in speedy way. AI researchers are always in progress to use this quality in the upcoming systems. In the upcoming years, AI systems will accomplish and go beyond the imagination of humans in doing complicated work. In next topic, the different kinds of artificial intelligence are discussed.

Reactive Machines AI

The initial types of artificial intelligence systems are entirely reactive having no storage. This system has the ability not to use past experiences to instruct present decisions and to manage memoirs. Its examples are IBM's chess-playing computer 'Deep Blue' beaten Garry Kasparov an international grandmaster in chess. Google's AlphaGo crushed the top human Go experts but it can't assess all the future coming moves. Its evaluation method is more interactive than Deep Blue's, using a neural network. The reactive system never gets tired and work effortlessly. This kind of AI has repetitive action plan. The interaction with outer world is very less.

Limited Memory AI

The type 2 AI is generally implemented in automatic / self-driving cars. These cars can sense the moves, speed and turns of other vehicles in the region round car continuously. The fixed data like track marks, traffic lights, traffic signals and turns on the roadway is preprogrammed in the AI machine. This helps automatic cars to run in a secure way in the presence of many vehicles. Time taken by type 2 systems to take careful directions / instructions in automatic vehicles is approximately 100 seconds.

Theory of Mind AI

The very advance mechanism of AI is theory of mind. In psychology, this term shows the wisdom of people and living creatures on the earth, having feelings which change their own nature. The existence of theory of mind is not as much in the world yet. The work is still in progress by making more advanced robots. This type of robot is capable to recognize physical movements and copy same to same. Many Hollywood and Bollybood directors had shown this idea in their movies.

Self-Aware AI

The alternate option of theory of mind is Self-aware AI. This type of AI is very superior and not came into existence fully yet, but when it comes, it can organize realization about themselves. In this type of AI systems machines are trained with some realization / awareness in it. As researchers did not get much success to create such machines but continues efforts are going on towards memory power, self realization in machines.

Artificial Narrow Intelligence (ANI)

ANI is the general mechanism, frequently used by mobile phones now days. In smart phones applications like Cortana and Siri facilitate users by responding their appeal or demand. But as the requirement of automation is increased day by day, artificial narrow intelligence is considered as 'weak AI'.

Artificial General Intelligence (AGI)

The other type of AI system is termed as 'strong AI' because of its operation more likely to human. Generally robots used are of ANI type, but some are AGI or beyond. For example "Pillo" robot which is meant for medical consultancy and help at home. It responds to all doubts related to the physical condition of the people. It gives instructions about the medicines taken by someone and provides guidance of their health. AGI robot plays a role of full-time doctor in our life.

Artificial Superhuman Intelligence (ASI)

The most advanced form of AI is super intelligence. Anything human can do or think to do can be achieved with this type of AI system. "Alpha 2" robot is the example of this category. This robot has the ability of handling a smart home in very efficient way. It can operate all the things of home as and when required. It can predict weather conditions and entertains you by music or some interesting stories. In near future, this type of robot becomes our family member.

As the different types of AI types are available, each is chosen according to their advantages and suitability to the system.

Advantages of Artificial Intelligence

Today, AI becomes the optimal choice for most of the applications because of its numerous advantages. Some of advantages of AI systems are discussed below.

Low Error The chances of error in AI systems are very low as compared to humans, if programmed carefully. The correctness, accuracy, truthfulness and speed are very high. It is not affected by unfriendly environments conditions. The environments which can be harmful for human like explore in space, mining high temperature conditions are easily handled by AI system. Robots are example of this which can perform dangerous task in harsh situations. It can take impartial decisions with less or no mistakes. Robots are emotionless and think logically.

Easy work capability AI systems are mostly human less systems. It works recurring, boring tasks in many painstaking places without human assistance. It makes the work easy in mining and digging fuels. It can even predict the working conditions before starts. AI systems act as trained assistant in various trials.

Security AI system acts as security agent in our homes, vehicles and in other systems. A card based systems and possibly other systems are prepared to find the defaulters. It helps in organizing and managing records in more secure manner.

Entertainment AI also relate with humans for entertainment. The 'jinnie' or robots playing many videogames with human on internet. The multimedia system lke alexa, Netflix and carvan fulfill all the requirements of audio and video on demand.

Medical help It is used for health checkup purposes, like disease diagnose and mentally state of any patient. It is used as assistance in operation theatres. Radio surgery and other critical surgeries can be done with more accuracy and efficiently.

Effortless task AI systems can works continuously without any gap. These systems never get fed up or exhausted.

Advantages and disadvantages are like two sides of a coin, both are opposite but still exist together. AI is also not spare from some limitations.

Disadvantages of Artificial Intelligence

The following are the disadvantages of AI which can't be ignored when AI is implemented in the system.

Cost Large amount of money and time is required to construct, reconstruct, and mantain the AI systems. Robotic cost more money and resources than any other system.

Storage and understanding AI is expansive, in terms of access and retrieval data in memory as humans can. It can understand and improve with tasks only if programmed properly otherwise not. It cannot work beyond a fixed program. Human technological perceptions and creativity can never be matched with AI.

Emotion less AI systems are senseless and emotionless. For simple work to be done one has to feed program in it. In medical science, it can never have emotions like a doctor or nurse has for their patients.

Increase unemployment AI systems like robots replacing human jobs, can lead to severe unemployment.

Misuse Smart phones and other technology made human dependent on AI and go down their natural understanding power. Automatic and powerful devices could be used to ruin the world by wrong people which could also be a fear for humanity.

Apart from advantages and disadvantages, system adopts AI according to its requirements / applications. The methods of AI made the system smarter according to their needs.

Computational Methods of AI

The followings are commonly used methods of AI in the various fields of industrial, medical and science.

- Machine learning, neural networks, hybrid neural network,
- Fuzzy systems
- Evolutionary algorithms, Genetic algorithm
- Bayesian network, Hidden Markov model
- Chaos theory

The various techniques used before AI came into existence in different fields. The wireless communication is also one of field where different traditional methods were used. In next topic various approaches for interface management were reviewed.

LITERATURE SURVEY

In this section, various traditional approaches used for interface management in wireless heterogeneous networks are discussed.

Mobile terminals are usually set with number of network interfaces. There are different attributes to choose randomly the best interface according to user preferences and/or application preferences. The Multiple Attributes Decision Making is an algorithmic approach was discussed for interface selection with multiple alternatives. The comparison was showed with other MADM techniques (Tran, P. N., & Boukhatem, N.,2008).

TOPSIS method was discussed for ranking and selecting alternatives. The ranking of alternatives was based on the relative similarity to the ideal solution. The proposed method gave the concept of positive ideal and negative ideal solutions to avoid the problem of the same similarity index. Proposed algorithm used for crisp and interval data (Roszkowska, E., 2011).

The various vertical handover techniques based on the multi criterions such as network based, terminal based, customer related or service based were compared. Multi handover decision techniques for vertical handoffs like decision function based strategies, user-centric strategies, multiple attribute decision strategies, fuzzy logic and neural networks based strategies and context-aware strategies were discussed. The context aware strategy was concluded the best in terms of efficiency, flexibility and implementation complexity (Chandavarkar, B. R., & Reddy, G. R. M. 2012).

The issue of security in various heterogeneous wireless networks architectures like Unified Cellular Ad hoc Network (UCAN), Integrated Cellular Ad hoc Relay (ICAR), Scalable Proxy Routing (SPR) and multiple hop cellular networks was raised. Author also discussed their security weaknesses in any attack using traditional approaches (Baseer, S.,2013).

A routing algorithm for interface selection in a network with multiple interfaces was proposed. Markov Decision Process (MDP) was applied in each node to find optimal policy. Proper path to the best access point was selected in a dynamic environment. Selection metrics in each node were interface load, link quality and destination condition. Proposed algorithm decreased interference and collision. Links were selected with better quality. The demerit of this approach was the increasing average delay (Jafari, A. H., & Shahhoseini, H. S., 2015).

With continuous change in technology in today's arena for better ease for the user there is a scope of improvement. There is a requirement of an intelligent system which have fast decision making capability along with physical tasks. A significant development has been found in these techniques; still, many challenges/ problems remain untouched.

AI is used in different fields like medical, business, science, communication and many more. AI is playing important role in our daily life and future. AI could be the best option resolution for the rising advanced communication system style. In next topic we will discuss the importance of AI in wireless heterogeneous networks.

IMPORTANCE OF AI IN WIRELESS HETEROGENEOUS NETWORKS (WHN)

The advancement of mobile technologies creates huge amount of traffic and require high data rate, as a result it needs additional bandwidth and superior quality of experience. To fulfill the upcoming requirement for data traffic, the wireless networks are switching towards heterogeneous networks.

The conventional network control and supervision activities are no longer effective. A talented method is to introduce artificial intelligence into the network heterogeneous environment, instead of physical optimization techniques. With artificial intelligence, the wireless networks are managed more separately and professionally, and the outcome is improved.

AI is a powerful tool with number of possible application areas, e.g., wireless signal processing, channel modeling, resource management and many more. Several smart techniques have been investigated and analyzed by the researchers. The developed techniques required by biological evolution process and a human brain, have a better potential to propose and better clarification. Fuzzy logic, machine learning and neural network are more related with these features. The importance of these AI methods in wireless communication is explained in next topic.

Fuzzy Logic in Wireless Heterogeneous Network

Fuzzy logic is an approach of many-valued logic which based on "degrees of truthness". It is different from true or false, 1 or 0 and Boolean logic on which the modern computer is based. It is utilized to carry the thought of half truth, wherever the real value might vary in between utterly true and utterly false. In contrast to the digital logic, Boolean logic has the truth values of variables may only be the number values one or zero. Fuzzy logic includes 0 and 1 as extreme cases of truth (or "the state of matters" or "fact") but also includes the various states of truth in between Fuzzy logic has been implemented to various areas, from control theory to artificial intelligence. While variables in arithmetic typically take numerical values, in fuzzy logic applications, non-numeric values are often used to facilitate the expression of rules and truth. In fuzzification, mathematical input values map into fuzzy membership functions. In opposite to it, de-fuzzification operations can be used to map a fuzzy output membership functions into a "crisp" output value that can be then used for decision or control purposes. The simple process of fuzzy logic is:

- **Fuzzification:** For fuzzy membership functions, fuzzify all input values into it.
- **Apply Rules:** To compute the fuzzy output functions, implement all possible rules in the rule base.
- **De-fuzzification:** Get "crisp" output values from the fuzzy output functions.

A fuzzy logic algorithm was proposed to improve the quality of transmission with low handover delay, less packet loss and reduced wrong handover (Prithviraj, A., Krishnamoorthy, K., & Vinothini, B., 2016). Combined fuzzy logic (FL) and genetic algorithms (GAs) for the access network selection was discussed (Alkhawlani, M., & Ayesh, A., 2008). and improved the QoE parameters: scalability, flexibility and simplicity. Fuzzy logic system for handover decision making for Global System for Mobile communication (GSM) network was proposed (Nyambati, E. T., & Oduol, V. K., 2017). Fuzzy Logic applications had been proposed for channel estimation, channel equalization and channel decoding in cellular networks. Traffic arrival rate and smallest reusable distance are the two fuzzy descriptors. These two attributes were taken to use the same channel in co-channel cells. Free channels within the reuse distance were considered (Nayak, P., Bhavani, V., & Shanthi, M, 2016).

The following are the main advantages of fuzzy logic which make it a preferred choice to be used:

- Fuzzy logic produces vague, imprecise and a qualitative output or data.
- Mathematics variables usually take numerical values; whereas in fuzzy logic linguistic variables are used.
- Linguistic variables are non-numeric values used to assist the appearance of rules and facts.
- It has well-built tool to reveal the problem solution space taken from its undecided input data.
- Able to stable the system as soon as possible or with the minimum time span.
- Fuzzy logic is a simple system and easy to design.
- Procedure to implement Fuzzy logic is very simple and easy to understand.

Neural Network in Wireless Heterogeneous Network

An Artificial Neural Network (ANN) is an information processing model motivated by biological nervous systems, like the brain process information. The main element is its general arrangement of information processing system. It comprises a huge number of well organized processing elements (neurons) working in accord to resolve particular problems. ANNs consist of many nodes, which imitate biological neurons of the human brain. The neurons are attached with links for the interaction between them. The nodes receive input data and simple operations on the data can be performed. The outcome is forwarded to next neurons and final output of every node is termed as its activation or node value. Each link is evaluated with weight. ANNs has the ability of learning which occurred by changing weight values. Two Artificial Neural Network topologies are – feed forward and feed backward. In feed forward the information flow is unidirectional. A unit sends information to other unit from which it does not get any information means one way communication. There is no back way. They are used in pattern generation / recognition / classification. They have predefined inputs and outputs.

A trained neural network can be act as an "expert" in the category of information. The expert can provide directions to unsolved situations of interest and answers "what-if" problems. An artificial neural network-based handoff decision algorithm was suggested to reduce the handoff latency in wireless heterogeneous infrastructures (Çalhan, A., & Çeken, C., 2013). The key parameters taken for this approach were data rate, monetary cost and RSSI information.

The following are the main advantages of the neural network:

- Particular hardware equipments are designed and manufactured for ANN assessment may be carried out in parallel which take advantage of real- time operation.
- It has the ability to learn operation depend upon information provided for training or prior experience.
- The feature or capability to retained information or redundant data keeps the system always on even any of its neurons fails.
- Neural network can develop its own organization (called self-organization)

Machine Learning in Wireless Heterogeneous Network

Machine learning (ML) is mainly based on algorithms and mathematical models. It makes the system capable to automatically learn and improve from its experiences. The process of learning starts with observations or input data. Machine learning algorithms make a mathematical model of trial data, known as "training data" in order to make predictions or decisions. Machine learning enables analysis of massive/large quantities of data. It generally gives faster and more accurate results in order to find profitable opportunities or dangerous risks. To fulfill such requirements, it may also require additional time and resources to train it properly. Combination of machine learning with AI and cognitive technologies can make it more effective in processing/analyzing large volumes of information. Some ML techniques are supervised, semi-supervised, unsupervised and reinforcement learning (RL). A machine learning approach named Q-learning was proposed to resolve the resource allotment problem in typical networks (Amiri, R., Mehrpouyan, H., Fridman, L., Mallik, R. K., Nallanathan, A., & Matolak, D., 2018). A cellular network was featured as a multiple agent network where each base station worked as an agent. Q-learning applied as an efficient approach to managing the resources of a multi-agent network. A context-aware mobility management (MM) approach was proposed for small cell networks, which uses reinforcement learning techniques and inter-cell coordination for optimizing the handover and throughput feature of user equipment (Simsek, M., Bennis, M., & Guvenc, I., 2015). Two common issues are over fitting and computation time.

The following are the main advantages of the machine learning:

- Developing low-complexity algorithms for wireless problems
- Overcoming the lack of network information/knowledge
- Facilitating self-organization capabilities
- Reducing signaling overhead
- Avoiding past faults
- Achieving better performance than traditional optimization

The above discussed computational methods are used in different way to achieve the good communication in wireless heterogeneous network. These methods are used to manage different aspects of communication like mobility management, network management, spectrum management, resource management and interference management etc. When there is heterogeneous network; the environment is of multi interface, multi channel. In this environment interface management plays a vital role to handle the situation.

ROLE OF INTERFACE MANAGEMENT IN WHN

An interface management system provides the interface between two pieces of equipments or protocol layers etc. The system has standardized function such as passing messages, connecting and disconnecting etc. Today, mobile terminals have several access interfaces of different network technologies. The combination of these access technologies needs an intelligent control to interface the best channel to best interface for best services. Interface management system manages the same situation of different nodes or networks equipped with different multi resources. The objective of the interface management process is to guarantee the proper functioning of a system composed of many interfacing sub-systems. An interface management protocol is designed to assign channels to interfaces in multichannel / multi interface environment. The function of interface management is the effective use of multiple channels when there are fewer interfaces per node than channels. Today's mobile devices are integrated with many devices like video camera, MP3 player, FM radio etc. All these devices required power to work. Three states of each interface were discussed; operating, standby, and turn-off state. An energy saving interface management was proposed to utilize the power consumption in this situation (Nguyen-Vuong, Q. T., Agoulmine, N., & Ghamri-Doudane, Y., 2008). A timescale methodology was used to assign channels to interfaces. It works on multi-channel and multi-interface environment. Channel pool management was used where channels are randomly selected from pool at a fixed interval of time. Time scheduler runs all the activity (Bhandari, V., & Vaidya, N. H., 2009). A discrete time modeling for multi-channel, multi-interface networks was proposed. A throughput optimal channel/interface allotment technique was used where many users were sharing the set of orthogonal channels. The stability region for certain simplified case was characterized (Halabian, H., Lambadaris, I., Lung, C. H., & Srinivasan, A., 2010). A device management interface in IoT was suggested as a Web service, which integrated the whole IoT network management aspects such as monitoring, manual device controls, and error reporting or data presentation (Ocak, M., 2014). Security is the prime concern in this approach for Short Messaging Service (SMS) over GSM network. A coded message from an android based transmitting terminal to field-programming gate array (FPGA) based receiving device. Messages were encrypted with rolling code to enhance the security of the interface (Wightwick, A., & Halak, B., 2016). A gateway was used in between application and platform component. It facilitates the user to access, modify and expand data at any time. With proper id and password one can easily sends its request .Gateway checks the authorization and authentication of user and forwarded the request (Nguyen, M. C., & Won, H. S., 2017).

Various approaches for interface management based on multiple criteria are classified into different classes shown in figure 1.

All these approaches use different traditional methods for interface management. Multiple parameters used by researchers to implement required functions. The pros and cons of these approaches are discussed in table 2.

Figure 1. Different interface management approaches

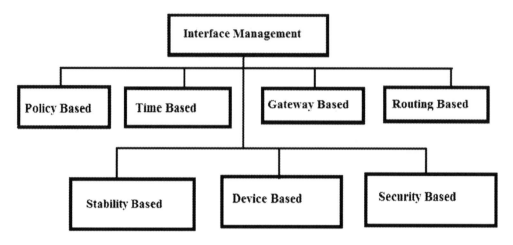

Table 2. Advantages and disadvantages of different interface management approaches

Approaches	Advantages	Disadvantages
Routing based	-Less packet drop rate -Less packet loss -Real- time support	-Cost -Complexity -Average delay increased
Policy- based	Smooth communication	- Decision speed
Time –based	- Flexible -Increase diversity gain	- Delay -Single path communication consideration -Low user consideration
Stability- based	Stay Stability criteria consideration	-Decision speed -Complexity
Device based	-Multi-criteria consideration -Reduces overhead -Low latency	-Slow operation -Not satisfactory in frequent updates - Complex
Security- based	-Simple operation -Security consideration	-One-way communication – Cost increased -Low user consideration
Gateway- based	-Easy accessibility of data -No extra hardware -Smooth communication	-Decision speed -Require the good knowledge of software

Any methods used in a system are measured by its performance. There are some performance parameters through which one can measure system efficiency, capability or performance.

QUALITY OF SERVICE AND QUALITY OF EXPERIENCE

Quality of Service (QoS) presents to the evaluation of a network. The measuring parameters like bandwidth, latency, error rates, and power etc. are very important in QoS. Quality of Service is a significant factor especially for internet services which has heavy data traffic, large volume of data. The multimedia applications, media streaming, online gaming and video-conferencing etc are some applications which dependent almost on a constant, stable and fast connection.

Quality of Experience (QoE) is a user-centered performance aspect for a given item for consumption or service. QoE is all about the experience the end users are receiving from a service and their satisfaction with it. in simple words, Quality of Experience of a service is the difference between customer's desired service experience and getting service in real. One of the important measures of QoS and QoE is reliability means whether system is reliable with any kind of implementation.

Reliability

Reliability is a performance measure parameter which presents the on the whole stability and regularity of a system. The ability of a system to consistently perform is termed as reliable system. This type of system required function task without degradation or failure. A system is termed as a high reliable if it generates same outputs in same environment. "It is the characteristic of a set of test scores that relates to the amount of random error from the measurement process that might be embedded in the scores. Scores that are highly reliable are accurate, reproducible, and consistent from one testing occasion to another.

AI for Reliability Prediction for Electronics and Communication

There is a constant challenge to provide reliable and high quality algorithm among communication, system and devices. The uses of real-time application on mobile devices are growing day by day. Reliability can be calculated at any point like on devices, network, on channel or at user 'end in the system. A trust based energy aware routing model was proposed for route discovery and node based on reliability. Route request from the source was accepted by a node only if its reliability was high. Otherwise, the route request was discarded. This approach formed a reliable route from source to destination thus increasing network life time, improving energy utilization and decreasing number of packet loss during transmission (Pushpalatha, M., Venkataraman, R., & Ramarao, T., 2009). A reliable routing algorithm based on fuzzy-logic (RRAF) for finding a reliable path in mobile ad hoc networks was proposed. In this approach for each node two parameters were used to calculate the lifetime of routes. Trust value and battery capacity were the two main parameters in this method that make the routing algorithm more reliable (Dana, A., Ghalavand, G., Ghalavand, A., & Farokhi, F., 2011).

CONCLUSION

Although there have been many algorithms for implementing interface management in wireless heterogeneous network. The problem is to automatically deal with the complexity of heterogeneous networks via the evolutionary methods. AI techniques could have been very useful for the interface management. Any technique is selected depending upon the various parameters, a tradeoff between these parameters, application, multimode device etc. To see these requirements and attractive features of fuzzy logic, neural network and machine learning or the hybrid of these techniques could be the best for interface management. Interface management works in multichannel and multi interfaces environment. As AI made the system smart and intelligent, it also needs a reliable system. Reliability plays a critical role in effective communication. Reliable systems are well-organized systems, and good organization is the key to succeeding in today's competitive market. These two terms describe different dealings of success, but both are equally important. Reliability refers to a system's ability to constantly perform fine, while efficiency measures that system's output in a given period. Fewer mistakes lead to greater efficiency over time. So AI methods can be used to calculate reliability to make system reliable as well as smart.

REFERENCES

Alkhawlani, M., & Ayesh, A. (2008). Access network selection based on fuzzy logic and genetic algorithms. *Advances in Artificial Intelligence*, 8(1), 1–12. doi:10.1155/2008/793058

Amiri, R., Mehrpouyan, H., Fridman, L., Mallik, R. K., Nallanathan, A., & Matolak, D. (2018). A machine learning approach for power allocation in HetNets considering QoS. In *2018 IEEE International Conference on Communications (ICC)* (pp. 1-7). IEEE. 10.1109/ICC.2018.8422864

Baseer, S. (2013). Heterogenous networks architectures and their security weaknesses. *International Journal of Computer and Communication Engineering*, 2(2), 90–93. doi:10.7763/IJCCE.2013.V2.145

Bhandari, V., & Vaidya, N. H. (2009). *Channel and interface management in a heterogeneous multi-channel multi-radio wireless network*. Illinois Univ at Urbana-Champaign. doi:10.21236/ADA555113

Çalhan, A., & Çeken, C. (2013). Artificial neural network based vertical handoff algorithm for reducing handoff latency. *Wireless Personal Communications*, 71(4), 2399–2415. doi:10.100711277-012-0944-4

Chandavarkar, B. R., & Reddy, G. R. M. (2012). Survey paper: Mobility management in heterogeneous wireless networks. *Procedia Engineering*, 30, 113–123. doi:10.1016/j.proeng.2012.01.841

Dana, A., Ghalavand, G., Ghalavand, A., & Farokhi, F. (2011). A Reliable routing algorithm for Mobile Adhoc Networks based on fuzzy logic. *International Journal of Computer Science Issues*, 8(3), 128–133.

Halabian, H., Lambadaris, I., Lung, C. H., & Srinivasan, A. (2010). Dynamic Channel and Interface Management in Multi-channel Multi-interface Wireless Access Networks. In *2010 IEEE Global Telecommunications Conference GLOBECOM 2010* (pp. 1-6). IEEE. 10.1109/GLOCOM.2010.5683566

Jafari, A. H., & Shahhoseini, H. S. (2015). A Reinforcement Routing Algorithm with Access Selection in the Multi–Hop Multi–Interface Networks. *Journal of Electrical Engineering*, 66(2), 70–78. doi:10.1515/jee-2015-0011

Nayak, P., Bhavani, V., & Shanthi, M. (2016). A Fuzzy Logic based Dynamic Channel allocation Scheme for wireless Cellular networks to optimize the frequency reuse. In 2016 IEEE Region 10 Conference (TENCON) (pp. 1111-1116). IEEE. doi:10.1109/TENCON.2016.7848181

Nguyen, M. C., & Won, H. S. (2017, February). Gateway-based access interface management in big data platform. In *2017 19th International Conference on Advanced Communication Technology (ICACT)* (pp. 447-450). IEEE. 10.23919/ICACT.2017.7890128

Nguyen-Vuong, Q. T., Agoulmine, N., & Ghamri-Doudane, Y. (2008). A user-centric and context-aware solution to interface management and access network selection in heterogeneous wireless environments. *Computer Networks*, 52(18), 3358–3372. doi:10.1016/j.comnet.2008.09.002

Nyambati, E. T., & Oduol, V. K. (2017). Analysis of The Impact of Fuzzy Logic Algorithm On Handover Decision in A Cellular Network. *International Journal of Innovation Education and Research*, *5*(5), 46–62.

Ocak, M. (2014). *Implementation of an internet of things device management interface.* Academic Press.

Prithviraj, A., Krishnamoorthy, K., & Vinothini, B. (2016). Fuzzy Logic-Based Decision-Making Algorithm to Optimize the Handover Performance in HetNets. *Circuits and Systems*, *7*(11), 3756–3777. doi:10.4236/cs.2016.711315

Pushpalatha, M., Venkataraman, R., & Ramarao, T. (2009). Trust based energy aware reliable reactive protocol in mobile ad hoc networks. *World Academy of Science, Engineering and Technology*, *56*(68), 356–359.

Roszkowska, E. (2011). Multi-criteria decision making models by applying the TOPSIS method to crisp and interval data. *Multiple Criteria Decision Making, 6*, 200-230.

Simsek, M., Bennis, M., & Guvenc, I. (2015). Mobility management in HetNets: A learning-based perspective. *EURASIP Journal on Wireless Communications and Networking*, *2015*(1), 26. doi:10.118613638-015-0244-2

Tran, P. N., & Boukhatem, N. (2008, September). Comparison of MADM decision algorithms for interface selection in heterogeneous wireless networks. In *2008 16th International Conference on Software, Telecommunications and Computer Networks* (pp. 119-124). IEEE.

Wightwick, A., & Halak, B. (2016). Secure communication interface design for IoT applications using the GSM network. In *2016 IEEE 59th International Midwest Symposium on Circuits and Systems (MWSCAS)* (pp. 1-4). IEEE. 10.1109/MWSCAS.2016.7870010

Chapter 13
PVT Variability Check on UCM Architectures at Extreme Temperature-Process Changes

Rajkumar Sarma
https://orcid.org/0000-0002-5551-1006
Lovely Professional University, India

Cherry Bhargava
https://orcid.org/0000-0003-0847-4780
Lovely Professional University, India

Shruti Jain
JUIT, India

ABSTRACT

The UCM (universal compressor-based multiplier) architecture promises to provide faster multiplication operation in supply voltage as low as 0.6 V. The basic component of UCM architecture is a universal compressor architecture that replaces the conventional Wallace tree algorithm. To extend the work further, in this chapter, a detailed PVT (process-voltage-temperature) analysis is performed using Cadence Virtuoso 90nm technology. The analysis shows that the delay of the UCM has reduced more significantly than the Wallace tree algorithm at extreme process, voltage, and temperature.

INTRODUCTION

Today's portable devices are capable of doing image filtering to face recognitions, an audio signal enhancement to voice recognition & gesture-based control to biometric authentication. All those functionalities are the applications of digital signal processing (DSP). A large number of mathematical operations are performed repeatedly and quickly on series of data samples by DSP algorithms. Most operating systems and general-purpose microprocessors can successfully execute DSP algorithms but

DOI: 10.4018/978-1-7998-1464-1.ch013

Figure 1. Basic multiplication operation

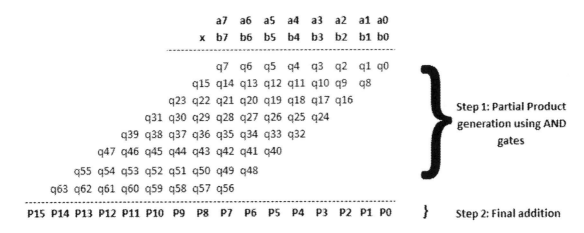

because of power efficiency constraints, they are not suitable for use in portable devices such as PDAs and mobile phones. However, the rapid growth of portable electronics has introduced the major challenges of high throughput for VLSI design engineers. Among the other digital blocks, multiplier plays a vital role while evaluating the performance of a DSP block. While performing convolution, filtering or any other DSP operations it is always desired to use an efficient multiplier unit. A basic design of a multiplier is as shown in the figure 1.

As shown in the figure 1, the multiplicand's & the multiplier's individual terms are ANDed to produce the partial products & positioned as per their weights. For example, in the figure 1, 'A2B0', 'A1B1' & 'A0B2' are aligned in a single column because the weight is two for all of the mentioned partial products. i.e. the summation of the bit location is any of 2+0, 1+1, 0+2, which are in all cases will be equal to 2. Hence, for the addition of partial products, its alignment is vital. At the next step, the partial product with same weights are added using full adder (in the case of 3 partial products), half adder (in the case of 2 partial products) or any compressor circuit (for adding 'n' number of partial products simultaneously).

In this research paper the novel UCM architecture as proposed in (Sarma, Bhargava & Jain, 2019), is further validated with the PVT analysis in cadence spectre tool in 90 nm CMOS technology. The UCM architecture uses a novel compressor-based multiplier algorithm which reduces the delay substantially.

The following sections are discussed as follows: in section 2, various different notable architectures related to multiplier are discussed in detail, in section 3, a quick review on the novel UCM architecture has been explained, in section 4, a detailed PVT analysis of the UCM architecture is discussed & in section 5, a detailed conclusion, future scopes & application of the UCM architecture is discussed.

VARIOUS MULTIPLIER ARCHITECTURES

As we know that the processing elements mainly involve the multiplication of two numbers. So, there is a need of multiplier in such type of processing systems. Various fast & efficient multipliers are described in the literature. Array multiplier (as shown in figure 2) is a basic multiplier which follows the principle of product generation & addition. But as the total number of addition levels increases, this architecture becomes bulkier with higher PDP.

Figure 2. Array multiplier architecture

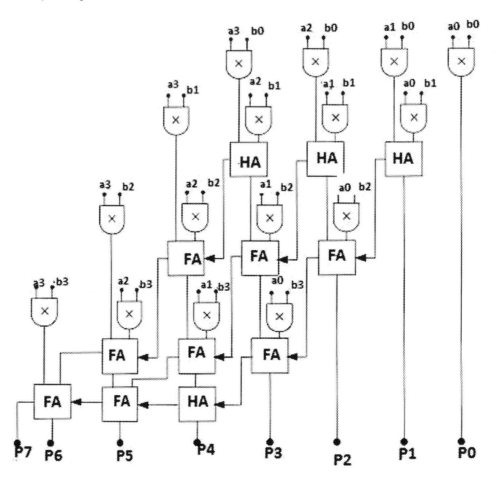

Solution for this problem can be Wallace tree multiplier based on Wallace tree structure. Here a multiplier is designed which generates the product of two numbers using purely combinational logic, i.e., in one gating step. Using straight forward diode-transistor logic, it appears presently possible to obtain products in under 1 micro sec, and quotients in 3 micro sec. A rapid square-root process is also outlined in the literature (Wallace, 1964). The figure for the same is shown in Figure 3.

However, in Wallace tree multiplier every partial product is added in a single direction from top to bottom so the number of adder increases. To overcome this problem a rectangular styled Wallace tree multiplier is proposed (Toh et al., 2001), in which the partial products are divided into two groups and added in the opposite direction. The partial products in the first group are added downward, and the partial products in the second group are added upward. On the other hand, in the literature (Onomi et al., 2001), a phase mode parallel multiplier is also proposed. The proposed multiplier has a Wallace-tree structure comprising trees of carry save adders for the addition of partial products. This structure has a regular layout; hence it is suitable for a pipeline scheme.

The conventional Wallace tree multiplier is based on carry save adder. In (Guevorkian et al., 2013) the speed of the multiplier has improved by introducing compressors instead of the carry save adder. 3-2 compressor, 4-2 compressor, 5-2 compressors and 7-2 compressors are used with Wallace tree multiplier.

Figure 3. Wallace tree multiplier (addition of partial product)

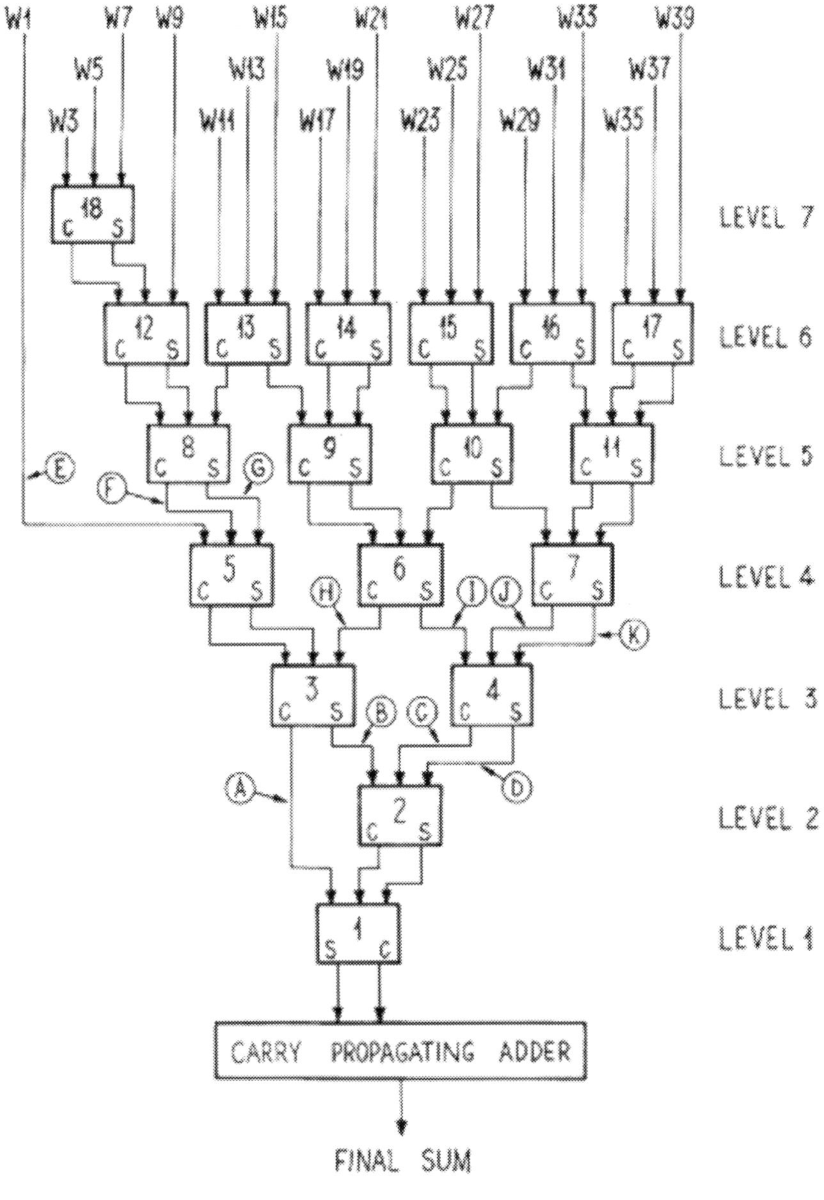

Higher order compressors have better performance compared with 3-2 compressor. So, the speed of the multiplier can be improved by introducing the higher order compressors.

A few architectures in the literature focused on the optimization of adder cell also. As adder is a basic cell in multiplier or divider, the optimization is mainly focused on adder part. A Carry-Select-Adder Optimization Technique is proposed in the literature (Liao et al., 2002) in which a carry-select-adder partitioning algorithm is for high-performance Booth-encoded Wallace-tree multipliers. By taking into various data arrival times, a branch-and-bound algorithm is proposed and a generalized technique to partition an n-bit carry-select adder into a number of adder blocks is proposed such that the overall delay of the design can be minimized. In a separate approach (Sousa, 2003), an improved algorithm for designing efficient modulo ($2^n + 1$) multipliers had been proposed. By manipulating the Booth tables and by applying a simple correction term, the proposed multiplier is the most efficient among all the known modulo ($2^n + 1$) multipliers and is almost as efficient as those for ordinary integer multiplication. On the other hand, (Singh, De & Maity, 2012) a comparative analysis is carried out for designing multiplier using complementary MOS (CMOS) logic style, complementary pass-transistor (CPL) logic style and double-pass transistor (DPL) logic style. Similarly a single precision reversible floating-point multiplier is proposed in the literature (Nachtigal, Thapliyal & Ranganathan, 2010). A 24-bit multiplier is proposed in this work by decomposing the whole 24-bit in three portions of 8 bit each.

The internal to the multiplier is adder. Therefore, an optimized adder can further enhance the performability of a multiplier. An adder (also referred to as summer) is a logic circuit which adds two or more variables. Adders are unavoidable part of logic circuits as they are not only used for addition but also to calculate the addresses, increment operations, table indices etc. Most common adders operate on binary numbers although they can be constructed for BCD, excess -3 formats etc. In the literature there are various full adder architectures are proposed. A novel low power hybrid full adder using MOSIS 90nm Technology (Khan, Kakde, & Suryawanshi, 2013) is proposed, which consumes very low power. The proposed design is compared with its conventional full adder which consists of 28 transistors. In a different approach, a hybrid 1-bit full adder is proposed by (Bhattacharyya et al., 2014) which uses CMOS as well as TG logic styles. The entire design was implemented in 90nm as well as 180nm technologies. At 1.8V supply voltage, the proposed design, offers very less power and moderately low delay. The adder proposed by (Bhattacharyya et al., 2014) is shown in figure 4.

Figure 4. Full adder design by (Bhattacharyya et al., 2014)

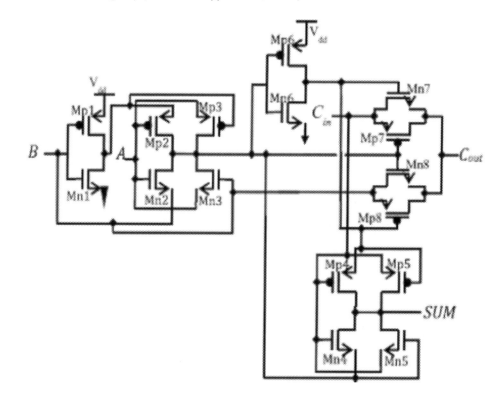

UCM ARCHITECTURE

The UCM architecture consists of three stages. The stage 1 & stage 3 remains the same for UCM architecture (as that of Wallace tree), because whether it is partial product generation or addition of intermediate sum or carry using fast adder, these can be chosen according to the requirement of the designer. Hence, it is more important to replace the stage 2 i.e. addition of partial product which creates sum & carry separately.

Addition of Partial Products

While adding partial products, the partial products are aligned in such a way that the summation of bit location of multiplicand & multiplier are equal. The summation of bit location can be called as `weight'

Figure 5. UCM architecture for 9 x 9 bit multiplication

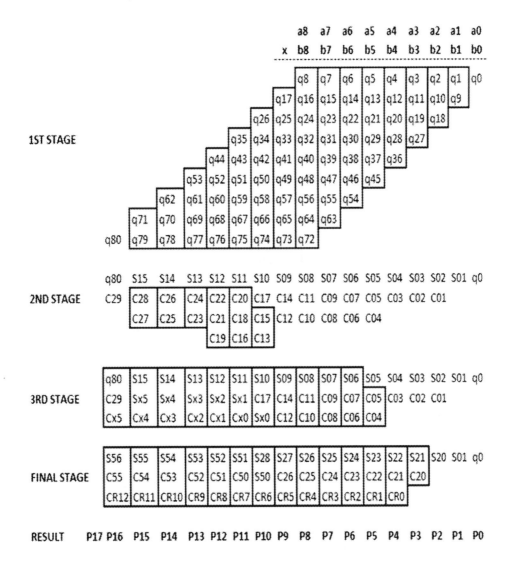

of a particular partial product. For example, in the figure 5, `q35', `q43', `q51', `q59', `q67' & `q75' are aligned in a single column because of the reason that the weight is eleven for all of the mentioned partial products, i.e. q35=a8b3, q43=a7b4, q51=a6b5 etc. So, the summation of the bit location is either of 8+3 or 7+4 or 6+5, which is in all cases are equal to 11. Hence, for the addition of partial products, its alignment is very important. Once the partial products are aligned the next step is to add all the partial product falling in that particular column. For adding a particular column firstly, the total number of stages & levels need to be identified. Each stage consists of an AND-XOR gate pair & the total number

Figure 6. AND-XOR gate arrangement with K stages & L levels having A0, A1, A2,....., AK partial products (with equal weights) for a particular column

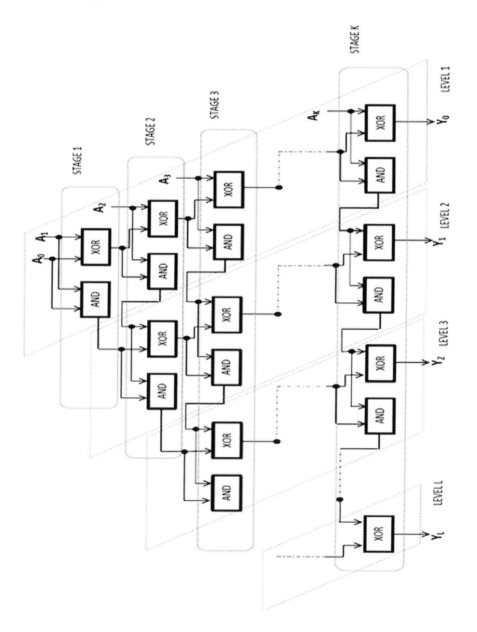

of stages in one level is counted from top to bottom. The total number of stages in the first level is 'i-1', where 'i' is the total number of partial products to be added in a particular column.

On the other hand, the horizontal count of AND-XOR pair is the total number of levels required for the design. In a different angle, we can say that the total number of levels required in a design is the total number of AND-XOR pair required in the bottom most stages. Basically, it is the count of AND-XOR pair from right to left. In each level, the total number of stages required will be decremented by one until it satisfies the formula:

where 'i' is the total number of the partial product to be added & 'n' is the total number of levels required. 'i' & 'n' are integers starting from 1, 2, 3, ..., ∞. For example, for adding 3 partial products in a column, the total number of levels will be: $2n-1 \geq 3$, so n=2. Similarly, if suppose i=8, i.e. $2n-1 \geq 8$, so n=4 & so on. The basic block diagram for K stages & L levels is shown in figure 5. In figure 6, A0, A1, A2 up to AK are the partial products; the term Y0 is the sum & Y1, Y2, Y3,....., YL are the carries. Therefore, in simple words, the algorithm shown in the figure 6 is a N-bit compressor circuit which generates sum of a particular column & single/multiple carries.

Special Cases

1.   In the last level, instead of AND-XOR pair, only XOR gate is to be used.
2.   If i=2, only one level is to be used to get the sum as well as carry. In this case, the output from the AND is the carry.
3.   For i=1, the input itself is the output (sum) & there is no carry output.

It is very important to note that the output through the level 1 is the sum of the partial products present in a particular column & the outputs of rest of the levels i.e. level 2 to level L are the corresponding carry bits. After getting the sum as well as carry bit of all columns, the next step is to add up the sum bits with the carry bit of the previous columns. For this any of the efficient algorithms such as dada algorithm, Wallace tree algorithm or even ripple carry adder can be used as the number of rows has reduced substantially. A detailed design is shown in figure 5.

PVT ANALYSIS

VLSI is an art of chip design, where specification is transformed to functional hardware. Cadence provides tools for front end as well as back end designs, where, after rigorous design steps, GDS-II file are finally sent for fabrication. But due to process complexity (i.e. pressure, supply voltage, temperature etc.) the YIELD of the fabricated designs is found very low. Major reason for yield loss is fabrication parameter variation among wafer to wafer. To improve the yield of design; the IC should be able to sustain extreme variation. Therefore, the design cycle must be validated through PVT and 3-sigma variation before fabrication.

The work proposed by (Sarma, Bhargava & Jain, 2019), provides a comparison of delays for 5x5 bit as well as for 9x9 bit operation for 0.6V, 0.7V, 0.8V & 0.9V. The same has been shown in the figure 7 & 8.

The comparison shows that the UCM architecture, as proposed by (Sarma, Bhargava & Jain, 2019), performs better than Wallace tree architecture at ultra-low supply voltages (less than 0.9V). Moreover, the UCM architecture performs even better for higher order bit multiplication. For example, the difference in delay of UCM & Wallace tree architecture for 9x9 bit operation is more than 5x5 bit operation (120 ps & 20 ps respectively). Therefore, (Sarma, Bhargava & Jain, 2019) has summarized that UCM architecture performs better than Wallace tree for higher order bit multiplication at ultra-low supply voltages (less than 0.9V).

To validate the performance of the UCM architecture further, a PVT analysis is carried out at different corners (Fast-Fast, Fast-Slow, Normal-Normal, Slow-Fast & Slow-Slow) & at three different extreme temperatures (-40°, 0° & +50° Celsius). Table 1 & 2 shows the delay comparison of UCM & Wallace tree 5x5 bit & 9x9 bit architecture respectively at 0.6 V & 0.9 V supply voltage in different corners along with variation in temperature (-40°, 0° & +50° Celsius)

Moreover, a graphical comparison of delay of UCM & Wallace tree 5x5 bit & 9x9 bit architectures at 0.6 V & 0.9 V supply voltage in different corners along with variation in temperature (-40°,0° & +50° Celsius) are shown in figure 9 & 10. The graphs in figure 9 & 10 clearly shows that there is a significant improvement in delay of UCM architecture in comparison to the Wallace tree architecture for 5x5 bit as well as 9x9 bit multiplication. Most important part is that, for 5x5 bit multiplication, at different corners & at extreme temperatures, the UCM architecture proves to be the better performer than Wallace tree architecture at ultra-low supply voltages. On the other hand, for 9x9 bit multiplication, the delay of UCM has a much more significant drop in comparison to the Wallace tree at 600 mV (at different corners & at extreme temperatures). Whereas, the delay of the UCM architecture is seems to be slightly higher than Wallace tree at slow-fast corner in -40°,0° & +50° Celsius for 9x9 bit multiplication at 900 mV. Moreover, as shown in the table 1, the minimum & maximum delay for 5x5 bit multiplication using UCM architecture at 600 mV are 2.665 ns & 2.937 ns respectively. Whereas the same for Wallace tree

Figure 7. Graphical comparison of 5x5 bit UCM & 5x5 bit Wallace tree multiplier at voltages below 1V

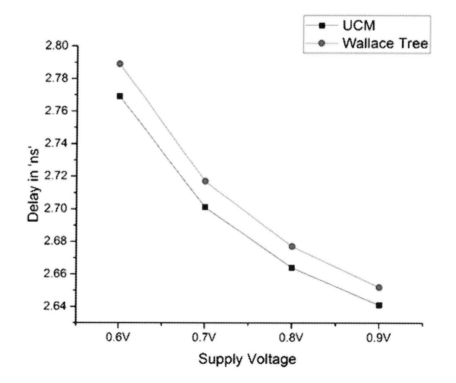

Figure 8. Graphical comparison of 9x9 bit UCM & 9x9 bit Wallace tree multiplier at voltages below 1V

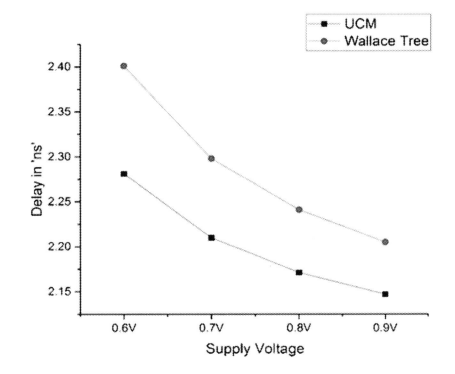

Table 1. Delay comparison of UCM & Wallace tree 5x5 bit architecture at 0.6 V & 0.9 V supply voltage in different corners along with variation in temperature (-40°,0° & +50° Celsius)

	UCM (in ns @ 600mV)	Wallace tree (in ns @ 600mV)	UCM (in ns @ 900mV)	Wallace tree (in ns @ 900mV)
Nominal (27)	2.769	2.789	2.641	2.652
FF_0 (-40)	2.665	2.677	2.59	2.597
FF_1 (0)	2.684	2.698	2.601	2.61
FF_2 (+50)	2.709	2.725	2.616	2.626
FS_0 (-40)	2.75	2.766	2.623	2.632
FS_1 (0)	2.782	2.801	2.64	2.651
FS_2 (+50)	2.822	2.845	2.663	2.676
NN_0 (-40)	2.72	2.735	2.613	2.622
NN_1 (0)	2.749	2.767	2.629	2.64
NN_2 (+50)	2.786	2.809	2.651	2.663
SF_0 (-40)	2.728	2.746	2.617	2.627
SF_1 (0)	2.76	2.782	2.635	2.647
SF_2 (+50)	2.802	2.829	2.658	2.673
SS_0 (-40)	2.826	2.849	2.656	2.668
SS_1 (0)	2.875	2.902	2.682	2.697
SS_2 (+50)	2.937	2.97	2.716	2.734

Table 2. Delay comparison of UCM & Wallace tree 9x9 bit architecture at 0.6 V & 0.9 V supply voltage in different corners along with variation in temperature (-40°,0° & +50° Celsius)

	UCM (in ns @ 600mV)	Wallace tree (in ns @ 600mV)	UCM (in ns @ 900mV)	Wallace tree (in ns @ 900mV)
Nominal (27)	2.281	2.401	2.147	2.205
FF_0 (-40)	2.171	2.239	1.138	1.195
FF_1 (0)	2.192	2.27	1.153	1.222
FF_2 (+50)	2.218	2.31	1.247	1.257
FS_0 (-40)	2.258	2.353	2.126	2.171
FS_1 (0)	2.291	2.402	2.145	2.198
FS_2 (+50)	2.334	2.463	2.169	2.233
NN_0 (-40)	2.228	2.322	1.235	1.252
NN_1 (0)	2.259	2.369	2.134	2.187
NN_2 (+50)	2.3	2.43	2.157	2.221
SF_0 (-40)	2.239	2.351	2.123	1.259
SF_1 (0)	2.274	2.406	1.421	1.289
SF_2 (+50)	2.32	2.479	2.168	1.439
SS_0 (-40)	2.339	2.484	2.162	2.227
SS_1 (0)	2.391	2.561	2.19	2.268
SS_2 (+50)	2.456	2.659	2.227	2.323

are 2.677 ns & 2.97 ns respectively. Similarly, the minimum & maximum delay for 5x5 bit multiplication using UCM architecture at 900 mV are 2.59 ns & 2.716 ns respectively. Whereas the same for Wallace tree are 2.597 ns & 2.734 ns respectively. Same thing if we observe for 9x9 bit multiplication using UCM architecture at 600 mV, the minimum & maximum delays are 2.171 ns & 2.456 ns respectively whereas for Wallace tree the values are 2.239 ns & 2.659 ns. On the other hand, for 9x9 bit multiplication using UCM architecture at 900 mV, the minimum & maximum delays are 1.138 ns & 2.227 ns respectively whereas for Wallace tree the values are 1.195 ns & 2.323 ns.

CONCLUSION

The UCM architecture has a wide range of acceptability in the field of digital system design. UCM architecture not only performs the best in a nominal Process, Voltage & Temperature but also in a wide range of variation in extreme temperature, process & ultra-low supply voltages. Especially, in the case of the higher order multiplication (9x9 bit) operation with supply voltage as low as 0.6 V, the delay has reduced by 5.05% (mean value) than Wallace tree multiplier architecture. Therefore, UCM multiplier will have a wide range of acceptability in the circuits where speed is the top most priority.

Figure 9. Graphical comparison of delay of UCM & Wallace tree 5x5 bit architecture at 0.6 V & 0.9 V supply voltage in different corners along with variation in temperature (-40°,0° & +50° Celsius)

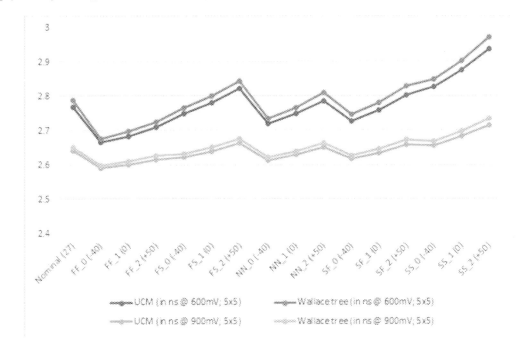

Figure 10. Graphical comparison of delay of UCM & Wallace tree 9x9 bit architecture at 0.6 V & 0.9 V supply voltage in different corners along with variation in temperature (-40°,0° & +50° Celsius)

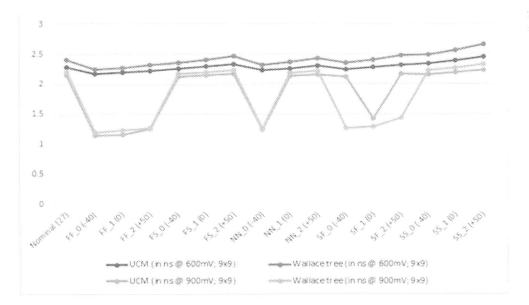

REFERENCES

Bhattacharyya, P., Kundu, B., Ghosh, S., Kumar, V., & Dandapat, A. (2014). Performance analysis of a low-power high-speed hybrid 1-bit full adder circuit. *IEEE Transactions on Very Large Scale Integration (VLSI) Systems, 23*(10), 2001-2008.

Guevorkian, D., Launiainen, A., Lappalainen, V., Liuha, P., & Punkka, K. (2005). A method for designing high-radix multiplier-based processing units for multimedia applications. *IEEE Transactions on Circuits and Systems for Video Technology, 15*(5), 716–725. doi:10.1109/TCSVT.2005.846436

Jaiswal, K. B., Seshadri, P., & Lakshminarayanan, G. (2015, March). Low power wallace tree multiplier using modified full adder. In *2015 3rd international conference on signal processing, communication and networking (ICSCN)* (pp. 1-4). IEEE. 10.1109/ICSCN.2015.7219880

Kataeva, I., Engseth, H., & Kidiyarova-Shevchenko, A. (2007). Scalable matrix multiplication with hybrid CMOS-RSFQ digital signal processor. *IEEE Transactions on Applied Superconductivity, 17*(2), 486–489. doi:10.1109/TASC.2007.901451

Khan, S., Kakde, S., & Suryawanshi, Y. (2013, December). VLSI implementation of reduced complexity wallace multiplier using energy efficient CMOS full adder. In *2013 IEEE International Conference on Computational Intelligence and Computing Research* (pp. 1-4). IEEE. 10.1109/ICCIC.2013.6724141

Krishna, K. G., Santhosh, B., & Sridhar, V. (2013). Design of wallace tree multiplier using compressors. *International journal of engineering sciences & research. Technology, 2*, 2249–2254.

Kshirsagar, R. D., Aishwarya, E. V., Vishwanath, A. S., & Jayakrishnan, P. (2013, December). Implementation of pipelined booth encoded wallace tree multiplier architecture. In *2013 International Conference on Green Computing, Communication and Conservation of Energy (ICGCE)* (pp. 199-204). IEEE. 10.1109/ICGCE.2013.6823428

Kuo, T. Y., & Wang, J. S. (2008, May). A low-voltage latch-adder based tree multiplier. In *2008 IEEE International Symposium on Circuits and Systems* (pp. 804-807). IEEE.

Liao, M. J., Su, C. F., Chang, C. Y., & Wu, A. H. (2002, May). A carry-select-adder optimization technique for high-performance Booth-encoded wallace-tree multipliers. In *2002 IEEE International Symposium on Circuits and Systems. Proceedings (Cat. No. 02CH37353)* (Vol. 1, pp. I-I). IEEE. 10.1109/ISCAS.2002.1009782

Luu, X. V., Hoang, T. T., Bui, T. T., & Dinh-Duc, A. V. (2014, October). A high-speed unsigned 32-bit multiplier based on booth-encoder and wallace-tree modifications. In *2014 International Conference on Advanced Technologies for Communications (ATC 2014)* (pp. 739-744). IEEE. 10.1109/ATC.2014.7043485

Nachtigal, M., Thapliyal, H., & Ranganathan, N. (2010, August). Design of a reversible single precision floating point multiplier based on operand decomposition. In *10th IEEE International Conference on Nanotechnology* (pp. 233-237). IEEE. 10.1109/NANO.2010.5697746

Onomi, T., Yanagisawa, K., Seki, M., & Nakajima, K. (2001). Phase-mode pipelined parallel multiplier. *IEEE Transactions on Applied Superconductivity, 11*(1), 541–544. doi:10.1109/77.919402

Paradhasaradhi, D., Prashanthi, M., & Vivek, N. (2014, March). Modified wallace tree multiplier using efficient square root carry select adder. In *2014 International Conference on Green Computing Communication and Electrical Engineering (ICGCCEE)* (pp. 1-5). IEEE. 10.1109/ICGCCEE.2014.6922214

Rao, M. J., & Dubey, S. (2012, December). A high speed and area efficient Booth recoded Wallace tree multiplier for Fast Arithmetic Circuits. In *2012 Asia Pacific Conference on Postgraduate Research in Microelectronics and Electronics* (pp. 220-223). IEEE. 10.1109/PrimeAsia.2012.6458658

Reddy, B. M., Sheshagiri, H. N., Vijayakumar, B. R., & Shanthala, S. (2014, December). Implementation of Low Power 8-Bit Multiplier Using Gate Diffusion Input Logic. In *2014 IEEE 17th International Conference on Computational Science and Engineering* (pp. 1868-1871). IEEE.

Sarma, R., Bhargava, C., Dhariwal, S., & Jain, S. (2019). UCM: A Novel Approach for Delay Optimization. *International Journal of Performability Engineering*, *15*(4).

Singh, A. K., De, B. P., & Maity, S. (2012). Design and Comparison of Multipliers Using Different Logic Styles. *International Journal of Soft Computing and Engineering*, *2*(2), 374–379.

Sousa, L. A. (2003). Algorithm for modulo ($2^n + 1$) multiplication. *Electronics Letters*, *39*(9), 752–754. doi:10.1049/el:20030467

Toh, N., Naemura, Y., Makino, H., Nakase, Y., Yoshihara, T., & Horiba, Y. (2001). A 600-MHz 54-bit multiplier with rectangular-styled Wallace tree. *IEEE Journal of Solid-State Circuits*, *36*(2), 249–257. doi:10.1109/4.902765

Wallace, C. S. (1964). A suggestion for a fast multiplier. *IEEE Transactions on Electronic Computers*, *EC-13*(1), 14–17. doi:10.1109/PGEC.1964.263830

Yi, Q., & Han, J. (2009, July). An improved design method for multi-bits reused booth multiplier. In *2009 4th International Conference on Computer Science & Education* (pp. 1914-1916). IEEE.

Chapter 14
Frequency–Based RO–PUF

Abhishek Kumar
Lovely Professional University, India

Jyotirmoy Pathak
https://orcid.org/0000-0002-9927-7231
Lovely Professional University, India

Suman Lata Tripathi
Lovely Professional University, India

ABSTRACT

Arbiter PUF and RO-PUF are two well-known architectures. Arbiter PUF is a simple architecture and easy to implement while RO-PUF require exponentially large hardware. As shown in this chapter, the digital design of RO-PUF response is 42.85% uniform and 46.25% unique.

INTRODUCTION

It is a primary requirement that random number must be truly random. Software algorithm of random number generator (RNG) does not produce truly random number (Beckmann & Potkonjak, 2009); the random number must originate from hardware feature. A random number should not be predicted even if every detail of the generator is known. There have been two approaches to random number generation: algorithmic (pseudorandom) and by a physical process (nondeterministic). PRNG is based on a mathematical formula (Devadas, Suh, Paral, Sowell, Ziola & Khandelwal, 2008) which produces a periodic sequence of number; can be predicted based on initial seed and upcoming sequence can predict with the previous pattern statically. Physical process-based random number generator generates each bit based on the physical variation of the circuit. Process variation of silicon IC manufacturing leads to a distant difference in each parameter. Physical unclonable function (PUF) (Pappu, Recht, Taylor & Gershenfeld, 2002) is a hardware security module extracts randomness from the process variation and environmental condition. A random number is the basic need of cryptography; a secret key is derived from a random number. A random number is a random string of binary 1 and 0 with a balanced number. Two well-known methods of random number generation are algorithmic and derive by physical process. Pseudo number

DOI: 10.4018/978-1-7998-1464-1.ch014

random number (PRN) generator is based on a mathematical algorithm, produces periodic sequence and repeats number after sequence length; not a completely random number (Karri & Potkonjak, 2014). A PRNG can be cryptanalysis upcoming number can be predicted statistically. So PRNG produces a deterministic random number. PRNG contains simple architecture and cost-effectivemethodsAbu we should cautious while used for security application. The alternative method to generate a random number is random number derivation from physical properties or hidden feature into the hardware. Hardware-based number nondeterministic, (Maes, 2013) consider as true random number generator (TRNG). According to Kerkchoff's principle, "A random number generator must generate completely unpredictable response even everything is known about generator" A hardware random number generator (HWRNG) requires an additional piece of hardware known as challenge-response pair (CRP) generator. Such a device is based on microscopic phenomena (Lao & Parhi, 2011) which are unpredictable such as thermal noise, photovoltaic effect or quantum effect .

MAIN FOCUS OF THE CHAPTER

A unique property of hardware element is identified to generate each bit of response whenever a challenge input applied (Suh, & Devadas, 2007). Physical unclonable function (PUF) is the widely studied CRP generator, whose response is a function of input challenge and unpredictable hardware feature. Figure1 (Reyneri, Corso, & Sacco, 1990) presents the expectation from a PUF is able to

1. Generate a unique response for each input challenge
2. Different response from different PUF for same input challenge
3. PUF response must be similar to the different operating condition.

(Becker, 2015) Silicon electronic circuit base PUF is preferred, with evidence of manufacturing variation of integrated circuit can't have similar properties. (Hori, Yoshida, Katashita & Satoh, 2010) PUF is a basic circuit intermix input challenge with intrinsic properties of silicon IC and produces non-linear responses. Their responses remain unpredictable even input is known and identical PUF (Kumar, Guajardo, Maes, Schrijen, & Tuyls, 2008). Security in the PUF circuit arises with inbuilt behavior of the electronic device/circuit. It is required to identify the hardware feature to generate a unique response (Rührmair, Sölter, Sehnke, 2009); 2 well-known structure of PUF is

(1) Delay-based arbiter PUF (Kim, Li, Markov & Koushanfar, 2013) consists of two parallel paths of a number of cascaded delay module implemented with multiplexer or tri-state buffer. A decision device (arbiter) at the far end produce high bit if the upper path is faster (Rajendran, Jyothi & Karri, 2011).

(2) Frequency variation ring oscillator PUF contains a large number of ring oscillators with different oscillation frequency. The multiplexer selects a random frequency and enables counter to start counting. At any time the present value of both counters is different. The comparator produces high response if counter1>counter2 (Majzoobi, Rostami, Koushanfar,Wallach & Devadas, 2012).

Figure 1. (a) Uniqueness of PUF (b) Reliability of PUF under the uneven condition

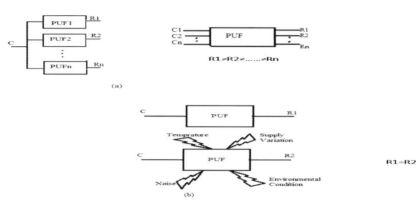

RING OSCILLATOR PUF (ROPUF)

Random number generation and cryptographic key generation (Cao, Zhang, Chang & Chen, 2015) is the niche application of PUF. Two popular architecture (Kacprzak, 1988) of PUF is delay based arbiter frequency variation based Ring oscillator (RO) PUF. In this work 32-bit response generator designed with Verilog HDL. The internal component of ROPUF presented in fig2 (Du & Bai, 2014) requires; ring oscillator (32 numbers), 16:1 Multiplexer (2 Numbers), counters (2 numbers) and a comparator. Ring oscillator generate different frequency when enabling signal goes high, LFSR runs independently on a given clock generate 16 different random number (Maiti & Schaumont, 2011) acts as a selection input for the multiplexer. Select a random frequency and enable counter ti counts upward. (Maes, Herrewege & Verbauwhede, 2012) Two different counters run continuously since operating frequency different its count value must be different, compare to generate a random response. It counter>counter2 response value is high else response is low (Xu & Potkonjak, 2016).

A ring oscillator is an odd number of cascaded connections of delay stage with the feedback connection. Each delay has own propagation delay. Fig presents a 3 stage delay-based ring oscillator; the delay stage is implemented with inverters with a logic gate NAND and NOT with finite delay and phase shift of 180. Frequency of RO is given as (Wang & Qu, 2018)

$$F_{oc} = \frac{1}{\tau_{pd}} = \frac{1}{2N\tau_p} \tag{1}$$

where tp is the sum of delay different delay and N is a number of states. RO PUF requires a large of ring oscillator each oscillator generates slight different frequency. To ensure frequency variation; (a) RO designed with a different number of delay stage (Schaub, Danger, Guilley & Rioul, 2018) 3, 5, 7,9,11, etc. (b) Delay of each stage should different is (a). The first stage of RO is NAND gate acts as start signs, to generate frequency next stage is a multiple logic gates with different delay and with a different number of stages. When enabling '0' RO do not oscillate and maintain constant value while enabling '1' it starts oscillation and frequency. Table1 listed below presents implemented RO with NAND and NOT with different stages, obtain frequency in the range of 33 kHz to 200KHz.

Figure 2. Ring Oscillator PUF (RO-PUF)

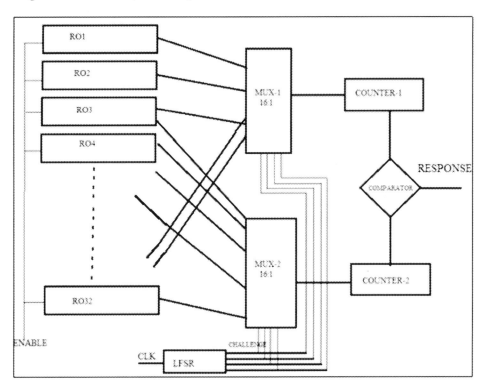

Figure 3. Stage Ring Oscillator

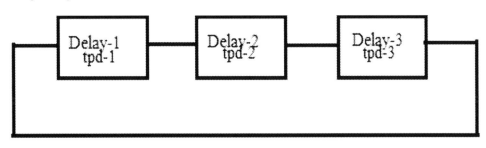

SIMULATION RESULT

In this work ROPUF implemented on a Xilinx Virtex5 FPGA (XC5VLX30) Xilinx ISE 9.2 and Xilinx XST used for logic synthesis. Fig4 presents the simulation result of ROPUF; for given time LFSR output as challenge input is 4'hE select two frequency of m1=62.5KHz and m2=50.5KHz. Each RO operate independently and LFSR randomly picks a frequency, it's is highly probable that selection rises in the timing window of RO frequency (Lao & Parhi, 2014) result in metastable condition as an output of the multiplexer. Metastable state states the when data input changes in setup time before the clock, or data changes in between hold time of clock; output is unpredictable presents as (X or Z). Next state of multiplexer output either high or low is completely random (Chaintoutis, Akriotou,Mesaritakis, Komnios,Karamitros,Fragkos & Syvridis, 2018). Whenever multiplexer selection input i.e. LFSR output

Table 1. Frequency in KHz with 32 ring oscillators

Ring Oscillator	Stage-1	Stage-2	Stage-3	Stage-4	Stage-5	Stage-6	Stage-7	Frequency (Khz)
RO1	NAND(2)	NOT(2)	NOT(2)					83
RO2	NAND(1)	NOT(1)	NOT(1)					200
RO3	NAND(2)	NOT(2)	NOT(2)	NOT(2)	NOT(2)			83.5
RO4	NAND(1)	NOT(2)	NOT(2)	NOT(2)	NOT(2)			85
RO5	NAND(1)	NOT(1)	NOT(2)	NOT(2)	NOT(2)			125
RO6	NAND(1)	NOT(1)	NOT(1)	NOT(2)	NOT(2)			120
RO7	NAND(1)	NOT(1)	NOT(1)	NOT(2)	NOT(2)			122
RO8	NAND(1)	NOT(1)	NOT(1)	NOT(1)	NOT(2)			121
RO9	NAND(1)	NOT(2)	NOT(2)	NOT(2)	NOT(2)	NOT(2)	NOT(2)	47.5
RO10	NAND(2)	NOT(2)	NOT(2)	NOT(2)	NOT(2)	NOT(2)	NOT(2)	83.33
RO11	NAND(1)	NOT(1)	NOT(1)	NOT(1)	NOT(1)	NOT(1)	NOT(1)	166.66
RO12	NAND(1)	NOT(2)	NOT(1)	NOT(2)	NOT(1)	NOT(2)	NOT(1)	100
RO13	NAND(2)	NOT(1)	NOT(2)	NOT(1)	NOT(2)	NOT(1)	NOT(2)	100.2
RO14	NAND(1)	NOT(1)	NOT(2)	NOT(2)	NOT(1)	NOT(1)	NOT(1)	100.5
RO15	NAND(3)	NOT(2)	NOT(2)	NOT(1)	NOT(1)	NOT(1)	NOT(3)	62.5
RO16	NAND(2)	NOT(3)	NOT(3)	NOT(3)	NOT(3)	NOT(3)	NOT(3)	94.5
RO17	NAND(4)	NOT(1)	NOT(2)	NOT(2)	NOT(4)	NOT(1)	NOT(2)	91.4
RO18	NAND(5)	NOT(3)	NOT(2)	NOT(5)	NOT(1)	NOT(2)	NOT(1)	50
RO19	NAND(3)	NOT(3)	NOT(3)	NOT(3)	NOT(3)	NOT(3)	NOT(3)	55.1
RO20	NAND(4)	NOT(4)	NOT(4)	NOT(4)	NOT(4)	NOT(4)	NOT(4)	41.66
RO21	NAND(5)	NOT(5)	NOT(5)	NOT(5)	NOT(5)	NOT(5)	NOT(5)	33.33
RO22	NAND(5)	NOT(1)	NOT(5)	NOT(5)	NOT(3)	NOT(1)	NOT(2)	62.5
RO23	NAND(1)	NOT(2)	NOT(3)	NOT(4)	NOT(5)	NOT(1)	NOT(1)	83.2
RO24	NAND(2)	NOT(3)	NOT(3)	NOT(4)	NOT(4)	NOT(4)	NOT(6)	62.5
RO25	NAND(7)	NOT(1)	NOT(2)	NOT(3)	NOT(3)	NOT(1)	NOT(4)	50.5
RO26	NAND(5)	NOT(1)	NOT(2)	NOT(3)	NOT(3)	NOT(1)	NOT(4)	62
RO27	NAND(2)	NOT(1)	NOT(2)	NOT(2)	NOT(1)	NOT(1)	NOT(1)	100
RO28	NAND(6)	NOT(6)	NOT(6)	NOT(6)	NOT(6)	NOT(6)	NOT(6)	55
RO29	NAND(2)	NOT(3)	NOT(2)	NOT(4)	NOT(5)	NOT(6)	NOT(1)	71.4
RO30	NAND(6)	NOT(6)	NOT(6)	NOT(1)	NOT(1)	NOT(1)	NOT(1)	55.6
RO31	NAND(6)	NOT(5)	NOT(4)	NOT(3)	NOT(2)	NOT(1)	NOT(6)	33
RO32	NAND(1)	NOT(2)	NOT(3)	NOT(4)	NOT(5)	NOT(6)	NOT(7)	45

changes in the timing window of multiplexer input i.e. RO frequency multiplexer output is X or Z. When multiplexer comes out of metastable region produces a stable output. Counters operate on the frequency selected by multiplexer counts rising edges; during undefined region counts stops and wait of next valid edge. Value of 2 counter states must be different. Counter operate for specific period stores counter output

Figure 4. Simulation result of RO-PUF

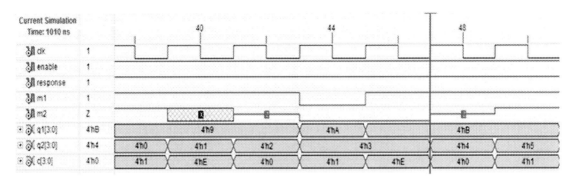

Figure 5. Synthesize result of ROPUF

Device Utilization Summary (estimated values)			
Logic Utilization	Used	Available	Utilization
Number of Slice Registers	31	18200	0%
Number of Slice LUTs	12	18200	0%
Number of fully used Bit Slices	6	37	16%
Number of bonded IOBs	33	220	15%
Number of BUFG/BUFGCTRLs	1	32	3%

q1=4'hB and q2=4'h3, generate response =1. Responses are stored in serial in the parallel output shift register (SIPO). Random number stores in SIPO are 32'h0007FFFF, 32'h 801FFE00, 32'h3FFC0000, 32'hE00001FF,32'h80016527,32'hEBB08679 etc. FPGA utilization report in fig5 shows that ROPUF requires6-bit slice, 1 BUFG, and 33 IOB.

PUF is characterized are validated with 3 properties (Rührmair, Devadas & Koushanfar, 2012) uniformity, uniqueness and randomness (Halak, 2018).

(i) *Uniformity* presents the uniform distribution of high and low bit (Cherkaoui, Bossuet & Marchand, 2016) of PUF response, Ideally PUF response contains 50% high and low bit (Baur & Boche, 2017). Designed ROPUF obtain average uniformity of 42.85%.

$$\text{Uniformity} = \frac{1}{n}\sum_{j=1}^{n} R_j \times 100\%$$

where n is the number of response bit and Rj is the response of jth PUF.

(ii) *Uniqueness* presents Response of PUF must unique for each challenge among a group of the chip. Uniqueness is measure with inter hamming distance (Avital, Levi, keren & Fish, 2016). Uniqueness measures the different response for same input challenge from different PUF (Kumar & Mishra, 2019).

$$\text{Inter Hamming Distance} = \frac{2}{k(k-1)} \sum_{i=1}^{k-1} \sum_{k=i+1}^{k} \frac{HD(Ri, Rj)}{n} \times 100\%$$

where k is number of PUF, n is a bit of response and responses of PUF are Ri and Rj for the same input challenge. When the same challenge applied to PUF1 and PUF2 output as R1 and R2; Hamming distance between R1 and R2 is calculated; ideally, uniqueness is approx to 50%. Designed Uniqueness is 46.25%

(iii) *Reliability* presents (Günlü, Kernetzky, İşcan, Sidorenko, Kramer & Schaefer, 2018) the ability to generate the same response for the same challenge under uneven environment (Kumar, Mishra & Kashwan, 2016). A PUF must produce a same response with the different even unfavorable operating condition; the response should not be effected by process variation, temperature vitiation, supply variation, thermal noise, inoperable environment condition. Variation (Idriss & Bayoumi, 2017) into its produces an error and reduces the reproducibility. Reliability is measured with intro hamming distance, a challenge is applied to a PUF (Khavari, Baur & Boche, 2017) under normal condition R3 and uneven condition (temperature, supply variation, noise, etc) R4. Hamming distance between R3 and R4 presents variation among response; ideally, its value should zero to have high reliability (Delvaux, 2019).

CONCLUSION

Software assisted security is not sufficient now-a-days, selected feature of hardware IC must include the computation to make the response difficult to learn for the adversary. PUF enhance the area little bit but prefer for hardware security module generate unique responses as a function of challenge and unique feature. Cryptographic key generation and random number generation is niche application of PUF. Delay based PUF require simple circuitry but slow rate of generation of random number while frequency variation based RO-PUF requires complex architecture can generate random number at higher speed. in this work 32-bit random number generator implemented with vertex5 FPGA, achieving uniformity of 42.85% and uniqueness 46.25%.

REFERENCES

Avital, M., Levi, I., Keren, O., & Fish, A. (2016). CMOS Based Gates for Blurring Power Information. *IEEE Transactions on Circuits and Systems. I, Regular Papers*, 63(7), 1033–1042. doi:10.1109/TCSI.2016.2546387

Baur, S., & Boche, H. (2017, December). Robust secure storage of data sources with perfect secrecy. In *2017 IEEE Workshop on Information Forensics and Security (WIFS)* (pp. 1-6). IEEE. 10.1109/WIFS.2017.8267669

Becker, G. T. (2015). On the pitfalls of using arbiter-PUFs as building blocks. *IEEE Transactions on Computer-Aided Design of Integrated Circuits and Systems*, 34(8), 1295–1307. doi:10.1109/TCAD.2015.2427259

Beckmann, N., & Potkonjak, M. (2009, June). Hardware-based public-key cryptography with public physically unclonable functions. In *International Workshop on Information Hiding* (pp. 206-220). Springer. 10.1007/978-3-642-04431-1_15

Cao, Y., Zhang, L., Chang, C. H., & Chen, S. (2015). A low-power hybrid RO PUF with improved thermal stability for lightweight applications. *IEEE Transactions on Computer-Aided Design of Integrated Circuits and Systems*, *34*(7), 1143–1147. doi:10.1109/TCAD.2015.2424955

Chaintoutis, C., Akriotou, M., Mesaritakis, C., Komnios, I., Karamitros, D., Fragkos, A., & Syvridis, D. (2018). Optical PUFs as physical root of trust for blockchain-driven applications. *IET Software*, *13*(3), 182–186. doi:10.1049/iet-sen.2018.5291

Cherkaoui, A., Bossuet, L., & Marchand, C. (2016). Design, evaluation, and optimization of physical unclonable functions based on transient effect ring oscillators. *IEEE Transactions on Information Forensics and Security*, *11*(6), 1291–1305. doi:10.1109/TIFS.2016.2524666

Delvaux, J. (2019). Machine-Learning Attacks on PolyPUFs, OB-PUFs, RPUFs, LHS-PUFs, and PUF–FSMs. *IEEE Transactions on Information Forensics and Security*, *14*(8), 2043–2058. doi:10.1109/TIFS.2019.2891223

Devadas, S., Suh, E., Paral, S., Sowell, R., Ziola, T., & Khandelwal, V. (2008, April). *Design and implementation of PUF-based" unclonable" RFID ICs for anti-counterfeiting and security applications. In 2008 IEEE international conference on RFID* (pp. 58–64). IEEE.

Du, C., & Bai, G. (2014, December). A novel relative frequency based ring oscillator physical unclonable function. In *2014 IEEE 17th International Conference on Computational Science and Engineering* (pp. 569-575). IEEE. 10.1109/CSE.2014.129

Günlü, O., Kernetzky, T., İşcan, O., Sidorenko, V., Kramer, G., & Schaefer, R. (2018). Secure and reliable key agreement with physical unclonable functions. *Entropy (Basel, Switzerland)*, *20*(5), 340. doi:10.3390/e20050340

Halak, B. (2018). Physically Unclonable Functions: Design Principles and Evaluation Metrics. In *Physically Unclonable Functions* (pp. 17–52). Cham: Springer. doi:10.1007/978-3-319-76804-5_2

Hori, Y., Yoshida, T., Katashita, T., & Satoh, A. (2010, December). Quantitative and statistical performance evaluation of arbiter physical unclonable functions on FPGAs. In *2010 International Conference on Reconfigurable Computing and FPGAs* (pp. 298-303). IEEE. 10.1109/ReConFig.2010.24

Idriss, T., & Bayoumi, M. (2017, September). Lightweight highly secure PUF protocol for mutual authentication and secret message exchange. In *2017 IEEE International Conference on RFID Technology & Application (RFID-TA)* (pp. 214-219). IEEE. 10.1109/RFID-TA.2017.8098893

Kacprzak, T. (1988). Analysis of oscillatory metastable operation of an RS flip-flop. *IEEE Journal of Solid-State Circuits*, *23*(1), 260–266. doi:10.1109/4.287

Karri, R., & Potkonjak, M. (2014). Special issue on emerging nanoscale architectures for hardware security, trust, and reliability: Part 1. *IEEE Transactions on Emerging Topics in Computing*, *2*(1), 2–3. doi:10.1109/TETC.2014.2318951

Khavari, M., Baur, S., & Boche, H. (2017, October). Optimal capacity region for PUF-based authentication with a constraint on the number of challenge-response pairs. In *2017 IEEE Conference on Communications and Network Security (CNS)* (pp. 575-579). IEEE. 10.1109/CNS.2017.8228679

Kumar, A., & Mishra, R. S. (2019). Challenge-Response Pair (CRP) Generator Using Schmitt Trigger Physical Unclonable Function. In *Advanced Computing and Communication Technologies* (pp. 213–223). Singapore: Springer. doi:10.1007/978-981-13-0680-8_20

Kumar, A., Mishra, R. S., & Kashwan, K. R. (2016, September). Challenge-response generation using RO-PUF with reduced hardware. In *2016 International Conference on Advances in Computing, Communications and Informatics (ICACCI)* (pp. 1305-1308). IEEE. 10.1109/ICACCI.2016.7732227

Kumar, S. S., Guajardo, J., Maes, R., Schrijen, G. J., & Tuyls, P. (2008, June). The butterfly PUF protecting IP on every FPGA. In *2008 IEEE International Workshop on Hardware-Oriented Security and Trust* (pp. 67-70). IEEE. 10.1109/HST.2008.4559053

Lao, Y., & Parhi, K. K. (2011, May). Reconfigurable architectures for silicon physical unclonable functions. In *2011 IEEE International Conference on Electro/Information Technology* (pp. 1-7). IEEE. 10.1109/EIT.2011.5978614

Lao, Y., & Parhi, K. K. (2014). Statistical analysis of MUX-based physical unclonable functions. *IEEE Transactions on Computer-Aided Design of Integrated Circuits and Systems, 33*(5), 649–662. doi:10.1109/TCAD.2013.2296525

Maes, R. (2013). Physically Unclonable Functions: Properties. In *Physically Unclonable Functions* (pp. 49–80). Berlin: Springer. doi:10.1007/978-3-642-41395-7_3

Maes, R., Van Herrewege, A., & Verbauwhede, I. (2012, September). PUFKY: A fully functional PUF-based cryptographic key generator. In *International Workshop on Cryptographic Hardware and Embedded Systems* (pp. 302-319). Springer. 10.1007/978-3-642-33027-8_18

Maiti, A., & Schaumont, P. (2011). Improved ring oscillator PUF: An FPGA-friendly secure primitive. *Journal of Cryptology, 24*(2), 375–397. doi:10.100700145-010-9088-4

Majzoobi, M., Rostami, M., Koushanfar, F., Wallach, D. S., & Devadas, S. (2012, May). Slender PUF protocol: A lightweight, robust, and secure authentication by substring matching. In *2012 IEEE Symposium on Security and Privacy Workshops* (pp. 33-44). IEEE. 10.1109/SPW.2012.30

Pappu, R., Recht, B., Taylor, J., & Gershenfeld, N. (2002). Physical one-wa functions. *Science, 297*(5589), 2026–2030. doi:10.1126cience.1074376 PMID:12242435

Rajendran, J., Jyothi, V., & Karri, R. (2011). Blue team red team approach to hardware trust assessment. *2011 IEEE 29th International Conference on Computer Design (ICCD)*.

Reyneri, L. M., Del Corso, D., & Sacco, B. (1990). Oscillatory metastability in homogeneous and inhomogeneous flip-flops. *IEEE Journal of Solid-State Circuits, 25*(1), 254–264. doi:10.1109/4.50312

Rührmair, U., Devadas, S., & Koushanfar, F. (2012). Security based on physical unclonability and disorder. In *Introduction to Hardware Security and Trust* (pp. 65–102). New York, NY: Springer. doi:10.1007/978-1-4419-8080-9_4

Rührmair, U., Sölter, J., & Sehnke, F. (2009). On the Foundations of Physical Unclonable Functions. *IACR Cryptology ePrint Archive, 2009*, 277.

Schaub, A., Danger, J. L., Guilley, S., & Rioul, O. (2018, August). An improved analysis of reliability and entropy for delay PUFs. In *2018 21st Euromicro Conference on Digital System Design (DSD)* (pp. 553-560). IEEE. 10.1109/DSD.2018.00096

Suh, G. E., & Devadas, S. (2007, June). Physical unclonable functions for device authentication and secret key generation. In *2007 44th ACM/IEEE Design Automation Conference* (pp. 9-14). IEEE.

Wang, Q., & Qu, G. (2018). A Silicon PUF Based Entropy Pump. *IEEE Transactions on Dependable and Secure Computing, 16*(3), 402–414. doi:10.1109/TDSC.2018.2881695

Xu, T., & Potkonjak, M. (2016). Digital bimodal functions and digital physical unclonable functions: architecture and applications. In *Secure System Design and Trustable Computing* (pp. 83–113). Cham: Springer. doi:10.1007/978-3-319-14971-4_3

Yao, Y., Kim, M., Li, J., Markov, I. L., & Koushanfar, F. (2013, March). ClockPUF: Physical Unclonable Functions based on clock networks. In *Proceedings of the Conference on Design, Automation and Test in Europe* (pp. 422-427). EDA Consortium. 10.7873/DATE.2013.095

Chapter 15

PID Plus Second Order Derivative Controller for Automatic Voltage Regulator Using Linear Quadratic Regulator

Shamik Chatterjee
Lovely Professional University, India

Vikram Kumar Kamboj
Lovely Professional University, India

Bhavana Jangid
Malaviya National Institute of Technology, Jaipur, India

ABSTRACT

This chapter presents linear quadratic regulator (LQR) for tuning the parameters of four-term proportional-integral-derivative plus second order derivative controller for controlling terminal voltage of alternator equipped with automatic voltage regulator (AVR) system. Different optimization techniques are considered for juxtaposition with the proposed controller on the basis of terminal voltage response profiles of the AVR system, and Bode plot analysis is carried out for comparing the frequency responses, and through root locus, the stability of the proposed controller is investigated. On-line responses are obtained by implementing a fast performing Sugeno fuzzy logic technique in the controller for working in off-nominal and on-line situations. The controller has undergone an investigation, while having changed system parameters, for the analysis of the robustness of the proposed controller. It is revealed that the performance of the proposed LQR-based controller exhibits a highly improved robust control system for controlling the AVR in power systems.

DOI: 10.4018/978-1-7998-1464-1.ch015

INTRODUCTION

An alternator or synchronous generator is accoutred with a voltage regulating device, called automatic voltage regulator (AVR), for perpetuating the resultant voltage of the alternator constant under quotidian conditions at variegated load levels.

GENERAL

The AVR, which is manoeuvred to modulate the output voltage of the synchronous generator or alternator, takes the fluctuating voltage and metamorphoses it to a constant voltage, under various load changes in the power system. This variation in voltage of the alternator leads to damage of the equipment associated with the power system. The low voltage of the system leads to increase in copper loss, poor power factor and low quality of power transfer capacity of the system. The equipment installed in power system are designed depending upon a particular terminal voltage level so if there is a decrement in the value of the terminal voltage, the system may become precarious, leads to change in the quality of the reactive power which is directly concomitant to the voltage of the system. Whereas if there is an increment of the terminal voltage over the rated voltage, the insulation of different components of power system equipments may breakdown and it may lead to poor performance and life of equipment associated with the system. Hence, to equilibrate the drawbacks of power system, a fast acting and high performance AVR is necessary. The reactive power flow is also controlled by the AVR which also clinches appropriate sharing of the reactive power amid all the alternators adhered in parallel. The AVR perpetuates the consistency of the terminal voltage on variation of the exciter voltage of the synchronous generator (Kundur *et al.*, 1994). The high inductance of field windings of the alternator and variation in load leads to difficulty in obtaining fast and stable response of the AVR. It is necessary to have improved performance of the AVR. Hence, the AVR system requires suitable controlling mechanism to execute as per desired accomplishment.

LITERATURE SURVEY

Researchers propounded disparate control strategies *viz.* robust control, optimal control and adaptive control, in the AVR system to scrutinize the system for procuring better ramification. In collation to modern control mechanisms, self-tuning adaptive control technique is easy to employ with variation in the parameters of the system. The traditional self-tuning control methods become sparse due to existence of more mathematical formulation in some working conditions because of non-linear load attributes and variable operating points. Since 2000, artificial intelligence and optimization methods based self-tuning control strategies are favored by the researchers.

The most preferable amidst controller by the researchers is proportional integral derivative (PID) controller which is employed for its structure's simplicity and robust performance over a large range of working conditions (Ang *et al.*, 2005). To improve the performance of the AVR system, the self tuning methods are employed at the starting, as per the reports displayed by Swidenbank *et al.*, in 1999, for controlling the turbine generator system (Swidenbank *et al.*, 1999). In the same year, Finch *et al.* (Finch *et al.*, 1999) have used the self tuning method for the AVR system. Evolution of new various types of heuristic optimization techniques take place for tuning the parameters of PID controller to have better improved voltage response of AVR. These popular optimization algorithms are accepted by the researchers for their research works across the world. The researchers have accepted the techniques due to their wide area of application in any level of difficulty of the problems and having less problems in computation, thus, proving it acceptability for different types of complex problems. Gaing, in 2004, proffered a particle swarm optimization (PSO) technique based PID controller for controlling the AVR system and further the author has compared the response with that obtained from genetic algorithm (GA) based PID controller (Gaing *et al.,* 2004). In 2006, Kim and Cho (Kim & Cho, 2006) have prospered a hybrid method, consisting of GA and bacterial foraging optimization algorithm, to ameliorate the execution of the PID controller for the AVR system. Gozde and Taplamacioglu (Gozde & Taplamacioglu, 2011) have tuned the parameters of the PID controller with artificial bee colony (ABC) algorithm for the AVR system. They have compared the response obtained from ABC with PSO and differential evolution algorithm (DEA) based results. Mukherjee and Ghoshal (Mukherjee & Ghoshal, 2007) employed crazy PSO (CRPSO) for tuning the PID controller parameters to control the AVR system. Kashki *et al.,* in 2007, have propounded continuous action of reinforcement learning automata (RLA) optimization method in order to optimize the PID controller's parameters for the AVR system. They have also collated their response with PSO and GA based PID controller (Kashki *et al.,* 2007). Zhu *et al.,* in 2009, propounded a chaotic ant swarm algorithm to tune the parameters of the PID controller for the AVR system (Zhu *et al.,* 2009). Based on lozi map, Coelho proffered chaotic optimization approach, in the same year, for the optimization of the PID controller's gains (Coelho *et al.,* 2009). Chatterjee *et al.* (Chatterjee *et al.,* 2009) have shown a collation between the responses obtained from velocity relaxed PSO based optimization technique and the CRPSO based optimization technique for the AVR system. In 2012, Panda *et al.* (Panda *et al.,* 2012) have proposed many optimizing liaisons (MOL), which is a paraphrased form of PSO, to obtain tuned PID controller gains for the AVR system and further they have analyzed the comparison of the responses with that obtained from the ABC, PSO and DEA for the AVR system. After two years, Mohanty *et al.* (Mohanty *et al.,* 2014) have proposed the local unimodal sampling optimization technique to optimize the parameters of the PID controller for the AVR system and they have also compared the responses with those achieved from PSO and DEA.

MOTIVATION

Most of the previous researchers, as per literature survey, have employed evolutionary algorithms to optimize the parameters of the PID controller to control the AVR system. The most appropriate response has not been reported by them, as observed from their simulation results, which have some drawbacks and may be overcome to have the desired response. The value of rise time (T_R), settling time (T_{SS}), overshoot (M_p) and steady state error (E_{SS}) should be less in the unit step responses as displayed in previous research works. The value of the stated parameters (*i.e.*, T_R, T_{SS}, M_p and E_{SS}) may become lesser than the values obtained from the given algorithms, *viz.* DEA (Gozde & Taplamacioglu, 2011), PSO (Gozde & Taplamacioglu, 2011) and ABC (Gozde & Taplamacioglu, 2011), for tuning the gains of the PID controller to control the AVR system. By minimizing the values of T_R, T_{SS}, M_p and E_{SS}, the recent optimization techniques based PID controller may provide better value of objective function by decreasing it. Hence, the terminal voltage transient response of the AVR system may become optimum.

As per the literature survey, there are few limitations in some evolutionary algorithms. PSO has different types of detailed studies which reports that there is a chance of divergence of the particle on defining the constants of acceleration and maximum velocity *i.e.*, it may reach infinity which is the phenomenon of the swarm which explodes (Clerc & Kennedy, 2002). The ABC technique has its own limitations such as (a) the nature of the function has to be optimized as important informations may lose, (b) calculations of the objective functions is large in number, (c) on the new optimization techniques, performance required to be improved, (d) slow achievement in the accuracy of the solution, (e) in local minima, solutions may ambush due to their achievement in higher accuracy, (f) the technique slows down on employment of sequential processing and (g) computational cost increases which leads to more required space for memory and more iterations. To obtain better performance than the traditional PID controller for the AVR system, a new controller has been implemented in the system, called proportional-integral-derivative plus second order derivative (PIDD^2) controller (Sahib, 2015). In comparison to the PID controller, the PIDD^2 controller delivers less settling time and rise time.

At minimum cost, dynamic system operation is the most perturbed matter for the optimal control's theory. The linear quadratic (LQ) problem is the case where a set of linear differential equations are used to describe the system dynamics as well as quadratic equation are used to denote the cost whose solution is procured by the LQ regulator (LQR). A mathematical algorithm *viz.* LQR algorithm is employed here to obtain the controller's parameters, which is controlling either a process or a machine or a system. It optimizes the cost function with weighing factors provided by the designer. The designer's work load decreases with the implementation of the LQR algorithm for the optimization of the controller's parameters. A pertinent state-feedback controller is obtained automatically by the LQR algorithm. In the LQR method, the only limitation of its application is the difficulty in obtaining the appropriate weighting factors. In this research work, the LQR algorithm is used to optimize the parameters of the studied traditional PID controller and as well as the proposed PIDD^2 controller to control the AVR system for the achievement of the desired terminal voltage response profiles.

In a variety of applications, Sugeno fuzzy logic (SFL) controller is used for controlling various plants. On-line parameters are adjusted as per the environment in which the SFL is operating and it furnishes good dynamic response in real time. The incremental variation in terminal voltage may be obtained by implementing the intelligent SFL technique in the system to determine the off-line optimum value of the controller's parameters.

CONTRIBUTION

The important contributions of this chapter are to:

1. tune the parameters of the PIDD^2 controller using the LQR method for the studied AVR system,
2. collate the voltage response profiles offered by the proposed LQR based PIDD^2 controller with those delivered by the traditional PID controller tuned by LQR, DEA, PSO and ABC for the control of the studied AVR system,
3. compare the larger bandwidth and better frequency response, by analyzing the Bode plots, of the proposed PIDD^2 controller with other optimization techniques employed to optimize PID controller gains for the AVR system,

4. investigate the stability of the proposed controller, compared with the other algorithms by analyzing the root locus of the control systems,
5. study the significance of the SFL tuned PIDD² controller's gains in real time atmosphere under variation in the working scenario of the AVR system and
6. study the robustness of the proposed LQR based PIDD² controller for the AVR system under uncertainties and variation in parameters of the studied system.

LAYOUT OF THE CHAPTER

The chapter is divided into nine sections in accordance with the sequence. In Section "AVR system modelling", the modeling of the AVR system is described with proper block diagram. The structure of the PIDD² controller is illustrated in Section "Structure of the PIDD² Controller". In the next section, *i.e.* Section "Mathematical Problem Formulation", the mathematical problem is formulated. Analysis of the AVR system controlled by the PID controller is reported in Section "Analysis of AVR System With PID Controller", while in Section "Analysis of the AVR System with PIDD² Controller", the PIDD² controller based AVR system is analyzed and reported with the transfer function block diagram. In Section "SFL for On-line Tuning of the PIDD² Controller", the on-line tuning of PIDD² controller using SFL technique is focused. The simulation results are displayed and discussed in Section "Result and Discussions". Finally, in Section "Conclusion", the concluding remarks of the research work are reported.

AVR SYSTEM MODELLING

The power system stability can be increased by controlling the excitation of the synchronous generator. The terminal voltage of the alternator can be maintained by implementing the AVR in the system, as it acts a feedback control loop which helps the alternator to maintain the voltage at a constant specified level. Automatic load frequency control (ALFC) loop is another control loop which maintains the frequency levels at rated value, conjugated in operation with the AVR. In operation, AVR is much faster than the ALFC considering the constituted small time constants of the units. Hence, the AVR's operation never affects the ALFC's operation. The AVR system consists of four major components which are generator, exciter, amplifier and sensor (Saadat, 2002). The sensor is to continuously sense the voltage at the terminal of the alternator. The signal is rectified and smoothened and then sent to the comparator for juxtaposition with a reference signal. The error voltage achieved from the comparator is then amplified and being sent to the exciter to control the field windings of the generator. In the present work, the four components' transfer function is assumed as linear for the modelling of the AVR system using mathematical expressions.

The components of the AVR system is displayed as a block diagram in Figure 1 (Mukherjee & Ghoshal, 2007; Chatterjee *et al.*, 2009). As per the block diagram, the terminal voltage of the synchronous generator (ΔV_t) is fed back through the voltage sensor, which senses the voltage (ΔV_s), to the comparator in which it will be compared with the reference voltage (ΔV_{ref}), which is the nominal voltage of value 1.0 p.u. If the terminal voltage changes then the error voltage (ΔV_e), *i.e.*, the output of the comparator also changes. The value of ΔV_e has to be minimized to achieve the nominal value as output. The system's transient response will give a steady state error with respect to the nominal value, if no controller

Figure 1. Block diagram of the studied AVR system

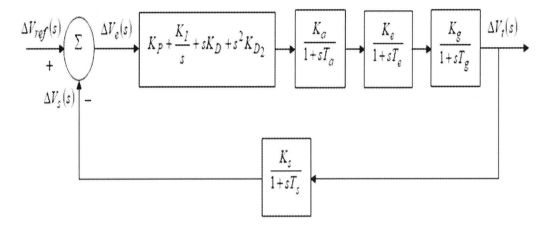

is connected. In Table 1, the transfer function of components of the AVR system is delineated (Gozde & Taplamacioglu, 2011; Chatterjee *et al.*, 2009). The values considered for the AVR system model's parameters are: $K_a = 10.0$, $T_a = 0.1$ s, $K_e = 1.0$, $T_e = 0.4$ s, $K_S = 1.0$, $T_S = 0.01$ s. The load decides the value of K_g and T_g (Gaing, 2004; Chatterjee *et al.*, 2009).

If the value of K_g and T_g is considered as 1.0 and 1.0 s, respectively, then the overall transfer function of the AVR system, $G_{AVR}(s)$, may be represented in (1).

$$G_{AVR}(s) = \frac{\Delta V_t(s)}{\Delta V_{ref}(s)} = \frac{0.1s + 10}{0.0004s^4 + 0.045s^3 + 0.555s^2 + 1.51s + 11}$$

$$= \frac{250(s + 100)}{(s + 98.82)(s + 12.63)(s^2 + 1.057s + 22.04)}$$

(1)

Table 1. Transfer function and parameter values of the AVR system.

AVR components	Transfer Function	Range of K	Range of T(s)	Value of K used	Value of T(s) used
Amplifier	$\dfrac{Ka}{1 + sT_a}$	$10 \leq K_a \leq 40$	$0.02 \leq T_a \leq 0.1$	10	0.1
Exciter	$\dfrac{Ke}{1 + sT_e}$	$1 \leq K_e \leq 10$	$0.4 \leq T_e \leq 1.0$	1	0.4
Generator	$\dfrac{Kg}{1 + sT_g}$	$0.7 \leq K_g \leq 1$	$1.0 \leq T_r \leq 2.0$	1	1
Sensor	$\dfrac{Ks}{1 + sT_s}$	$0.9 \leq K_s \leq 1.1$	$0.001 \leq T_z \leq 0.06$	1	0.01

It may be reported, from (1), that the transfer function has one zero at $Z=-100$ and two real poles at $S_1=-98.82$ and $S_2=-12.63$ and there are two complex poles which are at $S_{3,4}=-0.53\pm j4.66$. The value of $G_{AVR}(s)$ is approximated by terminating the zero at a value of -100 with the pole at -98.82 to achieve the approximated transfer function, $\tilde{G}_{AVR}(s)$, as expressed in (2).

$$\tilde{G}_{AVR}(s) = \frac{250}{(s+12.63)(s^2+1.057s+22.04)} \qquad (2)$$

The transient response of the voltage of the AVR system is displayed in Figure 2. The transient response exhibits the step parameters with high oscillations offering the value of M_p as 1.5064 p.u., T_R as 0.42 s, T_{SS} as 6.942s and E_{SS} as 0.0909. From this figure, it may be noted that the ΔV_t deviates from the nominal value. This deviation is not desirable and un-acceptable for the general operation of the generator in power system having such a high voltage in the unit of kilo-volt (kV). To avoid this drawback, a robust controller is required to install in the AVR system.

STRUCTURE OF THE PIDD² CONTROLLER

The most widely acceptable feedback controller in the industries is the PID controller which is being used, successfully, for over last 50 years. Despite having change in dynamic characteristics of the plant, the controller is easy to understand and robust enough to control it, which delivers outstanding control performance. The controller has three gain constants which are proportional, integral and derivative gains. The value of T_R can be minimized by the proportional controller but the controller is not able to terminate E_{SS}. If the gain of the proportional controller is made too high, then there is a chance of the

Figure 2. Voltage response of the AVR system without controller.

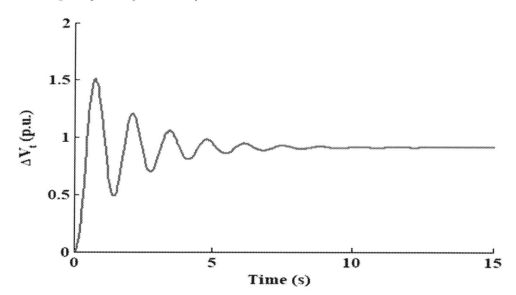

system to become unstable whereas if it is made too small, then it may end up to an output response of small magnitude with respect to a large input error and may become less sensitive controller. The value of E_{SS} can be terminated by the integral controller but it may result to a worst transient response. On making the value of integral gain too high, the system will result in higher value of M_P but a slow system response may be expected. The stability of the system may increase on increasing the value of the derivative gain which will reduce the value of M_P and, hence, the transient response will improve, but increasing the value of the derivative gain may lead to unstable system (Cominos & Munro 2002; Astrom & Hagglund, 1995). For designing the PID controller, three parameters *viz.* proportional gain (K_P), integral gain (K_I) and derivative gain (K_D) are required.

The transfer function of the PID controller, $G_{PID}(s)$, in Laplace mode, may be expressed as presented in (3) (Zhu *et al.*, 2009).

$$G_{PID}(s) = K_P + \frac{K_I}{s} + sK_D \tag{3}$$

For designing the PIDD^2 controller, in the traditional PID controller, a derivative controller of second order is implemented to achieve optimized M_P, T_R, T_S and E_{SS}. To design the controller perfectly, the four parameters *viz.* K_P, K_I, K_D and a second order derivative gain $\left(K_{D_2}\right)$ are required.

In Laplace mode, the transfer function of the PIDD^2 controller, $G_{PIDD^2}(s)$, may be presented in (4) (Sahib, 2015).

$$G_{PIDD^2}(s) = K_P + \frac{K_I}{s} + s\,K_D + s^2 K_{D_2} \tag{4}$$

The only difference between the traditional PID controller and the PIDD^2 controller is the extra derivative parameter of second order in the PIDD^2 controller (Sahib, 2015).

MATHEMATICAL PROBLEM FORMULATION

Design of Objective Function

The gains of the proposed controller for controlling the AVR system are optimized by LQR method in this research work. The system optimal state feedback matrix (K) is obtained by the LQR method which is implemented in a linear time invariant system, thus, minimizing a quadratic performance index J, as presented in (5).

$$J = \int_0^\infty \left(x^T Q x + u^T R u\right) dt \tag{5}$$

The symmetric positive semi definite weighting matrices (Q and R) have to be properly selected to achieve the minimum value of J. In the algebraic riccati equation (ARE), the design matrices Q and R are employed to obtain the minimum value of K which makes the system performance better. The ARE equation is presented in (6).

$$A^T P + P\left(A - BR^{-1}B^T P\right) + Q = 0 \tag{6}$$

In (6), the values of A and B are achieved by obtaining the AVR system's state space notation with the controller. The variable P may be solved by employing the optimized values of Q and R in (6), which requires,

Q is semi-definite, $i.e. Q \geq 0$

R is definite, $i.e. R > 0$ \hfill (7)

P is definite, $i.e.\ P > 0$

The value of K may be obtained by implementing the obtained value of P, as expressed in (8).

$$K = R^{-1}B^T P \tag{8}$$

Measure of Performance

The performance of the AVR system is analyzed by examining different frequency and time domain parameters. The performance indices (M_P, T_R and T_{SS}) are investigated to analyze the time domain performance. In addition to this, a mathematical analysis is also carried out for the investigation of the performance of time domain by evaluating the FOD value for each case of the AVR system, which depends upon the value of M_P, T_R and T_{SS}, as expressed in (9).

$$FOD = \left(1 - e^{-\beta}\right) * \left(M_P + E_{SS}\right) + e^{-\beta} * \left(T_{SS} - T_R\right) \tag{9}$$

In terms of phase margin (P_M), gain margin (G_M), phase crossover frequency (W_{CP}), gain crossover frequency (W_{CG}) and bandwidth (BW), the analysis of the frequency domain is carried out. For operation analysis purpose, investigation of the closed poles' location and damping ratio (ξ) for each system has to be done.

ANALYSIS OF AVR SYSTEM WITH PID CONTROLLER

To improve the ultimate response of the system, the PID controller is employed with the AVR system in the forward path for processing the voltage difference and providing a manoeuvred triggered signal ($V_e(s) = V_{ref}(s) - V_S(s)$).

Figure 3. Transfer function block diagram of PID controlled AVR system.

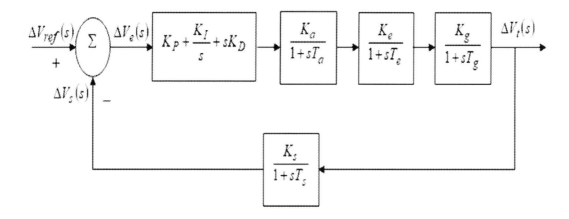

As per the transfer function presented in (3), three control actions with respect to the error signal are combined to form the PID controller (*viz.* proportional, integral and derivative). The block diagram representing the transfer function of the AVR system controlled by PID controller is displayed in Figure 3. If the notations of components of the AVR system are replaced by their respective values along with the load dependent parameters, K_g and T_g, as 1.0 and 1.0 s, respectively, then the overall transfer function of the PID controlled AVR system may be presented in (10).

$$G_{Total-PID}(s) = \frac{0.1K_D s^3 + (0.1K_P + 10K_D)s^2 + (0.1K_I + 10K_P)s + 10K_I}{0.0004s^5 + 0.0454s^4 + 0.555s^3 + (1.51 + 10K_D)s^2 + (1 + 10K_P)s + 10K_I} \quad (10)$$

The tuned parameters of the PID controller using LQR method for the AVR system are delineated in Table 2. On placing the values of K_P, K_I and K_D in (5), then the simplified transfer function, $G_{Total-PID}(s)$, of the LQR based PID controlled AVR system may be expressed in (11).

$$G_{Total-PID}(s) = \frac{0.09976s^3 + 10.429s^2 + 45.344s + 3.952}{0.0004s^5 + 0.0454s^4 + 0.555s^3 + 11.486s^2 + 46.305s + 3.952} \quad (11)$$

Table 2. Results of the transient response analysis.

Controller-method	K_P	K_I	K_D	K_{D2}	M_p (p.u.)	$T_R(s)$	$T_{ss}(s)$ (5% bant)	FOD
PID-DEA (Gozde & Taplamacioglu, 2011)	1.949	0.4430	0.3427	–	1.330	0.152	0.952	1.135
PID-PSO (Gozde & Taplamacioglu, 2011)	1.7774	0.3827	0.3184	–	1.300	0.161	1.000	1.130
PID-ABC (Gozde & Taplamacioglu, 2011)	1.6524	0.4083	0.3654	–	1.250	0.156	0.92	1.071
PID-LQR (Studied)	4.5305	0.3952	0.9976	–	1.029	0.499	1.3724	0.971
PIDD²-LQR (Proposed)	**32.362**	**0.3667**	**4.0737**	**0.0943**	**1**	**0.1744**	**0.3294**	**0.689**
An entry " – " means not applicable.								

The expression presented in (10), may be represented as pole-zero form in (12).

$$G_{Total-PID}(s) = \frac{249.4(s+99.99)(s+4.452)(s+0.089)}{(s+101.51)(s+0.09)(s+4.62)(s^2+6.28s+241.508)}$$

(12)

It may be observed from (12) that the PID controller lead to termination of the poles, s_1=-0.09 and s_2=-4.63 and zeroes of the studied AVR system on which the transient response of the system depends. For the improvement of the transient behaviour, the PID controller compensates the poles and zeroes of the system.

In Figure 4, the terminal voltage of the AVR system controlled by the LQR based PID controller is displayed. It may be observed that the transient response, shown in Figure 4, offers better performance index on being compared to the transient response shown in Figure 2, which is without any control actions. The performance index of the transient response of the LQR based PID controlled AVR system has the value of M_P=1.029 p.u., T_R = 0.499 s and T_{SS} = 1.3724 s. An improvement of about 47.74% and 80% is observed in the value of M_P and T_{SS} as collated to the studied AVR system without controller. There is hardly any change in the value of T_R which shows their almost equal speed of response. Furthermore, there is no steady state error in the response of the PID controlled system. The AVR system controlled by LQR based PID controller delivers better transient and steady state behaviour than the uncontrolled AVR system.

ANALYSIS OF THE AVR SYSTEM WITH PIDD² CONTROLLER

The transient performance of the AVR system with the proposed PIDD² controller is examined in this segment. In addition to the three controller gains of the PID controller, a second order derivative controller is employed. In (4), the transfer function of the PIDD² controller, in Laplace form, is expressed. As

Figure 4. Terminal voltage response profile of the AVR system with PID controller.

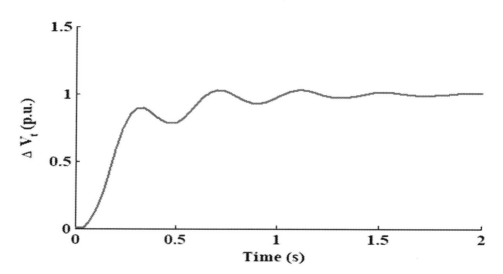

presented in (4), the PIDD^2 controller is formed by combining the K_P, K_I, K_D and K_{D_2}, with respect to the error signal. In Figure 5, the transfer function block diagram of the PIDD^2 controller employed in the AVR system is portrayed. If the variables of the AVR system parameters are substituted by their respective values and also replacing the variables K_g and T_g, which are load dependent, with 1.0 and 1.0 s respectively, the total transfer function, $G_{Total-PIDD^2}(s)$, of the AVR system controlled by the PIDD^2 controller may be expressed in (13).

$$G_{Total-PIDD^2}(s) = \frac{0.1K_{D_2}s^4 + \left(0.1K_D + 10K_{D_2}\right)s^3 + \left(10K_D + 0.1K_P\right)s^2 + \left(10K_P + 0.1K_I\right)s + 10K_I}{0.0004s^5 + 0.045s^4 + \left(10K_{D_2} + 0.55\right)s^3 + \left(10K_D + 1.51\right)s^2 + \left(10K_P + 1\right)s + 10K_I}$$

(13)

The controller gains of the PIDD^2 controller is optimized by the LQR technique for the AVR system. The algorithm of the LQR method is presented in Algorithm 1.

Algorithm 1: Main steps of LQR optimization technique for tuning the PIDD^2 controller gains

    Repeat

    Accept initial values of K_P, K_I, K_D and K_{D_2}

    Set $K_i = [K_P \ K_I \ K_D \ K_{D_2}]$

    Determine the state space representation of the system with the accepted values

    Reduce the system to a fourth order system by eliminating the relatively unstable states

    Initialize Q_i and R_i

    Repeat

    Determine the value of J_i

    Use the MATLAB command LQR to determine K_{i+1} matrix

    Set $K_i = K_{i+1}$

    Set $Q_i = Q_{i+1}$

    Set $R_i = R_{i+1}$

    Until (requirements are met)

    End

    Else (reset K_i)

    End

The tuned parameters of the PIDD^2 controller are depicted in Table 2, which replaces the variable in (13). Thus, the overall transfer, $G_{Total-PIDD^2}(s)$, function simplifies as presented in (14).

$$G_{Total-PIDD^2}(s) = \frac{0.00943s^4 + \left(1.35\right)s^3 + \left(43.97\right)s^2 + \left(323.66\right)s + 3.67}{0.0004s^5 + 0.045s^4 + \left(1.498\right)s^3 + \left(42.25\right)s^2 + \left(324.62\right)s + 3.67}$$

(14)

In pole zero form, the expression in (14) may be represented as in (15).

Figure 5. Transfer function block diagram of PIDD² controlled AVR system.

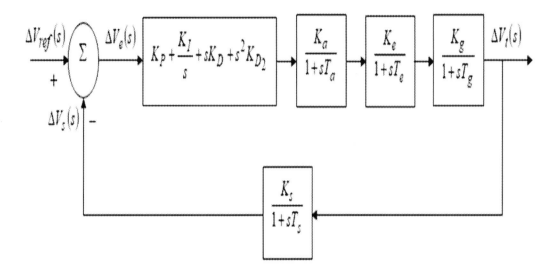

$$G_{Total-PIDD^2}(s) = \frac{23.575(s+0.012)(s+10.474)(s+32.740)(s+99.935)}{(s+0.012)(s+10.441)(s+80.787)(s^2+21.26s+960.68)} \tag{15}$$

It may be observed from (15) that the PIDD² controller cancels the pole, s_1=-0.012 and s_2=-10.441, and zero in the system, like the PID controller, to maintain the desired system's transient performance. They are accompanied with comparatively greater value of the time constants and thus offers greater influence on the system response as collated to the pole, s_3=-80.787, which has a lesser time constant with its effect on the transient behaviour sustaining for much less time. Thus an improved response is provided by the proposed controller on minimizing the transient disturbances due to the poles which are dominating the system's transient response. The zero employed by the second order derivative term mainly improves the transient response due to the proposed PIDD² controller in comparison to the conventional PID controller.

Table 2. Results of the transient response analysis

Controller-method	K_P	K_I	K_D	K_{D2}	M_P (p.u.)	$T_R(s)$	$T_{SS}(s)$ (5% bant)	FOD
PID-DEA (Gozde & Taplamacioglu, 2011)	1.949	0.4430	0.3427	_	1.330	0.152	0.952	1.135
PID-PSO (Gozde & Taplamacioglu, 2011)	1.7774	0.3827	0.3184	_	1.300	0.161	1.000	1.130
PID-ABC (Gozde & Taplamacioglu, 2011)	1.6524	0.4083	0.3654	_	1.250	0.156	0.92	1.071
PID-LQR (Studied)	4.5305	0.3952	0.9976	_	1.029	0.499	1.3724	0.971
PIDD²-LQR (Proposed)	**32.362**	**0.3667**	**4.0737**	**0.0943**	**1**	**0.1744**	**0.3294**	**0.689**
An entry " – " means not applicable.								

SFL FOR ON-LINE TUNING OF THE PIDD² CONTROLLER

The two inputs of the fuzzy logic controller (FLC) are incremental change in error in terminal voltage (Δe), which is the value based on the difference between the terminal voltage and reference voltage, and derivative of incremental change in error in terminal voltage ($\Delta \dot{e}$), which is the derivative value of the error *i.e.*, the rate of change of incremental change in error in terminal voltage. The FLC employs these two inputs to offer the control signal (u) (Mukherjee, 2009, Banerjee *et al..*, 2014, Mohagheghi *et al.*, 2004). The block diagram, representing PIDD² controller with the FLC, is portrayed in Figure 6. The error and the derivative of error can be expressed as presented in (16) and (17), respectively.

$$\Delta e = \left(\Delta V_{ref} - \Delta V_s \right) \tag{16}$$

$$\Delta \dot{e} = \frac{\Delta e_h - \Delta e_{h-1}}{t} \tag{17}$$

In (17), the variable Δe_h denotes the incremental change in error at time h whereas at time(h-1), the incremental change in error in terminal voltage (Δe_{h-1}) reported and the sample time is denoted by t (in s).

The seven fuzzy classes are used to divide the error and the derivative of error, which are the inputs of the FLC, as negative large (NL), negative medium (NM), negative small (NS), zero (ZE), positive small (PS), positive medium (PM) and positive large (PL). There are three membership functions, two inputs and one output, in the FLC as portrayed in Figure 7. In each membership functions contain two trapezoidal memberships and five triangular memberships out of seven memberships. The control rules of the two input membership functions show the output membership function based on the rules as delineated in Table 3. The method on which the control rules is executing is, *if* input 1 and input 2 *then* output 1. The control signal of the variation of the derivative of incremental change in error and incremental change in error may be presented as a surface plot which is displayed in (see Figure 8).

Figure 6. Block diagram of Fuzzy logic based PIDD2 controller.

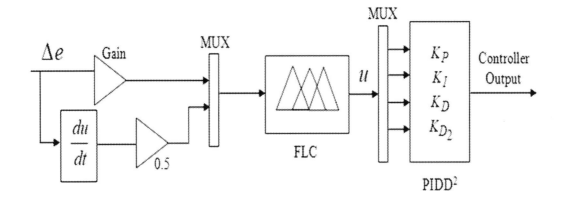

Table 3. Control rules of LQR-SFL based PIDD² controller

Input 1 / Input 2	NL	NM	NS	ZE	PS	PM	PL
NL	PL	PL	PL	PM	PM	PS	ZE
NM	PL	PM	PM	PM	PS	ZE	NS
NS	PL	PM	PS	PS	ZE	NS	NM
ZE	PM	PM	PS	ZE	NS	NM	NM
PS	PM	PS	ZE	NS	NS	NM	NL
PM	PS	ZE	NS	NM	NM	NM	NL
PL	ZE	NS	NM	NM	NL	NL	NL

Comparison of Sugeno FLC and Mamdani FLC systems

In the real world, the FLCs are implemented as an intelligent control method which leads to the increase in application. The Sugeno method is employed for its capability of adapting the environment and computationally, more compact as well as efficient compared to the Mamdani method. The adaptive techniques are considered for customizing the membership function which helps to design the FLC based on the specified data. The following points of Sugeno method are more advantageous over Mamdani system.

(i)   It is efficient, computationally.
(ii) It has great performance while formulating and calculating mathematical problems.
(iii) It can operate in linear methods.
(iv) It is compatible with the recent optimization techniques.

The following advantages are of Mamdani system.

1.   It is faster based on the user input.
2.   It is impetuous.
3.   It is accepted widely in the world.

If the system is static (*i.e.* the change in dynamics of the system is slow), the Mamdani FLC is preferable than Sugeno one whereas for a fast change in dynamics of the system, the Takagi-Sugeno method will work better than Mamdani system.

SOLUTIONS AND RECOMMENDATIONS

The PIDD² controller is employed to control the AVR system for the achievement of the desired response by optimizing the gains of the PIDD² controller by a tuning method *viz.* LQR. The analysis of the terminal voltage (ΔV_t) is carried out to investigate the performance for different cases. Using MATLAB/SIMULINK (version: 7.10) in 2.77 GHz, Intel Core™ $i7$ computer, the simulation of the system is done. The main observations of this research work are described below:

Figure 7. Membership function of lqr-sfl based pidd2 controller (a) input 1 (b) input 2 and (c) output

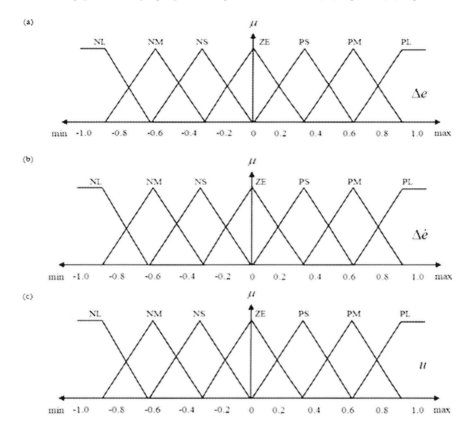

Figure 8. Surface plot of the controller's signal with the change in error and derivative of error signal

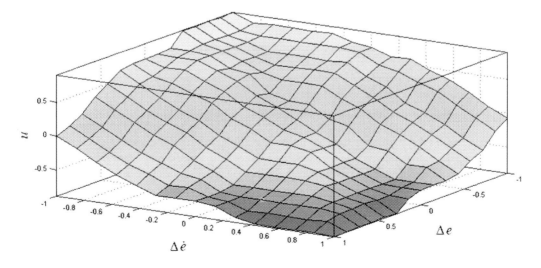

Transient Response Analysis

The analysis of the transient response investigates the performance of a system before reaching the steady state. The investigation may be carried out by analysing the system's oscillation, damping capability and speed which offers its characteristics. The transient response after reaching its steady state may also offer its steady state characteristics. The speed of the system may be reported by observing the value of T_R and the nature of oscillations of the system may be reported from the value of M_p and T_{SS} of the transient response. The performance index along with the FOD value of LQR based PID controller (Studied) and LQR based PIDD^2 controller (Proposed) along with optimizing techniques *viz.* DEA (Gozde & Taplamacioglu, 2011), PSO (Gozde & Taplamacioglu, 2011) and ABC (Gozde & Taplamacioglu, 2011) based PID controller for controlling the AVR system is delineated in Table 2. In Fig. 9, the comparison on transient voltage profiles offered by different evolutionary algorithms *viz.* DEA (Gozde & Taplamacioglu, 2011), PSO (Gozde & Taplamacioglu, 2011), ABC (Gozde & Taplamacioglu, 2011), PID-LQR (Studied) and PIDD^2-LQR (Proposed) is displayed with 1.0 p.u. perturbations in reference voltage for feigned input. It may be reported, from this figure, that the terminal voltage profile offered by the AVR system, controlled by the proposed LQR based PIDD^2 controller, manifests better response collated to those offered by different optimization technique based PID controller as per state-of-the-art literature. From Table 2, it may be observed that the studied LQR tuned PID controller has exhibited better performance index and FOD value in collation with DEA (Gozde & Taplamacioglu, 2011), PSO (Gozde & Taplamacioglu, 2011) and ABC (Gozde & Taplamacioglu, 2011) based PID controller for the AVR system. The FOD value offered by the LQR based PIDD^2 controller is minimum in comparison to other optimization techniques based PID controller for controlling the AVR system. Thus, the LQR based PIDD^2 controller manifests lesser number of oscillations, better damping capability (minimum T_{SS}) and most stable (minimum M_p) unit step transient response in juxtaposition with other optimization techniques tuned PID controlled AVR system.

Figure 9. Comparative terminal voltage response profile of the AVR system with different controllers.

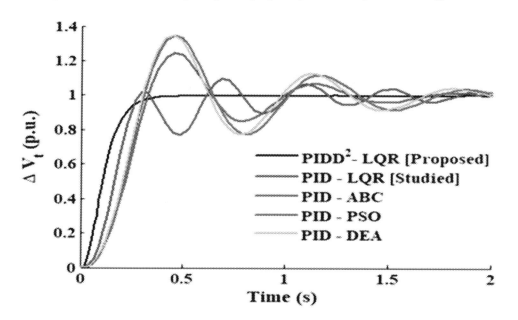

Bode Plot Analysis

The Bode plot analysis is carried out to obtain the frequency response of the AVR system. The magnitude and phase plots of each individual AVR system using controllers tuned by various algorithms are presented in Figure 10. Values of G_M, P_M, W_{CG}, W_{CP} and BW for these systems are represented in Table 4. From this Table 4, it may be observed, that in all the cases, the G_M and P_M for the AVR system is positive and W_{CP} is always greater than W_{CG}. These results indicate that all the considered systems are stable (Manke, 2011). But it may be observed that the highest values of P_M, W_{CG} and BW are obtained with PIDD^2 controlled AVR system (see Figure 11). Thus, the AVR system with PIDD^2 controller is relatively more stable than the other systems as per the investigation.

Table 4. Analysis of Bode plot based results

Controller-method	G_M (dB)	P_M (deg)	W_{CG} (rad/s)	W_{CP} (rad/s)	BW (rad/s)
PID-DEA (Gozde & Taplamacioglu, 2011)	∞	59	11.1	∞	12.800
PID-PSO (Gozde & Taplamacioglu, 2011)	∞	62.8	10.5	∞	12.182
PID-ABC (Gozde & Taplamacioglu, 2011)	∞	70.8	11.6	∞	12.879
PID-LQR (Studied)	∞	33.4	21.2	∞	23.5649
PIDD^2-LQR (Proposed)	∞	**88.8**	**47**	∞	**56.4787**

Figure 10. Bode plot for the AVR system with PID controller optimized by (a) DEA [Gozde & Taplamacioglu, 2011] (b) PSO [Gozde & Taplamacioglu, 2011] (c) ABC [Gozde & Taplamacioglu, 2011] and (d) LQR method [Studied].

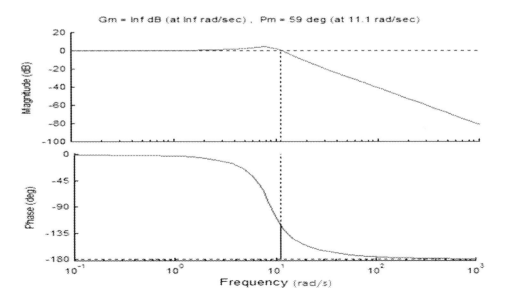

Figure 11. Bode plot for the AVR system with PIDD2 controller optimized by LQR method [Proposed].

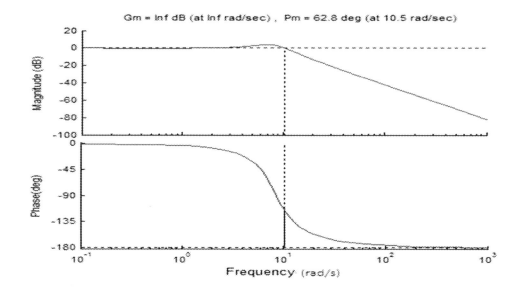

Root Locus Analysis

The root locus analysis helps to investigate the relative stability of a control system with respect to the other control systems. The transient behaviour of a system is closely related to the location of its closed loop poles (Ogata, 2010). Generally, in root locus method, by observing the locus of the closed loop poles of a system, comment on the system stability issue is, generally, made. The root loci of the AVR system using controllers optimized by different methods considered in this chapter are illustrated in Figure 12. Table 5 gives the closed loop poles and the corresponding ξ values for the various systems. It may be observed from Table 5 and Figure 13 that the conjugate poles for the AVR system using PIDD^2 controller is more to the left of the complex plane as compared to the other systems. Moreover, from Figure 13, it may be seen that, because of the additional zero introduced by the double derivative term of the PIDD^2 controller in the transfer function of the AVR system using PIDD^2 controller, the system root locus shows no tendency of intersecting the imaginary axis at any value of gain. The root locus deviates away from the imaginary axis. For other systems, the root locus asymptotes remain relatively close to the imaginary axis with the tendency of intersecting it at infinite value of gain. This signifies relatively more stability of the system with PIDD^2 controller. In addition, from Table 5 it may be observed that the conjugate poles of the PIDD^2 system have smaller ξ value than the systems tuned by other methods. This ensures a faster response of the proposed system while still maintaining zero overshoot and improved transient behaviour.

SFL Based Response

The on-line optimal parameters of the proposed controller are obtained by employing SFL based PIDD^2 model. The implementation of the SFL based PIDD^2 model in the system helps to determine the on-line and off-nominal input sets of parameters. Inevitably, the comparative analysis of the AVR system is

Figure 12. Root locus plot of the AVR system with PID controller tuned by (a) DEA [Gozde & Taplama-cioglu, 2011] (b) PSO [Gozde & Taplamacioglu, 2011] (c) ABC [Gozde & Taplamacioglu, 2011] and (d) LQR method [Studied].

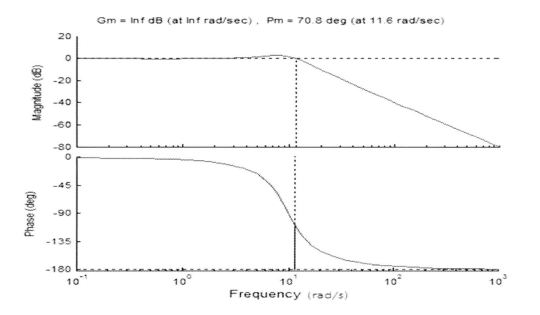

Figure 13. Root locus plot of the AVR system with PIDD2 controller tuned by LQR method [Proposed].

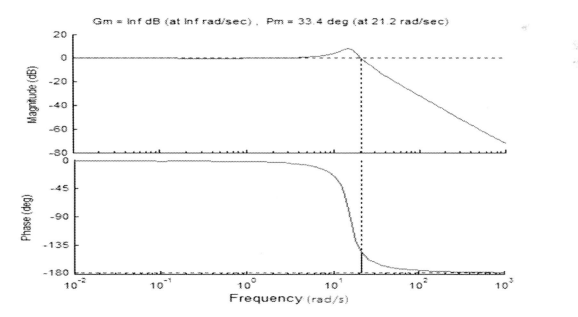

Table 5. Poles and damping ratios of the AVR system

Controller- method	Closed loop poles	Damping ratio
PID-DEA (Gozde & Taplamacioglu, 2011)	-99.7830	1
	-3.0767+j8.2010	0.351
	-3.0767-j8.2010	0.351
	-6.3353	1
	-0.2284	1
PID-PSO (Gozde & Taplamacioglu, 2011)	-99.7188	1
	-3.1337+j7.8084	0.372
	-3.1337-j7.8084	0.372
	-6.2986	1
	-0.2152	1
PID-ABC (Gozde & Taplamacioglu, 2011)	-99.9770	1
	-4.0929+j9.0723	0.411
	-4.0929-j9.0723	0.411
	-4.1129	1
	-0.2243	1
PID-LQR (Studied)	-101.51	1
	-3.14+j15.22	**0.202**
	-3.14-j15.22	**0.202**
	-4.62	1
	-0.09	1
PIDD²-LQR (Proposed)	-80.79	1
	-10.63+j29.12	0.343
	-10.63-j29.12	0.343
	-10.44	1
	-0.011	1

studied by wielding the gains of the PIDD^2 controller tuned by LQR-SFL technique. The terminal voltage response profile of LQR-SFL based PIDD^2 controlled AVR system is portrayed in Figure 14. It may be reported from this figure that, in respect of the tuning of on-line parameters of the PIDD^2 controller, the proposed LQR-SFL technique has endorsed its significance. Thus, with variation in operating conditions, the proposed LQR-SFL technique manifests adequate terminal voltage response profile.

Robustness Analysis

The variation in load leads to change in the operating points of the system due to flexibility of the power system parameters. Thus, a robust controller is essential for the AVR system for the uncertainties in the system. The behaviour of the system is investigated under system's parametric variation leading to analysis of the robustness of the proposed controller. In this chapter, the time constants of the AVR system's different blocks are varied, with a range of ±50% of the base value with steps of 25%, for introducing the system uncertainties.

Figure 14. SFL tuned PIDD² controlled AVR system's time-domain response based on the change in terminal voltage (p.u.).

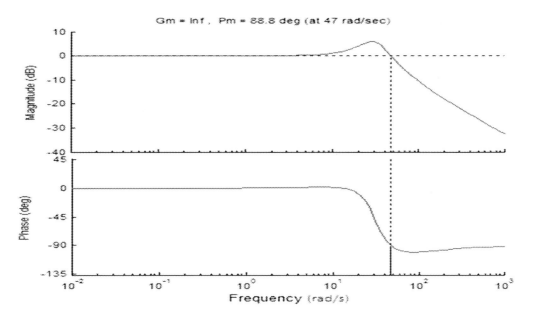

Table 6 .Results of robustness analysis for the AVR system tuned by the proposed PIDD2-LQR method

Parameter	Rate of change (%)	M_P (p.u)	T_{SS} (s) (5% bant)	T_R(s)
T_a	-50%	1.003	0.326	0.192
	-25%	1.002	0.324	0.183
	+25%	0.998	0.334	0.163
	+50%	0.998	0.351	0.143
T_e	-50%	1.001	0.342	0.194
	-25%	1.001	0.330	0.183
	+25%	0.998	0.318	0.162
	+50%	0.999	0.306	0.146
T_g	-50%	1.001	0.331	0.191
	-25%	1.003	0.330	0.182
	+25%	0.998	0.319	0.163
	+50%	0.999	0.307	0.149
T_s	-50%	0.996	0.320	0.174
	-25%	0.998	0.327	0.175
	+25%	1.001	0.335	0.176
	+50%	1.002	0.346	0.175

Table 7. Range of total deviation and percentage of maximum deviation for the AVR system tuned by PIDD²-LQR method

Parameter	Performance Parameters	Range of total deviation	Percentage of maximum deviation (%)
T_a	M_P (p.u.)	0.005	0.3
	T_{SS} (s)	0.027	6.56
	T_R (s)	0.049	0.09
T_e	M_P (p.u.)	0.003	**0.2**
	T_{SS} (s)	0.036	**3.83**
	T_R (s)	0.048	11.238
T_g	M_P (p.u.)	0.005	0.3
	T_{SS} (s)	0.024	6.8
	T_R (s)	0.042	6.28
T_s	M_P (p.u.)	0.006	0.4
	T_{SS} (s)	0.026	5.03
	T_R (s)	0.002	**0.917**

In Figure 15, the characteristics of the proposed LQR based PIDD² controlled AVR system with the variations in time constants are displayed. The performance index values *viz.* M_P, T_{SS} and T_R of the AVR system for different range of uncertainty under distinct levels of time constants variation is depicted in Table 6. In Table 7, the maximum percentage deviation and range of deviation of the terminal voltage response profile's parameters of the AVR system is delineated under variation of time constants of different blocks of the AVR system. It may be reported, from Table 7, that the percentage of the variation is very small and it may be accepted for its low value of deviation. The variation in T_e offers the least deviation in the value of M_P and T_{SS} which is 0.2% and 3.83%, respectively, of their respective base values. The value of T_R deviated, due to the variation in the value of T_S, 0.917% of the base value. It may be observed from the Table 7 that the average deviation of the performance index parameters *i.e.* M_P, T_{SS} and T_R are 0.3%, 5.55% and 9.631%, respectively. Thus, it may be reported that the proposed PIDD² controller proves to be a robust controller for controlling the AVR system under system's parametric variation.

CONCLUSION

In this chapter, analysis of the proposed PIDD² controller optimized by LQR method is carried out by comparing it with other optimization techniques based PID controller. The proposed controller is compared with other optimization techniques *viz.* DEA, PSO, ABC and LQR based PID controller for detailed analysis. The terminal voltage profiles procured by the proposed controller have shown the capability of the controller to control the studied AVR system. It may be observed from the transient response profiles that the performance indices and FOD value of the proposed controller based system manifests superior response of being fast, reduced oscillations and better damping over other optimization methods. Similarly, the frequency analysis of this system is found to have improved values of P_M, W_{CG} and BW as compared to the other systems showing its better stability characteristics. The root locus analysis shows

Figure 15. Profile of terminal voltage of PIDD2 controlled AVR system with variation in (a), (b), (c) and (d)

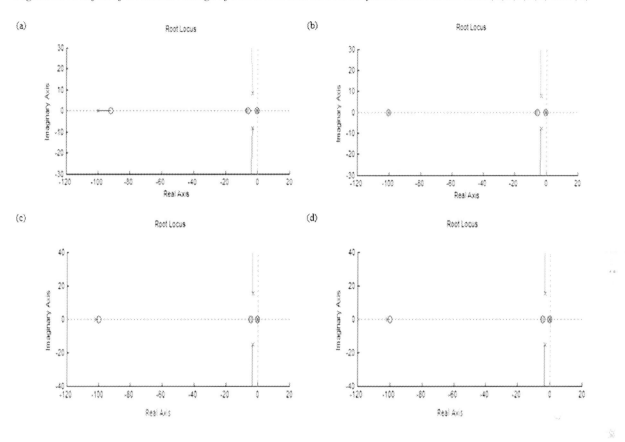

that the closed loop complex poles for the AVR system with PIDD^2 controller is located more to the left of imaginary axis on the complex plane as compared to other systems considered in this chapter. In addition, the root locus plot of the system with the proposed controller, unlike the other systems, shows no tendency to intersect the imaginary axis at any value of gain because of the introduction of an additional zero by the second order derivative term of the PIDD^2 controller in the system. To have better detailed knowledge about the proposed controller, the robustness analysis is also performed by changing the system parameters within a certain range. The analysis exhibits its robustness on controlling the studied system while varying the system parameters. The dynamic (on-line) response of the terminal voltage profile of the studied AVR system is achieved by implementing the fast acting SFL technique, along with the LQR method, in the system to tune the parameters of the PIDD^2 controller when operating in real time, on-line, off-nominal condition. It is observed that there is a low burden in obtaining on-line dynamic responses computationally. Hence the LQR–SFL technique may be employed to tune the PIDD^2 controller's gains, intelligently, for obtaining the on-line terminal voltage response profile of the studied AVR system. Thus in the concluding remarks, it may be stated that the proposed LQR based PIDD^2 controller is enough robust and capable to control the AVR system under specific system uncertainties.

REFERENCES

Ang, K. H., Chong, G., & Li, Y. (2005). PID control system analysis, design and technology. *IEEE Transactions on Control Systems Technology*, *13*(4), 559–576. doi:10.1109/TCST.2005.847331

Astrom, K. J., & Hagglund, T. (1995). *PID Controllers: Theory, Design and Tuning*. Instrument Society of America.

Banerjee, A., Mukherjee, V., & Ghoshal, S. P. (2014). Intelligent fuzzy-based reactive power compensation of an isolated hybrid power system. *International Journal of Electrical Power & Energy Systems*, *57*, 164–177. doi:10.1016/j.ijepes.2013.11.033

Chaterjee, A., Mukherjee, V., & Ghoshal, S. P. (2009). Velocity relaxed and craziness-based swarm optimized intelligent PID and PSS controlled AVR system. *International Journal of Electrical Power & Energy Systems*, *31*(8), 323–333. doi:10.1016/j.ijepes.2009.03.012

Clerc, M., & Kennedy, J. (2002). The particle swarm-explosion, stability and convergence in multidimensional complex space. *IEEE Transactions on Evolutionary Computation*, *6*(1), 58–73. doi:10.1109/4235.985692

Coelho, L. S. (2009). Tuning of PID controller for an automatic regulator voltage system using chaotic optimization approach. *Chaos Solutions Fract.*, *39*(4), 1504–1514. doi:10.1016/j.chaos.2007.06.018

Cominos, P., & Munro, N. (2002). PID controllers: recent tuning methods and design to specification. *Proc. IEE Control Theory and Applications*, *149*(1), 46-53. 10.1049/ip-cta:20020103

Finch, J. W., Zachariah, K. J., & Farsi, M. (1999). Turbo generator self-tuning automatic voltage regulator. *IEEE Transactions on Energy Conversion*, *14*(3), 843–848. doi:10.1109/60.790963

Gaing, Z. L. (2004). A particle swarm optimization approach for optimum design of PID controller in AVR system. *IEEE Transactions on Energy Conversion*, *19*(2), 384–391. doi:10.1109/TEC.2003.821821

Gozde, H., & Taplamacioglu, M. C. (2011). Comparative performance analysis of artificial bee colony algorithm for automatic voltage regulator (AVR) system. *Journal of the Franklin Institute*, *348*(8), 1927–1946. doi:10.1016/j.jfranklin.2011.05.012

Kashki, M., Abdel-Magid, Y. L., & Abido, M. L. (2008) A reinforcement learning automata optimization approach for optimum tuning of PID controller in AVR system. ICIC 2008, Advanced Intelligent Comput Theories and Appl. With Aspects of Artificial Intelligence, 684-692.

Kim, D. H., & Cho, J. H. (2006). A biologically inspired intelligent PID controller tuning for AVR systems. *International Journal of Control, Automation, and Systems*, *4*(5), 624–636.

Kundur, P., Balu, N. J., & Lauby, M. G. (1994). *Power system stability and control* (Vol. 7). New York: McGraw-Hill.

Manke, B. S. (2011). *Linear Control Systems*. New Delhi: Khanna Publishers.

Mohagheghi, S., Harley, R. G., & Venayagamoorthy, G. K. (2004). Modified Takagi–Sugeno fuzzy logic based controllers for a static compensator in a multi machine power system. *Proc. 39th IAS annual meeting conf.*, 2637-2642.

Mohanty, P. K., Sahu, B. K., & Panda, S. (2014). Tuning and assessment of proportional-integral-derivative controller for an automatic voltage regulator system employing local unimodal sampling algorithm. *Electric Power Components and Systems*, *42*(9), 959–969. doi:10.1080/15325008.2014.903546

Mukherjee, V. (2009). *Application of evolutionary optimization techniques for some selected power system problems* (Ph.D. dissertation). Durgapur, India: Dept Electrical Engg, National Institute of Technology.

Mukherjee, V., & Ghoshal, S. P. (2007). Intelligent particle swarm optimized fuzzy PID controller for AVR system. *Electric Power Systems Research*, *77*(12), 1689–1698. doi:10.1016/j.epsr.2006.12.004

Ogata, K. (2010). *Modern Control Engineering*. PHI Learning.

Panda, S., Sahu, B. K., & Mohanty, P. K. (2012). Design and performance analysis of PID controller for an automatic voltage regulator system using simplified particle swarm optimization. *Journal of the Franklin Institute*, *349*(8), 2609–2625. doi:10.1016/j.jfranklin.2012.06.008

Saadat, H. (2002). *Power System Analysis*. New Delhi: Tata McGraw Hill Ltd.

Sahib, M. A. (2015). A novel optimal PID plus second order derivative controller for AVR system. *J Eng Sc Technol.*, *18*(2), 194–206.

Swidenbank, E., Brown, M. D., & Flynn, D. (1999). Self-tuning turbine generator control for power plant. *Mechatronics*, *9*(5), 513–537. doi:10.1016/S0957-4158(99)00009-4

Zhu, H., Li, L., Zhao, Y., Guo, Y., & Yang, Y. (2009). CAS algorithm-based optimum design of PID controller in AVR system. *Chaos Solutions Fract.*, *42*(2), 792–800. doi:10.1016/j.chaos.2009.02.006

288

Chapter 16
40–GHz Inductor Less VCO

Abhishek Kumar

Lovely Professional University, India

ABSTRACT

In a modern communication system, voltage-controlled oscillator (VCO) acts as a basic building block for frequency generation. VCO with LC tank is preferred with passive inductor and varactor in radio frequency. Practical tuning range of VCO is low and unsuitable for wideband application. Switched capacitor and inductor can widen but at cost of chip area and complex system architecture. To overcome it, an equivalent circuit of the inductor is created. In this work, inductor-less VCO is implemented with CMOS 90nm technology that has center frequency 40GHz and frequency tuning range 37.7GHz to 41.9GHz.

INTRODUCTION

Communication between people found from the beginning of civilization; for communication, the message signal is modulated at the transmitted at a very high frequency; requires the use of a local oscillator. The local oscillator is a crucial component for frequency synthesizer in wireless communication; generate the same frequency for demodulation at the receiver end. Phase-locked loop (PLL) (Long, Foo, Weber, 2004) is an analog circuit can generate high frequency with the mechanism of feedback signal. The inherent block of PLL are (a) phase frequency detector (PFD) compare reference frequency with feedback, generate signal UP if feedback signal lags behind reference otherwise DOWN, it instructs (b) charge pump to produces high or low DC voltage (c) a low pass filter generates a constant voltage V_{const} (d) VCO takes this analog voltage and generate a clock signal of desired frequency. The oscillator is an electronic circuit can generate the sinusoidal, square, triangular, sawtooth wave, etc with positive feedback. There 2 types of oscillator LC tank oscillator and ring oscillator (Gordon & Voinigescu, 2004) . LC oscillator is preferred if the noise is a low and stringent requirement if accurate frequency. LC oscillators are extensively used in the RF circuit. Figure1 presents a block diagram of oscillator (Hess & Walter, 1993) whose oscillation frequency given as $f_{osc} = \dfrac{1}{2\pi\sqrt{LC}}$. Ring oscillator (RO) based VCO extensively uses the delay of each stage, RO (Liu, Chan,Wang & Su 2007) is known for an odd number

DOI: 10.4018/978-1-7998-1464-1.ch016

of cascaded delay stage usually implement an inverter along with the output of the last stage is feedback to the first stage. RO has the advantage of the ease of integration, small chip area and high frequency generation (Nagarajan, Seng, Mou & Kumar, 2011).

$$\frac{Y(s)}{X(s)} = \frac{H(s)}{1+\beta H(s)}$$

if βH(s) = -1 at s=jω, gain=∞ the circuit amplifies its own noise eventually begins to oscillate.

Figure 2 shows the output frequency of VCO maintains non-linear relations with a control voltage, $f_{out}=f_o+K_{VCO}.V_{ctrl}$. Where f0 is the frequency of oscillation, Kvco is the gain of the circuit. Frequency tuning is (Bunch & Raman, 2003) obligatory not to the complete application but also to reimburse for variation of the centre frequency of VCO.

LC-VCO

The active transistor (Mirajkar,Chand,Aniruddhan &Theertham, 2017) provides necessary negative trans-conductance 1/Gm to reimburse the tank losses. R is the resistance of the tank circuit. The stable

Figure 1. Block diagram of oscillator

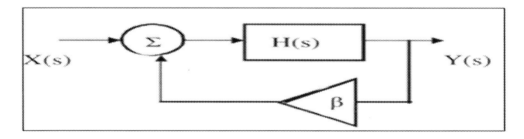

Figure 2. Transfer function of VCO

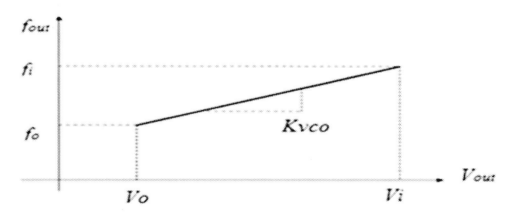

Figure 3. Block diagram of LCVCO

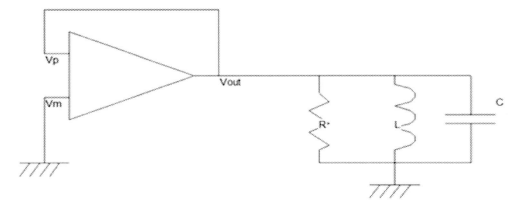

oscillator occurs when barkhaushen criterion is fulfilled, loop transfer functions exactly unity and phase gain of 360°.

2 elementary parameters of tunable VCO design is; constriction in resonator design tuning range and oscillation frequency (Trucco & Liberali, 2011) presented in figure3. Further, in both cases, the selection of the material depends on their contribution to noise performance for a wider tuning range (Jain, Gupta, Khatri & Banerjee, 2019) it is necessary for the inductor to have larger resistance ratio L/R and for varactor to have larger intrinsic maximum capacitance to minimum capacitance C_{max}/C_{min}

$$T_{loop}(s) = G_M \frac{sL}{1 + s\frac{L}{R} + s^2 LC}$$

$$Tuning = \frac{Cfix_{v,max}}{Cfix_{v,min}}.$$

The fundamental oscillation frequency of a basic LC-VCO is given as

$$f_{osc} = \frac{1}{2\pi \sqrt{L_{tank}(C_v + C_B)}}$$

where C_v is varactor capacitance and C_B isthe capacitance of switch capacitor bank.

$$C_B = C_p + \sum_{n=0}^{N} n.C_{unit}$$

C_p is the parasitic capacitance when all the capacitors in the bank are disconnected. The circuit is the capacitance of C_{Bank}.N is a maximum value of code n. The tuning range is an important aspect of LC oscillator where can generate frequency within a band, require a variable inductor and capacitor. This paper is focused on implementation on CMOS active inductor and switched capacitor bank to generate a range of oscillation frequency. LC VCO suffers from wide inductor area and narrow tuning range. Passive inductor (Haddad, Ghorbel & Rahajandraibe, 2017) shows poor integration and requires low characterization.

ACTIVE INDUCTOR

Spiral inductor using lossy silicon (Momen, & Kopru, 2015) could not easily attain high-quality operation (Dickson, LaCroix, Boret, Gloria, Beerkens, & Voinigescu, 2005) in CMOS technology. Moreover, it amplifies the area factor of the chip leading to the intrinsic mechanism. The alternative method is to exploit an active inductor which consists of two back to back connected trans-conductor generates inductive impedance and one of the terminals connected to capacitor call gyrator-C network. Basic gyrator-C topology stumpy inductance value and constricted tuning range (Fonte & Zito, 2009). Hence the structure can improve with cascaded transistor topology and active resistor in the feedback line. There is an inverse relationship between VI characteristics of the capacitor and inductor; the capacitive impedance has to be inverted in order to simulate inductance; deployment of the capacitor in a circuit to get desired phase relation to requiring gyrator circuit; a 2 port network. The active inductor is a special case of the gyrator. The active inductor has emerged as an alternative to the passive inductor due to compact size and capability to control over simulated inductance. Figure 4(a) presents a schematic of the active inductor (Racanelli & Kempf, 2006) and their small-signal model in figure4(b). M2 working in deep triode region acts as MOS resistor in addition to combining the output to the M1 transistor. To grant a vast bias current to M1 circuit is appended by an additional circuitry between M1 and M2. The supplementary circuitry is level shifter which consists of source follower transistor M3 and M4 as a current source (Absi, 2019). This configuration can work with even dwindling voltage supply without the want for any extra supply source other than VDD for greater voltage as in the case of NOS based active inductor. The PMOS based configuration of active inductor offers the advantages of low power consumption in contrast to the NMOS based design while demonstrating low resistance 50 ohms at the transmission. M1 doesn't endure body effect and has comparatively persistent V_{th} which further gives it consent to function with minimal voltage headroom. Because of body effect g_{m3}, g_{ds3}, g_{mb3} of M3 may fluctuate but the alternative will have an insignificant effect on the gain of the level shifter.

$$Gain_{levelshifter} = \frac{1}{\left(1 + \left(g_{ds4} + g_{ds3} + g_{mb3}\right)/g_{m3}\right)}$$

The inductive impedance of an active inductor is not conditional or Rs and can be amended by modifying V_{g2} which further alters the value of R2, the value of Rs is 50 ohm, The resonant frequency depends on the process technology active inductor overcome the disadvantage of magnetic coupling of the passive inductor. Moreover, are reducing in the contract of passive. The gain of the level shifter (Kananizadeh & Momeni, 2017) is taken as unity g_m, C_{gs}, $C_{ds,}$ and go_{ds} are the parameters of M1, R2 is on the resistance of M2. The small-signal model is depicted for the PMOS based active inductor and termination impedance

40-GHz Inductor Less VCO

Figure 4. Active inductor schematic and its small signal model

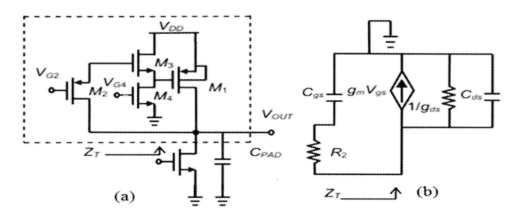

SWITCHED CAPACITOR BANK

Varactor diode is a variable capacitance which acts as a basic component for tuning the VCO frequency (Hu, Chou, Usami, Bowers & Blumenthal, 2004). Gate oxide capacitance in the accumulation and inversion region and total depletion capacitance is a series combination of depletion capacitance and Cox (Nandi,Pattanayak,Venkateswaran & Das, 2015).

Where WL is an area of varactor ε is the permittivity of silicon dioxide t_{ox} is the thickness of the oxide layer. Varactor diode preferred for tuning the circuit. Reverse bias PN junction varactor has low C_{max}/C_{min}. Another one is forward PN junction but it has larger amplitude swing leading to losses in the circuit; another commonly used varactor is MOS transistor. The ratio of capacitance to the area is higher for MOS varactor in comparison to the PN junction diode. Resulting in high capacitance tuning range, they also show string capacitance variation if the input voltage varies even in hundreds of mili-volts making a circuit to operate at low voltages. MOS varactor can works in a different region, depletion region, accumulation region and inversion. Inversion mode substrate of the varactor is separated and connected to the supply voltage. $V_{aractor}$ does not work in accumulation. PMOS base varactor used in the design, source, and drain of the PMOS transistor is shorted and the substrate is connected to V_{cc}, two identical PMOS transistor are shorted source and drain are connected in such a way that they form a mirror image at the junction and tuning voltage is applied at the junction. It displays in variable capacitance characteristic and displays wide tuning range; the invariability of small-signal plot inhibits tuning range deprivation by amplitude swing (Zheng, Arasu,Wong, Suan, Tran & Kwong, 2008). A MOS varactor in inversion mode known as IMOS varactor is less susceptible to latch-up. PMOS bulk is connected to body source and drain terminal are connected together shown in figure5. A tuning voltage applied to junction. Sinusoidal input of amplitude 0.5V and frequency 10GHz applied to the gate voltage. IMOS varactor implemented with PMOS length 100nm finger width 120nm employed 20fingers.

INDUCTOR LESS VCO

The schematic in figure 6 presents inductor less VCO; efficiently use for the generation of high frequency with low power consumption. Proposed architecture implemented with cadence virtuous schematic com-

Figure 5. Schematic of IMOS varactor diode

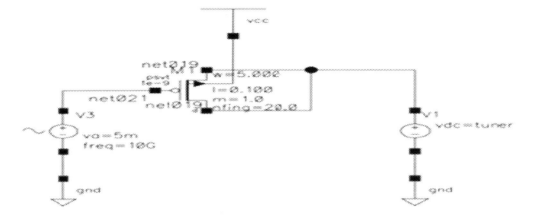

poser at CMOS 90nm technology, supply voltage 0.9 volt and simulation result is verified with cadence specter. The VCO consists of two complementary cross-coupled pairs, an active LC tank implemented by means of the active inductor and two p-MOSFET varicaps. -Gm from the NMOS cross-coupled pair is sufficient to compensate from the losses in the resonator circuit. Active inductor used for wide tuning range while varactor is used for fine-tuning range. VCO in figure 7 consist of 2 cross-coupled NMOS transistor provides negative trans-conductance. Proposed active inductor simulating the inductance of 401.75pH is used in place of conventional inductor (Gaoding & Bousquet, 2018). PMOS varactor provides the necessary capacitance value. The bias current source is looked upon as the major contributor of flicker noise. So, the PMOS transistor can be used to combat this problem. W/L of the PMOS must keep high in order to keep in working in the saturation region to keep voltage group across is lesser. The differential mode of operation is practiced in the proposed circuit by using 2 PMOS varactor by shorting their source and drain terminal and providing. The length and width of the NMOS transistor in the core are taken to be 0.2μm to 2.5μm respectively keeping in view the trade-off between tuning range (Xu, Saavedra & Chen, 2011) and phase noise. The capacitor C4 employed in parallel with the bias current imparts certain attenuation in terms of the high-frequency noise from the bias current source. The bias current course provides a considerable amount of flicker noise to the VCO. To combat this problem one PMOS transistor is used as a bias current source due to its characteristic of worse flicker noise upconversion compared to the NMOS of the identical dimension. The PMOS is operating saturation, if the W/L ratios of PMOS are kept high, there will be lesser voltage drop and hence improved the symmetry between two outputs of LC-VCO reducing the flicker noise upconversion. Interconnect of the source and VDD is provided to their substrate, the overall resonator circuit is formed

by connecting the gate node of 7 different pairs of PMOS varactor in parallel to provide a wide tuning range. Additional capacitances are used with the transistor in the NMOS core to eliminate high-frequency noise component while not contributing to the thermal noise. MOS varactor uses multiple fingers and the length is positioned at the minimum to get the good quality factor. Highest FOM can be obtained if the varactor has length <0.5μm and width 1-5μm.

2 PMOS varactor is connected in series through source/drain terminal/ Then 7 pairs of PMOS varactor are connected in parallel to form the digitally switchable capacitor bank. Biasing circuit consist of PMOS transistor with finger length 200nm and width 5μm. The capacitance value is 1.1fF, gate voltage

Figure 6. Inductor less VCO with tuning input

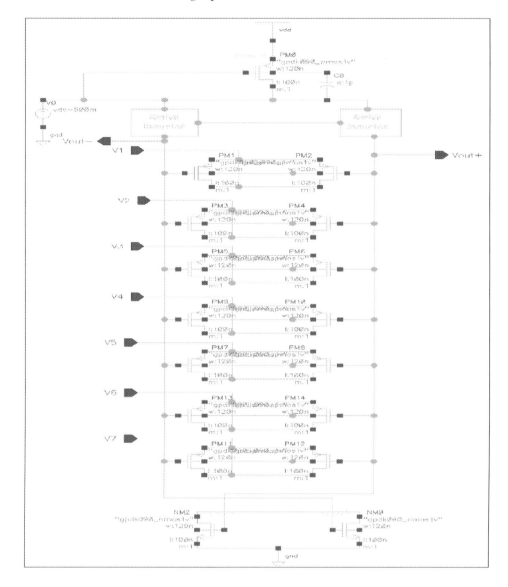

to the PMOS transistor is 0.5V. The VCO core consistsof 2 cross-coupled NMOS transistor each having length of 200nm, 20 fingers width of 150nm, total width is 3μm. The transient analysis was made of the proposed VCO from the analysis window, Time for analysis to be 50ns. Transient response in figure 7(a) presents first digital input 0 is given to all varactor leading the voltage, VCO takes 0.33ns to lock the frequency of 38.15GHz and for the frequency to settle. Complementary output of proposing inductor less VCO response figure7(b) presents named as out+ out- output waveform of VCO with capacitor bank select having value 0000000 is obtained to be 38.15GHz, when all 7-varactor pair connected to '1'. Oscillation frequency of VCO in figure8(b) increases and locking frequency achieved to 41.91GHz with settling time less than 300ps shown in figure8(a). Frequency increases with tuning voltage, figure9, its ranges between 38.15GHz to 41.91GHz.

Figure 7. Transient response for Vtune 0 and complementary output for Vtune 1

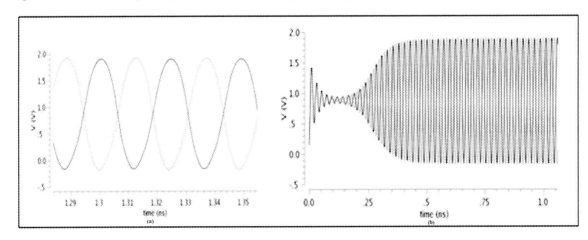

Figure 8. Transient response for Vtune 1 and complementary output for Vtune 0

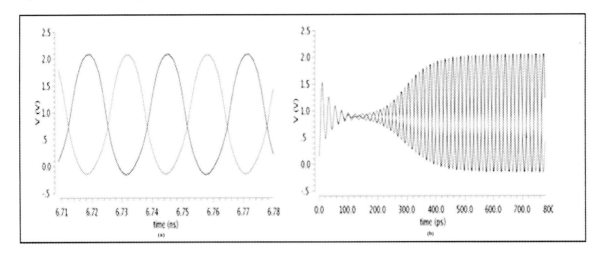

Figure 9. Frequency vs tuning

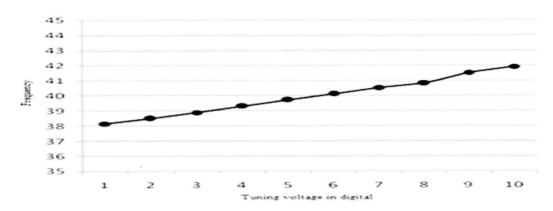

Table 1. Simulation result of presented work

Technology (nm)	90
Supply(v)	0.9
Power Dissipation (mw)	20
Frequency range (GHz)	38.15-41.9
Phase noise	-78.8 dBc/Hz at 1MHz offset

CONCLUSION

Due to the rapid development of various systems that have different operating frequencies, it becomes the need of the hour that a single oscillator is capable of tuning a wide range of carrier frequencies. At the same time oscillator must fulfill the requirement of having low phase noise low power and must permit full on-chip integration within the least possible area. Some of these requirement exhibit trades-offs. The active inductor has limited performance than passive LC oscillator but allows a high degree of circuit integration. Switched capacitor and switched inductor tuning range for wide frequency can be obtained. The disadvantage is enlarging the chip area and complexity of the control mechanism. By using inductor less circuit which tends to chip area we wish to fabricate the circuit with no physical inductor present on the chip or in a way using simulated inductor. Proposed VCO achieve a center frequency of 40GHz with a range of frequencies varies 38.15GHz to 41.9GHz; find application in Ka-band reduces chip area. Proposed VCO with an active inductor and IMOS varactor achieve the phase noise performance of -78.8dBc/Hz at 1 MHz offset.

REFERENCES

Al-Absi, M. A. (2019). Realization of a Large Values Floating and Tunable Active Inductor. *IEEE Access: Practical Innovations, Open Solutions, 7,* 42609–42613. doi:10.1109/ACCESS.2019.2907639

Bunch, R. L., & Raman, S. (2003). Large-signal analysis of MOS varactors in CMOS-G/sub m/LC VCOs. *IEEE Journal of Solid-State Circuits, 38*(8), 1325–1332. doi:10.1109/JSSC.2003.814416

Dickson, T. O., LaCroix, M. A., Boret, S., Gloria, D., Beerkens, R., & Voinigescu, S. P. (2005). 30-100-GHz inductors and transformers for millimeter-wave (Bi) CMOS integrated circuits. *IEEE Transactions on Microwave Theory and Techniques, 53*(1), 123–133. doi:10.1109/TMTT.2004.839329

Fonte, A., & Zito, D. (2009, July). Millimeter-wave high-Q CMOS active inductor. In 2009 Ph. D. Research in Microelectronics and Electronics (pp. 252-255). IEEE.

Gaoding, N., & Bousquet, J. F. (2018, December). A Fully Integrated Sub-GHz Inductor-less VCO with a Frequency Doubler. In *2018 25th IEEE International Conference on Electronics, Circuits and Systems (ICECS)* (pp. 469-472). IEEE. 10.1109/ICECS.2018.8617868

Gordon, M., & Voinigescu, S. P. (2004, September). An inductor-based 52-GHz 0.18/spl mu/m SiGe HBT cascode LNA with 22 dB gain. In *Proceedings of the 30th European Solid-State Circuits Conference* (pp. 287-290). IEEE. 10.1109/ESSCIR.2004.1356674

Haddad, F., Ghorbel, I., & Rahajandraibe, W. (2017, May). Multi-band inductor-less VCO for IoT applications. In *2017 IEEE International Symposium on Circuits and Systems (ISCAS)* (pp. 1-4). IEEE.

Hess, W., & Walter, R. (1993, September). A new K-band frequency divider by three using an injection locked oscillator in microstrip technique. In *1993 23rd European Microwave Conference* (pp. 391-393). IEEE. 10.1109/EUMA.1993.336571

Hu, Z., Nishimura, K., Chou, H., Rau, L., Usami, M., Bowers, J.E., & Blumenthal, D.J. (2004). *40-Gb / s Optical Packet Clock Recovery Using a Travelling-wave Electroabsorption Modulator-Based Ring Oscillator*. Academic Press.

Jain, V., Gupta, S. K., Khatri, V., & Banerjee, G. (2019, January). A 19.3-24.8 GHz Dual-Slope VCO in 65-nm CMOS for Automotive Radar Applications. In *2019 32nd International Conference on VLSI Design and 2019 18th International Conference on Embedded Systems (VLSID)* (pp. 118-123). IEEE.

Kananizadeh, R., & Momeni, O. (2017). A 190-GHz VCO with 20.7% tuning range employing an active mode switching block in a 130 nm SiGe BiCMOS. *IEEE Journal of Solid-State Circuits*, 52(8), 2094–2104. doi:10.1109/JSSC.2017.2689031

Liu, C. H., Chan, C. Y., Wang, R. L., & Su, Y. K. (2007, December). Low power current-reused voltage-controlled oscillator with optimum source damping resistors. In *2007 IEEE Conference on Electron Devices and Solid-State Circuits* (pp. 1017-1020). IEEE. 10.1109/EDSSC.2007.4450300

Long, J., Foo, J. Y., & Weber, R. J. (2004, February). A 2.4 GHz low-power low-phase-noise CMOS LC VCO. In *IEEE Computer Society Annual Symposium on VLSI* (pp. 213-214). IEEE. 10.1109/ISV-LSI.2004.1339533

Mirajkar, P., Chand, J., Aniruddhan, S., & Theertham, S. (2017). Low phase noise Ku-band VCO with optimal switched-capacitor bank design. *IEEE Transactions on Very Large Scale Integration (VLSI) Systems*, 26(3), 589–593.

Momen, H. G., Yazgi, M., & Kopru, R. (2015, November). Designing a new high Q fully CMOS tunable floating active inductor based on modified tunable grounded active inductor. In *2015 9th international conference on electrical and electronics engineering (ELECO)* (pp. 1-5). IEEE. 10.1109/ELECO.2015.7394542

Nagarajan, M., Ma, K., Seng, Y. K., Mou, S. X., & Kumar, T. B. (2011, September). A low power wide tuning range low phase noise VCO using coupled LC tanks. In *2011 Semiconductor Conference Dresden* (pp. 1-4). IEEE. doi:10.1109/SCD.2011.6068769

Nandi, R., Pattanayak, S., Venkateswaran, P., & Das, S. (2015). Electronically tunable differential integrator: linear voltage controlled quadrature oscillator. *International Scholarly Research Notices, 2015*.

Racanelli, M., & Kempf, P. (2006, January). Silicon foundry technology for RF products. In *Digest of Papers. 2006 Topical Meeting on Silicon Monolithic Integrated Circuits in RF Systems* (pp. 5-pp). IEEE.

Trucco, G., & Liberali, V. (2011). Analog design issues for mixed-signal CMOS integrated circuits. In *Advances in Analog Circuits* (pp. 165–180). London, UK: InTech. doi:10.5772/15176

Xu, J., Saavedra, C. E., & Chen, G. (2011). An active inductor-based VCO with wide tuning range and high DC-to-RF power efficiency. *IEEE Transactions on Circuits and Wystems. II, Express Briefs*, *58*(8), 462–466. doi:10.1109/TCSII.2011.2158713

Zheng, Y., Arasu, M. A., Wong, K. W., The, Y. J., Suan, A. P. H., Tran, D. D., . . . Kwong, D. L. (2008, February). A 0.18 m CMOS 802.15. 4a UWB transceiver for communication and localization. In 2008 IEEE International Solid-State Circuits Conference-Digest of Technical Papers (pp. 118-600). IEEE.

Compilation of References

Agarwal, M., Paul, B. C., Zhang, M., & Mitra, S. (2007). *Circuit failure prediction and its application to transistor aging.* Paper presented at the 25th IEEE VLSI Test Symposium, Berkeley, CA. 10.1109/VTS.2007.22

Aggarwal, J., & Bhargava, C. (2016). Reliability Prediction of Soil Humidity Sensor using Parts Count Analysis Method. *Indian Journal of Science and Technology*, *9*(47).

Al-Absi, M. A. (2019). Realization of a Large Values Floating and Tunable Active Inductor. *IEEE Access: Practical Innovations, Open Solutions*, *7*, 42609–42613. doi:10.1109/ACCESS.2019.2907639

Ali, A., & Majhi, S. (2010). PID controller tuning for integrating processes. *ISA Transactions*, *49*(1), 70–78. doi:10.1016/j.isatra.2009.09.001 PMID:19782358

Alkabani, Y., Koushanfar, F., & Potkonjak, M. (2007). Remote activation of ICs for piracy prevention and digital right management. *Proc. Int. Conf. Comput.-Aided Design*, 674-677. 10.1109/ICCAD.2007.4397343

Alkhawlani, M., & Ayesh, A. (2008). Access network selection based on fuzzy logic and genetic algorithms. *Advances in Artificial Intelligence*, *8*(1), 1–12. doi:10.1155/2008/793058

Allen, R., Billington, R., & Abdel-Gawad, N. (1986). The IEEE Reliability Test System -Extensions to and Evaluation of the Generating System. *IEEE Trans. on PWRS*, *1*(4), 1–7.

Al-Zubaidi, S., Ghani, J. A., & Haron, C. H. C. (2011). Application of ANN in milling process: A review. *Modelling and Simulation in Engineering*, *2011*, 9. doi:10.1155/2011/696275

Al-Zubaidi, S., Ghani, J. A., & Haron, C. H. C. (2013). Prediction of tool life in end milling of Ti-6Al-4V alloy using artificial neural network and multiple regression models. *Sains Malaysiana*, *42*(12), 1735–1741.

Aminian, K., Robert, P., Jéquier, E., & Schutz, Y. (1995). Incline, speed, and distance assessment during unconstrained walking. *Medicine and Science in Sports and Exercise*, *27*(2), 226–234. doi:10.1249/00005768-199502000-00012 PMID:7723646

Amiri, R., Mehrpouyan, H., Fridman, L., Mallik, R. K., Nallanathan, A., & Matolak, D. (2018). A machine learning approach for power allocation in HetNets considering QoS. In *2018 IEEE International Conference on Communications (ICC)* (pp. 1-7). IEEE. 10.1109/ICC.2018.8422864

Anderson, D. R. (2006). US patent infringement statute extends to the international market for copies of US software. *Journal of Intellectual Property Law & Practice*, *1*(4), 234–235.

Andriacchi, T. P., Ogle, J. A., & Galante, J. O. (1977). Walking speed as a basis for normal and abnormal gait measurements. *Journal of Biomechanics*, *10*(4), 261–268. doi:10.1016/0021-9290(77)90049-5 PMID:858732

Ang, K. H., Chong, G., & Li, Y. (2005). PID control system analysis, design and technology. *IEEE Transactions on Control Systems Technology*, *13*(4), 559–576. doi:10.1109/TCST.2005.847331

Antony, J., Bardhan Anand, R., Kumar, M., & Tiwari, M. (2006). Multiple response optimization using Taguchi methodology and neuro-fuzzy based model. *Journal of Manufacturing Technology Management*, *17*(7), 908–925. doi:10.1108/17410380610688232

Anwar, M. N., & Pan, S. (2013). Synthesis of the PID controller using desired closed-loop response. *IFAC Proc.*, *46*(32), 385-390.

Anwar, M., Shamsuzzoha, M., & Pan, S. (2015). A frequency domain PID controller design method using direct synthesis approach. *Arabian Journal for Science and Engineering*, *40*(4), 995–1004. doi:10.100713369-015-1582-4

Appenzeller, J. (2008). Carbon nanotubes for high-performance electronics-Progress and prospect. *Proceedings of the IEEE*, *96*(2), 201–211. doi:10.1109/JPROC.2007.911051

Appenzeller, J., Knoch, J., Tutuc, E., Reuter, M., & Guha, S. (2006, December). Dual-gate silicon nanowire transistors with nickel silicide contacts. In *2006 International Electron Devices Meeting* (pp. 1-4). IEEE. 10.1109/IEDM.2006.346842

Appenzeller, J., Lin, Y. M., Knoch, J., Chen, Z., & Avouris, P. (2005). Comparing carbon nanotube transistors-the ideal choice: A novel tunneling device design. *IEEE Transactions on Electron Devices*, *52*(12), 2568–2576. doi:10.1109/TED.2005.859654

Aronson, J. E., Liang, T.-P., & Turban, E. (2005). *Decision support systems and intelligent systems*. Pearson Prentice-Hall.

Astrom, K. J., & Hagglund, T. (1995). *PID Controllers Theory Design and Tuning* (2nd ed.). Instrument Society of America.

Astrom, K. J., & Hagglund, T. (1995). *PID Controllers: Theory, Design and Tuning*. Instrument Society of America.

Avital, M., Levi, I., Keren, O., & Fish, A. (2016). CMOS Based Gates for Blurring Power Information. *IEEE Transactions on Circuits and Systems. I, Regular Papers*, *63*(7), 1033–1042. doi:10.1109/TCSI.2016.2546387

Azadeh, A., Saberi, M., Anvari, M., Azaron, A., & Mohammadi, M. (2011). An adaptive-network-based fuzzy inference system-genetic algorithm clustering ensemble algorithm for performance assessment and improvement of conventional power plants. *Expert Systems with Applications*, *38*(3), 2224–2234. doi:10.1016/j.eswa.2010.08.010

Bachtold, A., Hadley, P., Nakanishi, T., & Dekker, C. (2001). Logic circuits with carbon nanotube transistors. *Science*, *294*(5545), 1317–1320. doi:10.1126cience.1065824 PMID:11588220

Baek, R. H., Baek, C. K., Jung, S. W., Yeoh, Y. Y., Kim, D. W., Lee, J. S., ... Jeong, Y. H. (2009). Characteristics of the Series Resistance Extracted From Si Nanowire FETs Using the $ Y $-Function Technique. *IEEE Transactions on Nanotechnology*, *9*(2), 212–217. doi:10.1109/TNANO.2009.2028024

Bailey, C., Lu, H., Stoyanov, S., Yin, C., Tilford, T., & Ridout, S. (2008). *Predictive reliability and prognostics for electronic components: Current capabilities and future challenges*. Paper presented at the 31st IEEE International Spring Seminar on Electronics Technology, ISSE'08, Budapest, Hungary.

Baker, R. (2007). The history of gait analysis before the advent of modern computers. *Gait & Posture*, *26*(3), 331–342. doi:10.1016/j.gaitpost.2006.10.014 PMID:17306979

Banerjee, A., Mukherjee, V., & Ghoshal, S. P. (2013). Modeling and seeker optimization based simulation for intelligent reactive power control of an isolated hybrid power system. *Swarm and Evolutionary Computation*, *13*, 85–100. doi:10.1016/j.swevo.2013.05.003

Banerjee, A., Mukherjee, V., & Ghoshal, S. P. (2014). Intelligent fuzzy-based reactive power compensation of an isolated hybrid power system. *International Journal of Electrical Power & Energy Systems, 57*, 164–177. doi:10.1016/j.ijepes.2013.11.033

Banerjee, I., & Das, P. (2012). Group technology based adaptive cell formation using a predator-prey genetic algorithm. *Applied Soft Computing, 12*(1), 559–572. doi:10.1016/j.asoc.2011.07.021

Bar-Cohen, Y. (2014). *High temperature materials and mechanisms*. CRC Press. doi:10.1201/b16545

Barlow, R. E., & Proschan, F. (1996). *Mathematical theory of reliability*. SIAM. doi:10.1137/1.9781611971194

Barnes, F. (1971). *Component Reliability*. Springer.

Baseer, S. (2013). Heterogenous networks architectures and their security weaknesses. *International Journal of Computer and Communication Engineering, 2*(2), 90–93. doi:10.7763/IJCCE.2013.V2.145

Baur, S., & Boche, H. (2017, December). Robust secure storage of data sources with perfect secrecy. In *2017 IEEE Workshop on Information Forensics and Security (WIFS)* (pp. 1-6). IEEE. 10.1109/WIFS.2017.8267669

Becker, G. T. (2015). On the pitfalls of using arbiter-PUFs as building blocks. *IEEE Transactions on Computer-Aided Design of Integrated Circuits and Systems, 34*(8), 1295–1307. doi:10.1109/TCAD.2015.2427259

Becker, H. I. (1957). *Low voltage electrolytic capacitor*. Google Patents.

Beckmann, N., & Potkonjak, M. (2009, June). Hardware-based public-key cryptography with public physically unclonable functions. In *International Workshop on Information Hiding* (pp. 206-220). Springer. 10.1007/978-3-642-04431-1_15

Bertani, A., Cappello, A., Benedetti, M. G., Simoncini, L., & Catani, F. (1999). Flat foot functional evaluation using pattern recognition of ground reaction data. *Clinical Biomechanics (Bristol, Avon), 14*(7), 484–493. doi:10.1016/S0268-0033(98)90099-7 PMID:10521632

Bhandari, V., & Vaidya, N. H. (2009). *Channel and interface management in a heterogeneous multi-channel multi-radio wireless network*. Illinois Univ at Urbana-Champaign. doi:10.21236/ADA555113

Bhargava, Banga, & Singh. (2014). *Failure prediction and health prognostics of electronic components: A review*. Paper presented at the IEEE Conference on Recent Advances in Engineering and Computational Sciences (RAECS), Chandigarh, India.

Bhargava, Banga, & Singh. (2017). Fabrication and Failure Prediction of Carbon-alum solid composite electrolyte based humidity sensor using ANN. *Science and Engineering of Composite Materials*. doi:10.1515ecm-2016-0272

Bhargava, Banga, & Singh. (2018). Failure prediction of humidity sensor DHT11 using various environmental testing techniques. *Journal of Materials and Environmental Sciences, 9*(7), 243-252.

Bhargava, C., Banga, V. K., & Singh, Y. (2014). *Failure prediction and health prognostics of electronic components: A review*. Paper presented at the IEEE Conference on Recent Advances in Engineering and Computational Sciences (RAECS), Chandigarh, India.

Bhargava, C., Banga, V., & Singh, Y. (2018). Mathematical Modelling and Residual Life Prediction of an Aluminium Electrolytic Capacitor. *Pertanika Journal of Science & Technology, 26*(2), 785–798.

Bhargava, C., & Handa, M. (2018). An Intelligent Reliability Assessment technique for Bipolar Junction Transistor using Artificial Intelligence Techniques. *Pertanika Journal of Science & Technology, 26*(4).

Bhattacharya, B., Mandal, K. K., & Chakravorty, N. (2012). Cultural Algorithm Based Constrained Optimization for Economic Load Dispatch of Units Considering Different Effects. *International Journal of Soft Computing and Engineering, 2*(2), 45–50.

Bhattacharyya, P., Kundu, B., Ghosh, S., Kumar, V., & Dandapat, A. (2014). Performance analysis of a low-power high-speed hybrid 1-bit full adder circuit. *IEEE Transactions on Very Large Scale Integration (VLSI) Systems, 23*(10), 2001-2008.

Billington, R., & Oteng-Adjei, J. (1991). Utilization Of Interrupted Energy Assessment Rates In Generation And Transmission System Planning. *IEEE Trans. on PWRS, 6*(3), 1245–1253.

Binti, Choong, Bin, Kamal, & Badal. (2017). Bit swapping linear feedback shift register for low power application using 130nm complementary metal oxide semiconductor technology. *IJE Transactions B. Applications, 30*, 1126–1133.

Borgovini, R., Pemberton, S., & Rossi, M. (1993). Failure mode, effects, and criticality analysis (FMECA). *Proceedings of the IEEE.*

Bradley, D. M., & Gupta, R. C. (2003). Limiting behaviour of the mean residual life. *Annals of the Institute of Statistical Mathematics, 55*(1), 217–226. doi:10.1007/BF02530495

Braunsteiner, E. E., & Mark, H. F. (1974). Aromatic polymers. *Journal of Polymer Science: Macromolecular Reviews, 9*(1), 83–126.

Breit, G. A., & Whalen, R. T. (1997). Prediction of human gait parameters from temporal measures of foot-ground contact. *Medicine and Science in Sports and Exercise, 29*(4), 540–547. doi:10.1097/00005768-199704000-00017 PMID:9107638

Bryllert, T., Wernersson, L. E., Froberg, L. E., & Samuelson, L. (2006). Vertical high-mobility wrap-gated InAs nanowire transistor. *IEEE Electron Device Letters, 27*(5), 323–325. doi:10.1109/LED.2006.873371

Bunch, R. L., & Raman, S. (2003). Large-signal analysis of MOS varactors in CMOS-G/sub m/LC VCOs. *IEEE Journal of Solid-State Circuits, 38*(8), 1325–1332. doi:10.1109/JSSC.2003.814416

Bushnell, M., & Agrawal, V. (2000). *Essentials of Electronic Testing for Digital, Memory and Mixed-Signal VLSI Circuits.* Boston: Kluwer Academic Publishers.

Çalhan, A., & Çeken, C. (2013). Artificial neural network based vertical handoff algorithm for reducing handoff latency. *Wireless Personal Communications, 71*(4), 2399–2415. doi:10.100711277-012-0944-4

Cantoro, M., Hofmann, S., Pisana, S., Scardaci, V., Parvez, A., Ducati, C., & Robertson, J. (2006). Catalytic chemical vapor deposition of single-wall carbon nanotubes at low temperatures. *Nano Letters, 6*(6), 1107–1112. doi:10.1021/nl060068y PMID:16771562

Cao, Y., Zhang, L., Chang, C. H., & Chen, S. (2015). A low-power hybrid RO PUF with improved thermal stability for lightweight applications. *IEEE Transactions on Computer-Aided Design of Integrated Circuits and Systems, 34*(7), 1143–1147. doi:10.1109/TCAD.2015.2424955

Cavagna, G. A., & Kaneko, M. (1977). Mechanical work and efficiency in level walking and running. *The Journal of Physiology, 268*(2), 467–481. doi:10.1113/jphysiol.1977.sp011866 PMID:874922

Cavanagh, P. R., & Lafortune, M. A. (1980). Ground reaction forces in distance running. *Journal of Biomechanics, 13*(5), 397–406. doi:10.1016/0021-9290(80)90033-0 PMID:7400169

Chaintoutis, C., Akriotou, M., Mesaritakis, C., Komnios, I., Karamitros, D., Fragkos, A., & Syvridis, D. (2018). Optical PUFs as physical root of trust for blockchain-driven applications. *IET Software, 13*(3), 182–186. doi:10.1049/iet-sen.2018.5291

Chakraborty, R. S., & Bhunia, S. (2009). HARPOON: An obfuscation-based SoC design methodology for hardware protection. *IEEE Transactions on Computer-Aided Design of Integrated Circuits and Systems, 28*(10), 1493–1502. doi:10.1109/TCAD.2009.2028166

Chandavarkar, B. R., & Reddy, G. R. M. (2012). Survey paper: Mobility management in heterogeneous wireless networks. *Procedia Engineering, 30*, 113–123. doi:10.1016/j.proeng.2012.01.841

Chaterjee, A., Mukherjee, V., & Ghoshal, S. P. (2009). Velocity relaxed and craziness-based swarm optimized intelligent PID and PSS controlled AVR system. *International Journal of Electrical Power & Energy Systems, 31*(8), 323–333. doi:10.1016/j.ijepes.2009.03.012

Chen, C., Zhang, B., Vachtsevanos, G., & Orchard, M. (2011). Machine condition prediction based on adaptive neuro–fuzzy and high-order particle filtering. *IEEE Transactions on Industrial Electronics, 58*(9), 4353–4364. doi:10.1109/TIE.2010.2098369

Chen, D., & Seborg, D. E. (2002). PI/PID controller design based on direct synthesis and disturbance rejection. *Industrial & Engineering Chemistry Research, 41*(19), 4807–4822. doi:10.1021/ie010756m

Chen, F., Tang, B., & Chen, R. (2013). A novel fault diagnosis model for gearbox based on wavelet support vector machine with immune genetic algorithm. *Measurement: Journal of the International Measurement Confederation, 46*(1), 220–232. doi:10.1016/j.measurement.2012.06.009

Chen, L., Kiong, S., Dong, X., & Mei, X. (2004). Modeling and optimization of fed-batch fermentation processes using dynamic neural networks and genetic algorithms. *Biochemical Engineering Journal, 22*(1), 51–61. doi:10.1016/j.bej.2004.07.012

Chen, P., Zhang, W., & Zhu, L. (2006) Design and tuning method of PID controller for a class of inverse response processes. *Proceedings of the 2006 American Control Conference.* 10.1109/ACC.2006.1655367

Chen, T., Li, J., Jin, P., & Cai, G. (2013). Reusable rocket engine preventive maintenance scheduling using a genetic algorithm. *Reliability Engineering & System Safety, 114*, 52–60. doi:10.1016/j.ress.2012.12.020

Cherkaoui, A., Bossuet, L., & Marchand, C. (2016). Design, evaluation, and optimization of physical unclonable functions based on transient effect ring oscillators. *IEEE Transactions on Information Forensics and Security, 11*(6), 1291–1305. doi:10.1109/TIFS.2016.2524666

Chidambaram, M., & Sree, R. P. (2003). A simple method of tuning of PID controller for integrating/dead time processes. *Computers & Chemical Engineering, 27*(2), 211–215. doi:10.1016/S0098-1354(02)00178-3

Chien, I. L., Chung, Y. C., Chen, B. S., & Chuang, C. Y. (2003). Simple PID controller tuning method for processes with inverse response plus dead time or large overshoot response plus dead time. *Industrial & Engineering Chemistry Research, 42*(20), 4461–4477. doi:10.1021/ie020726z

Chrisey, D. B., & Hubler, G. K. (Eds.). (1994). Pulsed laser deposition of thin films. Academic Press.

Clerc, M., & Kennedy, J. (2002). The particle swarm-explosion, stability and convergence in multidimensional complex space. *IEEE Transactions on Evolutionary Computation, 6*(1), 58–73. doi:10.1109/4235.985692

Cleves, M. (2008). *An introduction to survival analysis using Stata.* Stata Press.

Clifford, J. P., John, D. L., Castro, L. C., & Pulfrey, D. L. (2004). Electrostatics of partially gated carbon nanotube FETs. *IEEE Transactions on Nanotechnology, 3*(2), 281–286. doi:10.1109/TNANO.2004.828539

Cocchi, R. P., Baukus, J. P., Chow, L. W., & Wang, B. J. (2014, June). Circuit camouflage integration for hardware IP protection. *Proceedings of the 51st Annual Design Automation Conference*, 1-5 10.1145/2593069.2602554

Coelho, L. S. (2009). Tuning of PID controller for an automatic regulator voltage system using chaotic optimization approach. *Chaos Solutions Fract.*, *39*(4), 1504–1514. doi:10.1016/j.chaos.2007.06.018

Cohen, G. M., Rooks, M. J., Chu, J. O., Laux, S. E., Solomon, P. M., Ott, J. A., & Haensch, W. (2007). Nanowire metal-oxide-semiconductor field effect transistor with doped epitaxial contacts for source and drain. *Applied Physics Letters*, *90*(23), 233110. doi:10.1063/1.2746946

Cohen, G., & Coon, G. (1953). Theoretical consideration of retarded control. *Transaction of ASME*, *75*(1), 827-834.

Cominos, P., & Munro, N. (2002). PID controllers: recent tuning methods and design to specification. *Proc. IEE Control Theory and Appl.*, *149*(1), 46-53. 10.1049/ip-cta:20020103

Conradie, D. G., Morison, L. E., & Joubert, J. W. (2008). Scheduling at coal handling facilities using Simulated Annealing. *Mathematical Methods of Operations Research*, *68*(2), 277–293. doi:10.100700186-008-0221-1

Crowe, A., Samson, M. M., Hoitsma, M. J., & van Ginkel, A. A. (1996). The influence of walking speed on parameters of gait symmetry determined from ground reaction forces. *Human Movement Science*, *15*(3), 347–367. doi:10.1016/0167-9457(96)00005-X

Czanderna, A. W. (2012). *Methods of surface analysis* (Vol. 1). Elsevier.

Dana, A., Ghalavand, G., Ghalavand, A., & Farokhi, F. (2011). A Reliable routing algorithm for Mobile Adhoc Networks based on fuzzy logic. *International Journal of Computer Science Issues*, *8*(3), 128–133.

Das, K., Lashkari, R. S., & Sengupta, S. (2006). Reliability considerations in the design of cellular manufacturing systems. *International Journal of Quality & Reliability Management*, *23*(7), 880–904. doi:10.1108/02656710610679851

De Informa, C., & De Marrocos, P. (1999). Global optimization of energy and production in process industries : A genetic algorithm application. *Control Engineering Practice*, *7*(4), 549–554. doi:10.1016/S0967-0661(98)00194-4

Delvaux, J. (2019). Machine-Learning Attacks on PolyPUFs, OB-PUFs, RPUFs, LHS-PUFs, and PUF–FSMs. *IEEE Transactions on Information Forensics and Security*, *14*(8), 2043–2058. doi:10.1109/TIFS.2019.2891223

Derycke, V., Martel, R., Appenzeller, J., & Avouris, P. (2002). Controlling doping and carrier injection in carbon nanotube transistors. *Applied Physics Letters*, *80*(15), 2773–2775. doi:10.1063/1.1467702

Devadas, S., Suh, E., Paral, S., Sowell, R., Ziola, T., & Khandelwal, V. (2008, April). *Design and implementation of PUF-based" unclonable" RFID ICs for anti-counterfeiting and security applications. In 2008 IEEE international conference on RFID* (pp. 58–64). IEEE.

Dickson, T. O., LaCroix, M. A., Boret, S., Gloria, D., Beerkens, R., & Voinigescu, S. P. (2005). 30-100-GHz inductors and transformers for millimeter-wave (Bi) CMOS integrated circuits. *IEEE Transactions on Microwave Theory and Techniques*, *53*(1), 123–133. doi:10.1109/TMTT.2004.839329

Dislis,, C., Ambler,, A.,, & Dick,, J. (1991). Economic Effects in Design and Test. *IEEE Design & Test of Computers*, *8*, 64-77. doi:10.1109/54.107206

Doggett, A. M. (2005). Root cause analysis: A framework for tool selection. *The Quality Management Journal*, *12*(4), 34–45. doi:10.1080/10686967.2005.11919269

Dotoli, M., Fanti, M. P., Mangini, A. M., & Ukovich, W. (2009). On-line fault detection in discrete event by petri nuts and integer linear programming. *Automatica*, *45*(11), 2665–2672. doi:10.1016/j.automatica.2009.07.021

Du, C., & Bai, G. (2014, December). A novel relative frequency based ring oscillator physical unclonable function. In *2014 IEEE 17th International Conference on Computational Science and Engineering* (pp. 569-575). IEEE. 10.1109/CSE.2014.129

Duffuaa, S. O., & Sultan, K. S. (1999). A stochastic programming model for scheduling maintenance personnel. *Applied Mathematical Modelling, 25*(5), 385–397. doi:10.1016/S0307-904X(98)10009-4

Dutt, S., & Li, L. (2009). Trust-based design and check of FPGA circuits using two-level randomize ECC structure. *ACM Transactions on Reconfigurable Technology and Systems, 2*(1), 1–36. doi:10.1145/1502781.1508209

Dylis, D. D., & Priore, M. G. (2001). *A comprehensive reliability assessment tool for electronic systems.* Paper presented at the IEEE Annual Symposium on Reliability and Maintainability, Philadelphia, PA. 10.1109/RAMS.2001.902485

E, J., Ames, K., & Ames, K. (2007). A simulated annealing algorithm for system cost minimization subject to reliability constraints. *Communications in Statistics-Simulation and Computation,* 37–41.

El-Amin, I., Duffuaa, S., & Abbas, M. (2000). A tabu search algorithm for maintenance scheduling of generating units. *Electric Power Systems Research, 54*(2), 91–99. doi:10.1016/S0378-7796(99)00079-6

Erdem, O., & Serdar, S. (2018). A fast digit based Montgomery multiplier designed for FPGAs with DSP resources. *Microprocessor and Microsystem, Elsevier, 62,* 12–19. doi:10.1016/j.micpro.2018.06.015

Fard, N., & Li, C. (2009). Optimal simple step stress accelerated life test design for reliability prediction. *Journal of Statistical Planning and Inference, 139*(5), 1799–1808. doi:10.1016/j.jspi.2008.05.046

Favre, J., Hayoz, M., Erhart-Hledik, J. C., & Andriacchi, T. P. (2012). A neural network model to predict knee adduction moment during walking based on ground reaction force and anthropometric measurements. *Journal of Biomechanics, 45*(4), 692–698. doi:10.1016/j.jbiomech.2011.11.057 PMID:22257888

Fei, T., Jiang, K., Liu, S., & Zhang, T. (2014). Humidity sensors based on Li-loaded nanoporous polymers. *Sensors and Actuators. B, Chemical, 190,* 523–528. doi:10.1016/j.snb.2013.09.013

Fenton, N. E., & Ohlsson, N. (2000). Quantitative analysis of faults and failures in a complex software system. *IEEE Transactions on Software Engineering, 26*(8), 797–814. doi:10.1109/32.879815

Finch, J. W., Zachariah, K. J., & Farsi, M. (1999). Turbo generator self-tuning automatic voltage regulator. *IEEE Transactions on Energy Conversion, 14*(3), 843–848. doi:10.1109/60.790963

Fong & Maier. (2008). Power plant maintenance scheduling using Ant Colony Optimization – An improved formulation. *Engineering Optimization, 404*(4), 309–329.

Fonte, A., & Zito, D. (2009, July). Millimeter-wave high-Q CMOS active inductor. In 2009 Ph. D. Research in Microelectronics and Electronics (pp. 252-255). IEEE.

Fountain, J. H. (1965). *A general computer simulation technique for assessments and testing requirements.* Paper presented at the ACM Conference on the SHARE design automation project, New York, NY. 10.1145/800266.810757

Frank, K. D., Rich, C., & Longcore, T. (2006). Effects of artificial night lighting on moths. Ecological Consequences of Artificial Night Lighting, 305-344.

Fried, J. R. (1982). Polymer Technology. 1. The Polymers of Commercial Plastics. *Plastics Engineering, 38*(6), 49–55.

Fruehauf, P. S., Chien, I. L., & Lauritsen, M. D. (1994). Simplified IMC-PID tuning rules. *ISA Transactions, 33*(1), 43–59. doi:10.1016/0019-0578(94)90035-3

Gaing, Z. L. (2004). A particle swarm optimization approach for optimum design of PID controller in AVR system. *IEEE Transactions on Energy Conversion, 19*(2), 384–391. doi:10.1109/TEC.2003.821821

Gaoding, N., & Bousquet, J. F. (2018, December). A Fully Integrated Sub-GHz Inductor-less VCO with a Frequency Doubler. In *2018 25th IEEE International Conference on Electronics, Circuits and Systems (ICECS)* (pp. 469-472). IEEE. 10.1109/ICECS.2018.8617868

Garg, H., & Sharma, S. P. (2011). Multi-Objective Optimization of Crystallization Unit in a Fertilizer Plant Using Particle Swarm Optimization. *International Journal of Applied Science and Engineering, 9*(4), 261–276.

Gaston, K. J., Bennie, J., Davies, T. W., & Hopkins, J. (2013). The ecological impacts of nighttime light pollution: A mechanistic appraisal. *Biological Reviews of the Cambridge Philosophical Society, 88*(4), 912–927. doi:10.1111/brv.12036 PMID:23565807

George, W. E. (2016). Array Multipliers for High Throughput in Xilinx FPGAs with 6-Input LUTs. *Computers MDPI, 5*, 1–25.

Giakas, G., & Baltzopoulos, V. (1997). Time and frequency domain analysis of ground reaction forces during walking: An investigation of variability and symmetry. *Gait & Posture, 5*(3), 189–197. doi:10.1016/S0966-6362(96)01083-1

Gioftsos, G., & Grieve, D. W. (1996). The use of artificial neural networks to identify patients with chronic low-back pain conditions from patterns of sit-to-stand manoeuvres. *Clinical Biomechanics (Bristol, Avon), 11*(5), 275–280. doi:10.1016/0268-0033(96)00013-7 PMID:11415632

Glaser, R. E. (1980). Bathtub and related failure rate characterizations. *Journal of the American Statistical Association, 75*(371), 667–672. doi:10.1080/01621459.1980.10477530

Gnani, E., Gnudi, A., Reggiani, S., Luisier, M., & Baccarani, G. (2008). Band effects on the transport characteristics of ultrascaled snw-fets. *IEEE Transactions on Nanotechnology, 7*(6), 700–709. doi:10.1109/TNANO.2008.2005777

Gokulachandran, J., & Mohandas, K. (2015). Comparative study of two soft computing techniques for the prediction of remaining useful life of cutting tools. *Journal of Intelligent Manufacturing, 26*(2), 255–268. doi:10.100710845-013-0778-2

Goran, M. I., & Sun, M. (1998). Total energy expenditure and physical activity in prepubertal children: Recent advances based on the application of the doubly labeled water method. *The American Journal of Clinical Nutrition, 68*(4), 944S–949S. doi:10.1093/ajcn/68.4.944S PMID:9771877

Gordon, M., & Voinigescu, S. P. (2004, September). An inductor-based 52-GHz 0.18/spl mu/m SiGe HBT cascode LNA with 22 dB gain. In *Proceedings of the 30th European Solid-State Circuits Conference* (pp. 287-290). IEEE. 10.1109/ESSCIR.2004.1356674

Gozde, H., & Taplamacioglu, M. C. (2011). Comparative performance analysis of artificial bee colony algorithm for automatic voltage regulator (AVR) system. *Journal of the Franklin Institute, 348*(8), 1927–1946. doi:10.1016/j.jfranklin.2011.05.012

Grasso, R., Bianchi, L., & Lacquaniti, F. (1998). Motor patterns for human gait: Backward versus forward locomotion. *Journal of Neurophysiology, 80*(4), 1868–1885. doi:10.1152/jn.1998.80.4.1868 PMID:9772246

Greason, W. D., & Castle, G. P. (1984). The Effects of Electrostatic Discharge on Microelectronic Devices A Review. *IEEE Transactions on Industry Applications, IA-20*(2), 247–252. doi:10.1109/TIA.1984.4504404

Gu, J., & Pecht, M. (2008). *Prognostics and health management using physics-of-failure.* Paper presented at the IEEE Annual Symposium on Reliability and Maintainability Symposium (RAMS 2008), Las Vegas, NV 10.1109/RAMS.2008.4925843

Gualous, H., Bouquain, D., Berthon, A., & Kauffmann, J. (2003). Experimental study of supercapacitor serial resistance and capacitance variations with temperature. *Journal of Power Sources, 123*(1), 86–93. doi:10.1016/S0378-7753(03)00527-5

Guevorkian, D., Launiainen, A., Lappalainen, V., Liuha, P., & Punkka, K. (2005). A method for designing high-radix multiplier-based processing units for multimedia applications. *IEEE Transactions on Circuits and Systems for Video Technology, 15*(5), 716–725. doi:10.1109/TCSVT.2005.846436

Gu, J., Qu, G., & Zhou, Q. (2019), Information hiding for trusted system design. *Proceedings of the 46th Design Automation Conference (DAC '09)*, 698–701.

Günlü, O., Kernetzky, T., İşcan, O., Sidorenko, V., Kramer, G., & Schaefer, R. (2018). Secure and reliable key agreement with physical unclonable functions. *Entropy (Basel, Switzerland), 20*(5), 340. doi:10.3390/e20050340

Habtour, E., Drake, G. S., & Davies, C. (2011). *Modeling damage in large and heavy electronic components due to dynamic loading.* Paper presented at the IEEE Annual Symposium on Reliability and Maintainability Symposium (RAMS), Lake Buena Vista.

Haddad, F., Ghorbel, I., & Rahajandraibe, W. (2017, May). Multi-band inductor-less VCO for IoT applications. In *2017 IEEE International Symposium on Circuits and Systems (ISCAS)* (pp. 1-4). IEEE.

Hafez, A. L., Zawbaa, H. M., Hassanien, A. E., & Fahmy, A. A. (2014). Networks community detection using artificial bee colony swarm optimization. In *Proc. of the Fifth Int. Conf. on Innov. in Bio-Inspired Comp. and Appl. (IBICA).* Ostrava, Czech Republic: Springer.

Halabian, H., Lambadaris, I., Lung, C. H., & Srinivasan, A. (2010). Dynamic Channel and Interface Management in Multi-channel Multi-interface Wireless Access Networks. In *2010 IEEE Global Telecommunications Conference GLOBECOM 2010* (pp. 1-6). IEEE. 10.1109/GLOCOM.2010.5683566

Halak, B. (2018). Physically Unclonable Functions: Design Principles and Evaluation Metrics. In *Physically Unclonable Functions* (pp. 17–52). Cham: Springer. doi:10.1007/978-3-319-76804-5_2

Handbook, M. S. (1995). MIL-HDBK-217F. Department of Defense, US.

Hao, Z., Kefa, C., & Jianbo, M. (2001). Combining neural network and genetic algorithms to optimize low NOx pulverized coal combustion. *Fuel, 80*(15), 2163–2169. doi:10.1016/S0016-2361(01)00104-1

Harada, K., Katsuki, A., & Fujiwara, M. (1993). Use of ESR for deterioration diagnosis of electrolytic capacitor. *IEEE Transactions on Power Electronics, 8*(4), 355–361. doi:10.1109/63.261004

Haringa, G., Jordan, G., & Garver, L. (1991). Application of Monte Carlo Simulation to Multi-Area Reliability Evaluations. *IEEE Computer Applications in Power, 4*(1), 21–25. doi:10.1109/67.65031

Hashemian, H. M., & Bean, W. C. (2011). State-of-the-art predictive maintenance techniques. *IEEE Transactions on Instrumentation and Measurement, 60*(10), 3480–3492. doi:10.1109/TIM.2009.2036347

Hayward, G., & Davidson, V. (2003). Fuzzy logic applications. *Analyst (London), 128*(11), 1304–1306. doi:10.1039/b312701j PMID:14700220

Hess, W., & Walter, R. (1993, September). A new K-band frequency divider by three using an injection locked oscillator in microstrip technique. In *1993 23rd European Microwave Conference* (pp. 391-393). IEEE. 10.1109/EUMA.1993.336571

Hori, Y., Yoshida, T., Katashita, T., & Satoh, A. (2010, December). Quantitative and statistical performance evaluation of arbiter physical unclonable functions on FPGAs. In *2010 International Conference on Reconfigurable Computing and FPGAs* (pp. 298-303). IEEE. 10.1109/ReConFig.2010.24

Ho, W. K., Hang, C. C., & Cao, L. S. (1995). Tuning of PID controllers based on gain and phase margin specification. *Automatica, 31*(3), 497–502. doi:10.1016/0005-1098(94)00130-B

Høyland, A., & Rausand, M. (1994). *System reliability theory: models and statistical methods*. Wiley.

Hu, Z., Nishimura, K., Chou, H., Rau, L., Usami, M., Bowers, J.E., & Blumenthal, D.J. (2004). *40-Gb/s Optical Packet Clock Recovery Using a Travelling-wave Electroabsorption Modulator-Based Ring Oscillator*. Academic Press.

Huang, X., Denprasert, P. M., Zhou, L., Vest, A. N., Kohan, S., & Loeb, G. E. (2017). Accelerated life-test methods and results for implantable electronic devices with adhesive encapsulation. *Biomedical Microdevices, 19*(3), 46. doi:10.100710544-017-0189-9 PMID:28536859

Hussain & Padma. (2013). Test Vector Generator (TPG) for Low Power Logic Built-In Self-Test (BIST). International Journal of Advanced Research in Electrical. *Electronics and Instrumentation Engineering, 2*, 1634–1640.

Hussin, S. M., Hassan, M. Y., Wu, L., Abdullah, M. P., Rosmin, N., & Ahmad, M. A. (2018). Mixed Integer Linear Programming for Maintenance Scheduling in Power System Planning. *Indonesian Journal of Electrical Engineering and Computer Science, 11*(2), 607–613. doi:10.11591/ijeecs.v11.i2.pp607-613

Huston, H. H., & Clarke, C. P. (1992). *Reliability defect detection and screening during processing-theory and implementation*. Paper presented at the IEEE 30th Annual Symposium on International Reliability Physics, San Diego, CA.

Idriss, T., & Bayoumi, M. (2017, September). Lightweight highly secure PUF protocol for mutual authentication and secret message exchange. In *2017 IEEE International Conference on RFID Technology & Application (RFID-TA)* (pp. 214-219). IEEE. 10.1109/RFID-TA.2017.8098893

Inoue, T., Henmi, H., Yoshikawa, Y., & Ichihara, H. (2011). High-Level Synthesis for Multi-Cycle transient fault Tolerant Datapaths. *Proc. 17 the IEEE International On-Line Testing Symposium*, 13-18. 10.1109/IOLTS.2011.5993804

Iqbal, Z., Potkonjak, M., Dey, S., & Parker, A. C. (1993). Critical Path Minimization Using Retiming and Algebraic Speed-Up. *30th ACM/IEEE Design Automation Conference*, 573-577. 10.1145/157485.165046

Issa, S. (2008). Bit-swapping LFSR for low-power BIST. *Electronics Letters, 44*(6), 401–402. doi:10.1049/el:20083481

Jafari, A. H., & Shahhoseini, H. S. (2015). A Reinforcement Routing Algorithm with Access Selection in the Multi–Hop Multi–Interface Networks. *Journal of Electrical Engineering, 66*(2), 70–78. doi:10.1515/jee-2015-0011

Jain, V., Gupta, S. K., Khatri, V., & Banerjee, G. (2019, January). A 19.3-24.8 GHz Dual-Slope VCO in 65-nm CMOS for Automotive Radar Applications. In *2019 32nd International Conference on VLSI Design and 2019 18th International Conference on Embedded Systems (VLSID)* (pp. 118-123). IEEE.

Jaiswal, K. B., Seshadri, P., & Lakshminarayanan, G. (2015, March). Low power wallace tree multiplier using modified full adder. In *2015 3rd international conference on signal processing, communication and networking (ICSCN)* (pp. 1-4). IEEE. 10.1109/ICSCN.2015.7219880

Jánó, R., & Pitică, D. (2011). *Parameter monitoring of electronic circuits for reliability prediction and failure analysis*. Paper presented at the IEEE 34th International Spring Seminar on Electronics Technology (ISSE), Tratanska Lomnica, Slovakia.

Jeng, J. C., & Lin, S. W. (2012). Robust proportional–integral–derivative controller design for stable/integrating processes with inverse response and time delay. *Industrial & Engineering Chemistry Research, 51*(6), 2652–2665. doi:10.1021/ie201449m

Jiao, Y., Lei, S., Pei, Z., & Lee, E. (2004). Fuzzy adaptive networks in machining process modeling: Surface roughness prediction for turning operations. *International Journal of Machine Tools & Manufacture, 44*(15), 1643–1651. doi:10.1016/j.ijmachtools.2004.06.004

Jing, Z., & Zhan, J. (2008). Fabrication and gas-sensing properties of porous ZnO nanoplates. *Advanced Materials, 20*(23), 4547–4551. doi:10.1002/adma.200800243

Jinno, M., Bandow, S., & Ando, Y. (2004). Multiwalled carbon nanotubes produced by direct current arc discharge in hydrogen gas. *Chemical Physics Letters, 398*(1-3), 256–259. doi:10.1016/j.cplett.2004.09.064

Joo, S. B., Oh, S. E., Sim, T., Kim, H., Choi, C. H., Koo, H., & Mun, J. H. (2014). Prediction of gait speed from plantar pressure using artificial neural networks. *Expert Systems with Applications, 41*(16), 7398–7405. doi:10.1016/j.eswa.2014.06.002

Jose, S., Voogt, F., van der Schaar, C., Nath, S., Nenadović, N., Vanhelmont, F., . . . Šakić, A. (2017). *Reliability tests for modelling of relative humidity sensor drifts.* Paper presented at the IEEE International Symposium on Reliability Physics (IRPS'17), Monterey, CA

Juang, C.F. (2004). A hybrid of genetic algorithm and particle swarm optimization for recurrent network design. *IEEE Trans. Syst. Man. Cybern., B, 34*(2), 997-1006.

Kacprzak, T. (1988). Analysis of oscillatory metastable operation of an RS flip-flop. *IEEE Journal of Solid-State Circuits, 23*(1), 260–266. doi:10.1109/4.287

Kajal, S., Tewari, P. C., & Saini, P. (2013). Availability optimization for coal handling system using a genetic algorithm. *International Journal of Performability Engineering, 9*(1), 109–116.

Kalbfleisch, J. D., & Prentice, R. L. (2011). *The statistical analysis of failure time data* (Vol. 360). John Wiley & Sons.

Kamboj, Bhardwaj, Bhullar, Arora, & Kaur. (2012). Mathematical model of reliability assessment for generation system. *2012 IEEE International Power Engineering and Optimization Conference*, 258-262. 10.1109/PEOCO.2012.6231118

Kananizadeh, R., & Momeni, O. (2017). A 190-GHz VCO with 20.7% tuning range employing an active mode switching block in a 130 nm SiGe BiCMOS. *IEEE Journal of Solid-State Circuits, 52*(8), 2094–2104. doi:10.1109/JSSC.2017.2689031

Kančev, D., Gjorgiev, B., & Čepin, M. (2011). Optimization of test interval for aging equipment: A multi-objective genetic algorithm approach. *Journal of Loss Prevention in the Process Industries, 24*(4), 397–404. doi:10.1016/j.jlp.2011.02.003

Karimi, A., Garcia, D., & Longchamp, R. (2003). PID controller tuning using Bode's integrals, IEEE Transaction. *Control System Technol, 11*(6), 812–821. doi:10.1109/TCST.2003.815541

Karri, R., & Potkonjak, M. (2014). Special issue on emerging nanoscale architectures for hardware security, trust, and reliability: Part 1. *IEEE Transactions on Emerging Topics in Computing, 2*(1), 2–3. doi:10.1109/TETC.2014.2318951

Karuppusami, G., & Gandhinathan, R. (2006). Pareto analysis of critical success factors of total quality management: A literature review and analysis. *The TQM Magazine, 18*(4), 372–385. doi:10.1108/09544780610671048

Kashki, M., Abdel-Magid, Y. L., & Abido, M. L. (2008) A reinforcement learning automata optimization approach for optimum tuning of PID controller in AVR system. ICIC 2008, Advanced Intelligent Comput Theories and Appl. With Aspects of Artificial Intelligence, 684-692.

Kataeva, I., Engseth, H., & Kidiyarova-Shevchenko, A. (2007). Scalable matrix multiplication with hybrid CMOS-RSFQ digital signal processor. *IEEE Transactions on Applied Superconductivity, 17*(2), 486–489. doi:10.1109/TASC.2007.901451

Kaviri, A. G., Jaafar, M. N. M., & Lazim, T. M. (2012). Modeling and multi-objective exergy-based optimization of a combined cycle power plant using a genetic algorithm. *Energy Conversion and Management*, *58*, 94–103. doi:10.1016/j.enconman.2012.01.002

Kececioglu, D., & Tian, X. (1998). Reliability education: A historical perspective. *IEEE Transactions on Reliability*, *47*(3), SP390–SP398. doi:10.1109/24.740556

Keller, T. S., Weisberger, A. M., Ray, J. L., Hasan, S. S., Shiavi, R. G., & Spengler, D. M. (1996). Relationship between vertical ground reaction force and speed during walking, slow jogging, and running. *Clinical Biomechanics (Bristol, Avon)*, *11*(5), 253–259. doi:10.1016/0268-0033(95)00068-2 PMID:11415629

Khan, S., Kakde, S., & Suryawanshi, Y. (2013, December). VLSI implementation of reduced complexity wallace multiplier using energy efficient CMOS full adder. In *2013 IEEE International Conference on Computational Intelligence and Computing Research* (pp. 1-4). IEEE. 10.1109/ICCIC.2013.6724141

Kharchenko, V. A. (2015). Problems of reliability of electronic components. *Modern Electronic Materials*, *1*(3), 88–92. doi:10.1016/j.moem.2016.03.002

Khavari, M., Baur, S., & Boche, H. (2017, October). Optimal capacity region for PUF-based authentication with a constraint on the number of challenge-response pairs. In *2017 IEEE Conference on Communications and Network Security (CNS)* (pp. 575-579). IEEE. 10.1109/CNS.2017.8228679

Khayer, M. A., & Lake, R. K. (2009). The quantum and classical capacitance limits of InSb and InAs nanowire FETs. *IEEE Transactions on Electron Devices*, *56*(10), 2215–2223. doi:10.1109/TED.2009.2028401

Khodr, H. M., Gomez, J. F., Barnique, L., Vivas, J. H., Paiva, P., Yusta, J. M., & Urdaneta, A. J. (2002). A Linear Programming Methodology for the Optimization of Electric Power – Generation Schemes. *IEEE Transactions on Power Systems*, *17*(3), 864–869. doi:10.1109/TPWRS.2002.800982

Kim, D. H., & Cho, J. H. (2006). A biologically inspired intelligent PID controller tuning for AVR systems. *International Journal of Control, Automation, and Systems*, *4*(5), 624–636.

Kim, H., Hong, J., Park, K. Y., Kim, H., Kim, S. W., & Kang, K. (2014). Aqueous rechargeable Li and Na ion batteries. *Chemical Reviews*, *114*(23), 11788–11827. doi:10.1021/cr500232y PMID:25211308

Kim, R., & Lundstrom, M. S. (2008). Characteristic features of 1-D ballistic transport in nanowire MOSFETs. *IEEE Transactions on Nanotechnology*, *7*(6), 787–794. doi:10.1109/TNANO.2008.920196

Kim, Y. A., Muramatsu, H., Hayashi, T., & Endo, M. (2012). Catalytic metal-free formation of multi-walled carbon nanotubes in atmospheric arc discharge. *Carbon*, *50*(12), 4588–4595. doi:10.1016/j.carbon.2012.05.044

Kirby, E. D., & Chen, J. C. (2007). Development of a fuzzy-nets-based surface roughness prediction system in turning operations. *Computers & Industrial Engineering*, *53*(1), 30–42. doi:10.1016/j.cie.2006.06.018

Kirkpatrick, S., & Gelatt, C.D., M. P. V. (1983). Optimization by Simulated Annealing. *Science*, *220*(4598), 671–680. PMID:17813860

Klass, B. K. J. C., & Van, P. (2006). System Reliability: Concepts and Applications. Edward Arnold.

Knoch, J., Riess, W., & Appenzeller, J. (2008). Outperforming the conventional scaling rules in the quantum-capacitance limit. *IEEE Electron Device Letters*, *29*(4), 372–374. doi:10.1109/LED.2008.917816

Korotcenkov, G. (2013). Chemical Sensors: Simulation and Modeling: Vol. 5. *Electrochemical Sensors*. Momentum Press.

Kötz, R., Hahn, M., & Gallay, R. (2006). Temperature behavior and impedance fundamentals of supercapacitors. *Journal of Power Sources*, *154*(2), 550–555. doi:10.1016/j.jpowsour.2005.10.048

Koushanfar, F. (2012). *Hardware metering: A survey. In Introduction to Hardware Security and Trust* (pp. 103–122). New York, NY: Springer. doi:10.1007/978-1-4419-8080-9_5

Kram, R., & Taylor, C. R. (1990). Energetics of running: A new perspective. *Nature*, *346*(6281), 265–267. doi:10.1038/346265a0 PMID:2374590

Krishna, K. G., Santhosh, B., & Sridhar, V. (2013). Design of wallace tree multiplier using compressors. *International journal of engineering sciences & research. Technology*, *2*, 2249–2254.

Kroft, D. (1974). Comments on A Twos Complement Parallel Array Multiplication Algorithm. *IEEE Transactions on Computers*, *C-23*(12), 1327–1328. doi:10.1109/T-C.1974.223863

Kshirsagar, R. D., Aishwarya, E. V., Vishwanath, A. S., & Jayakrishnan, P. (2013, December). Implementation of pipelined booth encoded wallace tree multiplier architecture. In *2013 International Conference on Green Computing, Communication and Conservation of Energy (ICGCE)* (pp. 199-204). IEEE. 10.1109/ICGCE.2013.6823428

Kullstam, P. A. (1981). Availability, MTBF and MTTR for repairable M out of N system. *IEEE Transactions on Reliability*, *30*(4), 393–394. doi:10.1109/TR.1981.5221134

Kulshreshtha, D. C., & Chauhan, D. S. (2009). *Electronics Engineering*. New Age Publications.

Kumar, A., & Mishra, R. S. (2019). Challenge-Response Pair (CRP) Generator Using Schmitt Trigger Physical Unclonable Function. In *Advanced Computing and Communication Technologies* (pp. 213–223). Singapore: Springer. doi:10.1007/978-981-13-0680-8_20

Kumar, A., Mishra, R. S., & Kashwan, K. R. (2016, September). Challenge-response generation using RO-PUF with reduced hardware. In *2016 International Conference on Advances in Computing, Communications and Informatics (ICACCI)* (pp. 1305-1308). IEEE. 10.1109/ICACCI.2016.7732227

Kumar, P., & Tewari, P. C. (2017). Performance analysis and optimization for CSDGB filling system of a beverage plant using particle swarm optimization. *International Journal of Industrial Engineering Computations*, *8*, 303–314. doi:10.5267/j.ijiec.2017.1.002

Kumar, R. (2017). Redundancy optimization of a coal-fired power plant using a simulated annealing technique. *International Journal of Intelligent Enterprise*, *4*(3), 191–203. doi:10.1504/IJIE.2017.087625

Kumar, S. S., Guajardo, J., Maes, R., Schrijen, G. J., & Tuyls, P. (2008, June). The butterfly PUF protecting IP on every FPGA. In *2008 IEEE International Workshop on Hardware-Oriented Security and Trust* (pp. 67-70). IEEE. 10.1109/HST.2008.4559053

Kumganty, S. (1994). Effect of HVDC Component Enhancement on the Overall Reliability Performance. *IEEE Transactions on Power Delivery*, *9*(1).

Kundu, K., Rossini, M., & Portioli-staudacher, A. (2019). A study of kanban assembly line feeding system through integration of simulation and particle swarm optimization. *International Journal of Industrial Engineering and Computations*, *10*, 421–442. doi:10.5267/j.ijiec.2018.12.001

Kundur, P., Balu, N. J., & Lauby, M. G. (1994). *Power system stability and control* (Vol. 7). New York: McGraw-Hill.

Kuo, T. Y., & Wang, J. S. (2008, May). A low-voltage latch-adder based tree multiplier. In *2008 IEEE International Symposium on Circuits and Systems* (pp. 804-807). IEEE.

Lange, H., Sioda, M., Huczko, A., Zhu, Y. Q., Kroto, H. W., & Walton, D. R. M. (2003). Nanocarbon production by arc discharge in water. *Carbon*, *41*(8), 1617–1623. doi:10.1016/S0008-6223(03)00111-8

Lao, Y., & Parhi, K. K. (2011, May). Reconfigurable architectures for silicon physical unclonable functions. In *2011 IEEE International Conference on Electro/Information Technology* (pp. 1-7). IEEE. 10.1109/EIT.2011.5978614

Lao, Y., & Parhi, K. K. (2014). Protecting DSP circuits through obfuscation. *IEEE International Symposium on Circuits and Systems (ISCAS)*, 798-801. 10.1109/ISCAS.2014.6865256

Lao, Y., & Parhi, K. K. (2014). Statistical analysis of MUX-based physical unclonable functions. *IEEE Transactions on Computer-Aided Design of Integrated Circuits and Systems*, *33*(5), 649–662. doi:10.1109/TCAD.2013.2296525

Lao, Y., & Parhi, K. K. (2015). Obfuscating DSP Circuits via High-Level Transformations. *IEEE Transactions on Very Large Scale Integration (VLSI) Systems*, *23*(5), 819–830.

Lapa, C. M. F., Pereira, C. M. N. A., & De Barros, M. P. (2006). A model for preventive maintenance planning by genetic algorithms based on cost and reliability. *Reliability Engineering & System Safety*, *91*(2), 233–240. doi:10.1016/j.ress.2005.01.004

Le May, I. (1998). Product liability and failure analysis. *Technology Law and Insurance*, *3*(2), 163–171. doi:10.1080/135993798349550

Lebby, M. S., Blair, T. H., & Witting, G. F. (1996). *Electronic book*. Google Patents.

Lee, S.-B., Kim, I., & Park, T.-S. (2008). Fatigue and fracture assessment for reliability in electronics packaging. *International Journal of Fracture*, *150*(1-2), 91–104. doi:10.100710704-008-9224-4

Lee, Y., Park, S., Lee, M., & Brosilow, C. (1998). PID controller tuning for desired closed-loop responses for SI/SO systems. *AIChE Journal. American Institute of Chemical Engineers*, *44*(1), 106–115. doi:10.1002/aic.690440112

Lefik, M. (2013). Some aspects of application of artificial neural network for numerical modeling in civil engineering. *Bulletin of the Polish Academy of Sciences. Technical Sciences*, *61*(1), 39–50. doi:10.2478/bpasts-2013-0003

Liao, M. J., Su, C. F., Chang, C. Y., & Wu, A. H. (2002, May). A carry-select-adder optimization technique for high-performance Booth-encoded wallace-tree multipliers. In *2002 IEEE International Symposium on Circuits and Systems. Proceedings (Cat. No. 02CH37353)* (Vol. 1, pp. I-I). IEEE. 10.1109/ISCAS.2002.1009782

Li, J., & Lach, J. (2018). At-speed delay characterization for IC authentication and Trojan horse detection. *Proceedings of the IEEE International Workshop on Hardware-Oriented Security and Trust (HOST '08)*, 8–14.

Li, J., Zhang, Q., Yang, D., & Tian, J. (2004). Fabrication of carbon nanotube field effect transistors by AC dielectrophoresis method. *Carbon*, *42*(11), 2263–2267. doi:10.1016/j.carbon.2004.05.002

Li, J., Zhang, Q., Yan, Y., Li, S., & Chen, L. (2007). Fabrication of carbon nanotube field-effect transistors by fluidic alignment technique. *IEEE Transactions on Nanotechnology*, *6*(4), 481–484. doi:10.1109/TNANO.2007.897868

Liu, C. H., Chan, C. Y., Wang, R. L., & Su, Y. K. (2007, December). Low power current-reused voltage-controlled oscillator with optimum source damping resistors. In *2007 IEEE Conference on Electron Devices and Solid-State Circuits* (pp. 1017-1020). IEEE. 10.1109/EDSSC.2007.4450300

Long, J., Foo, J. Y., & Weber, R. J. (2004, February). A 2.4 GHz low-power low-phase-noise CMOS LC VCO. In *IEEE Computer Society Annual Symposium on VLSI* (pp. 213-214). IEEE. 10.1109/ISVLSI.2004.1339533

Lu, C. J., & Meeker, W. O. (1993). Using degradation measures to estimate a time-to-failure distribution. *Technometrics*, *35*(2), 161–174. doi:10.1080/00401706.1993.10485038

Luu, X. V., Hoang, T. T., Bui, T. T., & Dinh-Duc, A. V. (2014, October). A high-speed unsigned 32-bit multiplier based on booth-encoder and wallace-tree modifications. In *2014 International Conference on Advanced Technologies for Communications (ATC 2014)* (pp. 739-744). IEEE. 10.1109/ATC.2014.7043485

Lu, Y., & Christou, A. (2017). Lifetime Estimation of Insulated Gate Bipolar Transistor Modules Using Two-step Bayesian Estimation. *IEEE Transactions on Device and Materials Reliability*, *17*(2), 414–421. doi:10.1109/TDMR.2017.2694158

Ma, H., & Wang, L. (2005). *Fault diagnosis and failure prediction of aluminum electrolytic capacitors in power electronic converters.* Paper presented at the IEEE 31st Annual Conference on Industrial Electronics Society (IECON 2005), Raleigh, NC.

Macin, V., Tormos, B., Sala, A., & Ramirez, J. (2006). Fuzzy logic-based expert system for diesel engine oil analysis diagnosis. *Insight-Non-Destructive Testing and Condition Monitoring*, *48*(8), 462–469. doi:10.1784/insi.2006.48.8.462

Maes, R. (2013). Physically Unclonable Functions: Properties. In *Physically Unclonable Functions* (pp. 49–80). Berlin: Springer. doi:10.1007/978-3-642-41395-7_3

Maes, R., Van Herrewege, A., & Verbauwhede, I. (2012, September). PUFKY: A fully functional PUF-based cryptographic key generator. In *International Workshop on Cryptographic Hardware and Embedded Systems* (pp. 302-319). Springer. 10.1007/978-3-642-33027-8_18

Maiti, A., & Schaumont, P. (2011). Improved ring oscillator PUF: An FPGA-friendly secure primitive. *Journal of Cryptology*, *24*(2), 375–397. doi:10.100700145-010-9088-4

Majzoobi, M., Rostami, M., Koushanfar, F., Wallach, D. S., & Devadas, S. (2012, May). Slender PUF protocol: A lightweight, robust, and secure authentication by substring matching. In *2012 IEEE Symposium on Security and Privacy Workshops*(pp. 33-44). IEEE. 10.1109/SPW.2012.30

Manke, B. S. (2011). *Linear Control Systems*. New Delhi: Khanna Publishers.

Mann, N. R., Singpurwalla, N. D., & Schafer, R. E. (1974). *Methods for statistical analysis of reliability and life data*. Wiley.

Manogaran, G., Varatharajan, R., & Priyan, M. (2018). Hybrid recommendation system for heart disease diagnosis based on multiple kernel learning with adaptive neuro-fuzzy inference system. *Multimedia Tools and Applications*, *77*(4), 4379–4399. doi:10.100711042-017-5515-y

Manoravi, P., Selvaraj, I. I., Chandrasekhar, V., & Shahi, K. (1993). Conductivity studies of new polymer electrolytes based on the poly (ethylene glycol)/sodium iodide system. *Polymer*, *34*(6), 1339–1341. doi:10.1016/0032-3861(93)90799-G

Mansouri, I., Gholampour, A., Kisi, O., & Ozbakkaloglu, T. (2018). Evaluation of peak and residual conditions of actively confined concrete using neuro-fuzzy and neural computing techniques. *Neural Computing & Applications*, *29*(3), 873–888. doi:10.100700521-016-2492-4

Ma, R., Yu, N., & Hu, J. (2013). Application of Particle Swarm Optimization Algorithm in the Heating System Planning Problem. *The Scientific World Journal*, 1–13. PMID:23935429

Martin, P. L. (1999). *Electronic failure analysis handbook: techniques and applications for electronic and electrical packages, components, and assemblies*. McGraw-Hill Professional Publishing.

Mazhar, M., Kara, S., & Kaebernick, H. (2007). Remaining life estimation of used components in consumer products: Life cycle data analysis by Weibull and artificial neural networks. *Journal of Operations Management*, *25*(6), 1184–1193. doi:10.1016/j.jom.2007.01.021

McMahon, T. A., Valiant, G., & Frederick, E. C. (1987). Groucho running. *Journal of Applied Physiology, 62*(6), 2326–2337. doi:10.1152/jappl.1987.62.6.2326 PMID:3610929

McMillan, G. K. (2011). *Industrial applications of PID control, PID control in the third Millennium: Lessons learned and new approaches.* Springer.

Miller, R., & Nelson, W. (1983). Optimum simple step-stress plans for accelerated life testing. *IEEE Transactions on Reliability, 32*(1), 59–65. doi:10.1109/TR.1983.5221475

Mirajkar, P., Chand, J., Aniruddhan, S., & Theertham, S. (2017). Low phase noise Ku-band VCO with optimal switched-capacitor bank design. *IEEE Transactions on Very Large Scale Integration (VLSI) Systems, 26*(3), 589–593.

Mirjalili, S. (2015). Moth-flame optimization algorithm: A novel nature-inspired heuristic paradigm. *Knowledge Syst., 89*, 228–249. doi:10.1016/j.knosys.2015.07.006

Misra, K. B. (2008). *Handbook of performability engineering.* Springer Science & Business Media. doi:10.1007/978-1-84800-131-2

Mohagheghi, S., Harley, R. G., & Venayagamoorthy, G. K. (2004). Modified Takagi–Sugeno fuzzy logic based controllers for a static compensator in a multi machine power system. *Proc. 39th IAS annual meeting conf.*, 2637-2642.

Mohamed, A. W., Sabry, H. Z., & Abd-Elaziz, T. (2013). Real parameter optimization by an effective differential evolution algorithm. *Egyptian Informatics Journal, 14*(1), 37–53. doi:10.1016/j.eij.2013.01.001

Mohanta, D. K., Sadhu, P. K., & Chakrabarti, R. (2007). A deterministic and stochastic approach for safety and reliability optimization of captive power plant maintenance scheduling using GA/SA-based hybrid techniques: A comparison of results. *Reliability Engineering & System Safety, 92*(2), 187–199. doi:10.1016/j.ress.2005.11.062

Mohanty, P. K., Sahu, B. K., & Panda, S. (2014). Tuning and assessment of proportional-integral-derivative controller for an automatic voltage regulator system employing local unimodal sampling algorithm. *Electric Power Components and Systems, 42*(9), 959–969. doi:10.1080/15325008.2014.903546

Momen, H. G., Yazgi, M., & Kopru, R. (2015, November). Designing a new high Q fully CMOS tunable floating active inductor based on modified tunable grounded active inductor. In *2015 9th international conference on electrical and electronics engineering (ELECO)* (pp. 1-5). IEEE. 10.1109/ELECO.2015.7394542

Moore, G. (1998). Cramming More Components Onto Integrated Circuits. *Proceedings of the IEEE, 86*(1), 82–85. doi:10.1109/JPROC.1998.658762

Moore, G. E. (2006). Cramming more components onto integrated circuits. *IEEE Solid-State Circuits Society Newsletter, 20*(3), 33–35. doi:10.1109/N-SSC.2006.4785860

Mukherjee, V. (2009). *Application of evolutionary optimization techniques for some selected power system problems* (Ph.D. dissertation). Durgapur, India: Dept Electrical Engg, National Institute of Technology.

Mukherjee, V., & Ghoshal, S. P. (2007). Intelligent particle swarm optimized fuzzy PID controller for AVR system. *Electric Power Systems Research, 77*(12), 1689–1698. doi:10.1016/j.epsr.2006.12.004

Munoz-Estrada. Perez-Ruiz, & Barquin. (2004). *Including Combined- Cycle Power Plants in Generation System Reliability Studies.* In The 8th International Conference on Probabilistic Methods Applied to Power Systems, Iowa State University, Ames, IA. doi:10.1007/978-94-015-7949-0

Munro, C. F., Miller, D. I., & Fuglevand, A. J. (1987). Ground reaction forces in running: A reexamination. *Journal of Biomechanics, 20*(2), 147–155. doi:10.1016/0021-9290(87)90306-X PMID:3571295

314

Murthy, D. (2007). Product reliability and warranty: An overview and future research. *Production, 17*(3), 426–434. doi:10.1590/S0103-65132007000300003

Nachtigal, M., Thapliyal, H., & Ranganathan, N. (2010, August). Design of a reversible single precision floating point multiplier based on operand decomposition. In *10th IEEE International Conference on Nanotechnology* (pp. 233-237). IEEE. 10.1109/NANO.2010.5697746

Nagarajan, M., Ma, K., Seng, Y. K., Mou, S. X., & Kumar, T. B. (2011, September). A low power wide tuning range low phase noise VCO using coupled LC tanks. In *2011 Semiconductor Conference Dresden* (pp. 1-4). IEEE. doi:10.1109/SCD.2011.6068769

Nakamura, S., Harada, Y., & Seno, M. (1991). Novel metalorganic chemical vapor deposition system for GaN growth. *Applied Physics Letters, 58*(18), 2021–2023. doi:10.1063/1.105239

Nandi, R., Pattanayak, S., Venkateswaran, P., & Das, S. (2015). Electronically tunable differential integrator: linear voltage controlled quadrature oscillator. *International Scholarly Research Notices, 2015*.

Nayak, P., Bhavani, V., & Shanthi, M. (2016). A Fuzzy Logic based Dynamic Channel allocation Scheme for wireless Cellular networks to optimize the frequency reuse. In *2016 IEEE Region 10 Conference (TENCON)* (pp. 1111-1116). IEEE. doi:10.1109/TENCON.2016.7848181

Nazari, M., Kashanian, S., Moradipour, P., & Maleki, N. (2018). A novel fabrication of sensor using ZnO-Al2O3 ceramic nanofibers to simultaneously detect catechol and hydroquinone. *Journal of Electroanalytical Chemistry, 812*, 122–131. doi:10.1016/j.jelechem.2018.01.058

Neophytou, N., & Kosina, H. (2012). Numerical study of the thermoelectric power factor in ultra-thin Si nanowires. *Journal of Computational Electronics, 11*(1), 29–44. doi:10.100710825-012-0383-1

Neri, L., Allen, V., & Anderson, R. (1979). Reliability based quality (RBQ) technique for evaluating the degradation of reliability during manufacturing. *Microelectronics and Reliability, 19*(1-2), 117–126. doi:10.1016/0026-2714(79)90369-X

Nguyen, M. C., & Won, H. S. (2017, February). Gateway-based access interface management in big data platform. In *2017 19th International Conference on Advanced Communication Technology (ICACT)* (pp. 447-450). IEEE. 10.23919/ICACT.2017.7890128

Nguyen, D., & Bagajewicz, M. (2008). Optimization of Preventive Maintenance Scheduling in Processing Plants. *European Symposium on Computer Aided Process Engineering*, 319–324. 10.1016/S1570-7946(08)80058-2

Nguyen, D., & Bagajewicz, M. (2010). Optimization of Preventive Maintenance in Chemical Process Plants. *Industrial & Engineering Chemistry Research, 49*(9), 4329–4339. doi:10.1021/ie901433b

Nguyen-Vuong, Q. T., Agoulmine, N., & Ghamri-Doudane, Y. (2008). A user-centric and context-aware solution to interface management and access network selection in heterogeneous wireless environments. *Computer Networks, 52*(18), 3358–3372. doi:10.1016/j.comnet.2008.09.002

Nie, T., & Toyonaga, M. (2007). An efficient and reliable watermarking system for IP Protection. *IEICE Transactions on Fundamentals of Electronics, Communications and Computer Science, E90-A*(9), 1932–1939. doi:10.1093/ietfec/e90-a.9.1932

Nyambati, E. T., & Oduol, V. K. (2017). Analysis of The Impact of Fuzzy Logic Algorithm On Handover Decision in A Cellular Network. *International Journal of Innovation Education and Research, 5*(5), 46–62.

O'Connor, P. D., O'Connor, P., & Kleyner, A. (2012). *Practical reliability engineering*. John Wiley & Sons.

Ocak, M. (2014). *Implementation of an internet of things device management interface.* Academic Press.

Ogata, K. (2010). *Modern Control Engineering.* PHI Learning.

Oh, S. E., Choi, A., & Mun, J. H. (2013). Prediction of ground reaction forces during gait based on kinematics and a neural network model. *Journal of Biomechanics, 46*(14), 2372–2380. doi:10.1016/j.jbiomech.2013.07.036 PMID:23962528

Oh, W. C., & Chen, M. L. (2008). Synthesis and characterization of CNT/TiO$_2$ composites thermally derived from MWCNT and titanium (IV) n-butoxide. *Bulletin of the Korean Chemical Society, 29*(1), 159–164. doi:10.5012/bkcs.2008.29.1.159

Oliveira, G., Cunha, S., & Pereira, M. (1987). A Direct Method for Multi-Area Reliability Evaluation. *IEEE Trans. on PWRS, 2*(4), 934–942.

Onomi, T., Yanagisawa, K., Seki, M., & Nakajima, K. (2001). Phase-mode pipelined parallel multiplier. *IEEE Transactions on Applied Superconductivity, 11*(1), 541–544. doi:10.1109/77.919402

Onwunalu, J. E., & Durlofsky, L. J. (2010). Application of a particle swarm optimization algorithm for determining optimum well location and type. *Computers & Geosciences, 14*(1), 183–198. doi:10.100710596-009-9142-1

Ortiz, M., Leroy, Y., & Needleman, A. (1987). A finite element method for localized failure analysis. *Computer Methods in Applied Mechanics and Engineering, 61*(2), 189–214. doi:10.1016/0045-7825(87)90004-1

Panda, R. C. (2008). Synthesis of PID tuning rule using the desired closed loop response. *Industrial & Engineering Chemistry Research, 47*(22), 8684–8692. doi:10.1021/ie800258c

Panda, R. C. (2009). Synthesis of PID controller for unstable and integrating processes. *Chemical Engineering Science, 64*(12), 2807–2816. doi:10.1016/j.ces.2009.02.051

Panda, S., Sahu, B. K., & Mohanty, P. K. (2012). Design and performance analysis of PID controller for an automatic voltage regulator system using simplified particle swarm optimization. *Journal of the Franklin Institute, 349*(8), 2609–2625. doi:10.1016/j.jfranklin.2012.06.008

Pan, S., & Anwar, N. M. (2013). A frequency response matching method for PID controller design for industrial processes with time delay, ICAC3 2013. *CCIS, 361*, 636–646.

Pant, S., Anand, D., Kishor, A., & Singh, B. (2015). A Particle Swarm Algorithm for Optimization of Complex System Reliability. *International Journal of Performability Engineering, 11*(1), 33–42.

Pappu, R., Recht, B., Taylor, J., & Gershenfeld, N. (2002). Physical one-wa functions. *Science, 297*(5589), 2026–2030. doi:10.1126cience.1074376 PMID:12242435

Paradhasaradhi, D., Prashanthi, M., & Vivek, N. (2014, March). Modified wallace tree multiplier using efficient square root carry select adder. In *2014 International Conference on Green Computing Communication and Electrical Engineering (ICGCCEE)* (pp. 1-5). IEEE. 10.1109/ICGCCEE.2014.6922214

Parhi, K. (1989). Algorithm transformation techniques for concurrent processors. *Proceedings of the IEEE, 77*(12), 1879–1895. doi:10.1109/5.48830

Parhi, K., & Messerschmitt, D. (1989). Pipeline interleaving and parallelism in recursive digital filters. I. Pipelining using scattered look-ahead and decomposition. *IEEE Transactions on Acoustics, Speech, and Signal Processing, 37*(7), 1099–1117. doi:10.1109/29.32286

Parhi, K., & Messerschmitt, D. (1991). Static rate-optimal scheduling of iterative data-flow programs via optimum unfolding. *IEEE Transactions on Computers, 40*(2), 178–195. doi:10.1109/12.73588

Parhi, K., Wang, C.-Y., & Brown, A. (1992). Synthesis of control circuits in folded pipelined DSP architectures. *IEEE Journal of Solid-State Circuits, 27*(1), 29–43. doi:10.1109/4.109555

Parler, S. G. (1999). *Thermal modeling of aluminum electrolytic capacitors.* Paper presented at the IEEE 34th Annual Meeting on Industry Applications Conference, Phoenix, AZ. 10.1109/IAS.1999.799180

Pascoe, N. (2011). *Essential Reliability Technology Disciplines in Development.* Northern Telecomm Europe Ltd. doi:10.1002/9780470980101

Peacock, A. J., & Calhoun, A. (2012). *Polymer Chemistry: Properties and Application.* Carl Hanser Verlag GmbH Co KG.

Pecht, M. (2009). *Product reliability, maintainability, and supportability handbook.* CRC Press. doi:10.1201/9781420009897

Pecht, M., & Gu, J. (2009). Physics-of-failure-based prognostics for electronic products. *Transactions of the Institute of Measurement and Control, 31*(3-4), 309–322. doi:10.1177/0142331208092031

Peng, S., & Vayenas, N. (2014). *Maintainability Analysis of Underground Mining Equipment Using Genetic Algorithms: Case Studies with an LHD Vehicle.* Academic Press.

Peng, H., & Ling, X. (2008). Optimal design approach for the plate-fin heat exchangers using neural networks cooperated with genetic algorithms. *Applied Thermal Engineering, 28*(5–6), 642–650. doi:10.1016/j.applthermaleng.2007.03.032

Persson, P., & Astrom, K. J. (1992). Dominant pole design – A unified view of PID controller tuning. *IFAC Proc., 25*(14), 377-382.

Potkonjak. (2010). Synthesis of trustable ICs using untrusted CAD tools. *Proceedings of the Design Automation Conference*, 633–634.

Potkonjak, M., Nahapetian, A., Nelson, M., & Massey, T. (2009). Hardware trojan horse detection using gate-level characterization. *Proceedings of the 46th Design Automation Conference (DAC '09)*, 688–693. 10.1145/1629911.1630091

Pradhan, D. K., Samantaray, B. K., Choudhary, R. N. P., & Thakur, A. K. (2005). Effect of plasticizer on microstructure and electrical properties of a sodium ion conducting composite polymer electrolyte. *Ionics, 11*(1-2), 95–102. doi:10.1007/BF02430407

Praveen, & Shiva, & Kurian. (2013). Improved Design of Low Power TPG Using LPLFSR. *International Journal of Computer & Organization Trends, 3*, 101–106.

Praveen, & Swamy, & Shanmukha. (2018). Design of BIST with Low Power Test Vector Generator. *Journal of Circuits, Systems, and Computers, 27*, 1–18. doi:10.1142/S0218126618500780

Prithviraj, A., Krishnamoorthy, K., & Vinothini, B. (2016). Fuzzy Logic-Based Decision-Making Algorithm to Optimize the Handover Performance in HetNets. *Circuits and Systems, 7*(11), 3756–3777. doi:10.4236/cs.2016.711315

Puente, J., Pino, R., Priore, P., & de la Fuente, D. (2002). A decision support system for applying failure mode and effects analysis. *International Journal of Quality & Reliability Management, 19*(2), 137–150. doi:10.1108/02656710210413480

Pushpalatha, M., Venkataraman, R., & Ramarao, T. (2009). Trust based energy aware reliable reactive protocol in mobile ad hoc networks. *World Academy of Science, Engineering and Technology, 56*(68), 356–359.

Qian, R., Wu, Y., Duan, X., Kong, G., & Long, H. (2018). SVM multi-classification optimization research based on multi-chromosome genetic algorithm. *International Journal of Performability Engineering, 14*(4), 631–638.

Qin, Y., Chung, H. S., Lin, D., & Hui, S. (2008). *Current source ballast for high power lighting emitting diodes without electrolytic capacitor.* Paper presented at the IEEE 34th Annual Conference on Industrial Electronics (IECON'08), Orlando, FL.

Racanelli, M., & Kempf, P. (2006, January). Silicon foundry technology for RF products. In *Digest of Papers. 2006 Topical Meeting on Silicon Monolithic Integrated Circuits in RF Systems* (pp. 5-pp). IEEE.

Raheja, D. G., & Gullo, L. J. (2012). *Design for reliability.* John Wiley & Sons. doi:10.1002/9781118310052

Rajeev, D., Dinakaran, D., & Singh, S. (2017). Artificial neural network based tool wear estimation on dry hard turning processes of AISI4140 steel using coated carbide tool. *Bulletin of the Polish Academy of Sciences. Technical Sciences, 65*(4), 553–559. doi:10.1515/bpasts-2017-0060

Rajendran, J., Jyothi, V., & Karri, R. (2011). Blue team red team approach to hardware trust assessment. *2011 IEEE 29th International Conference on Computer Design (ICCD).*

Ramakrishnan, A., Syrus, T., & Pecht, M. (2000). Electronic Hardware Reliability. In The RF and Microwave Handbook (pp. 102-110). Academic Press. doi:10.1201/9781420036879.ch22

Rao, A. S., Rao, V. S. R., & Chidambaram, M. (2009). Direct synthesis-based controller design for integrating processes with time delay. *Journal of the Franklin Institute, 346*(1), 38–56. doi:10.1016/j.jfranklin.2008.06.004

Rao, B. (1996). *Handbook of condition monitoring.* Elsevier.

Rao, M. J., & Dubey, S. (2012, December). A high speed and area efficient Booth recoded Wallace tree multiplier for Fast Arithmetic Circuits. In *2012 Asia Pacific Conference on Postgraduate Research in Microelectronics and Electronics* (pp. 220-223). IEEE. 10.1109/PrimeAsia.2012.6458658

Rausand, M., & Arnljot, H. (2004). *System reliability theory: models, statistical methods, and applications* (Vol. 396). John Wiley & Sons.

Razavieh, A., Singh, N., Paul, A., Klimeck, G., Janes, D., & Appenzeller, J. (2011, June). A new method to achieve RF linearity in SOI nanowire MOSFETs. In *2011 IEEE Radio Frequency Integrated Circuits Symposium* (pp. 1-4). IEEE. 10.1109/RFIC.2011.5940626

Reddy, B. M., Sheshagiri, H. N., Vijayakumar, B. R., & Shanthala, S. (2014, December). Implementation of Low Power 8-Bit Multiplier Using Gate Diffusion Input Logic. In *2014 IEEE 17th International Conference on Computational Science and Engineering* (pp. 1868-1871). IEEE.

Reiter, J., Velická, J., & Míka, M. (2008). Proton-conducting polymer electrolytes based on methacrylates. *Electrochimica Acta, 53*(26), 7769–7774. doi:10.1016/j.electacta.2008.05.066

ResearchGate. (n.d.). Retrieved from https://www.researchgate.net/figure/The-gait-cycle-A-schematic-representation-of-gait-cycle-with-stance-red-and-swing_fig1_249968026

Reyneri, L. M., Del Corso, D., & Sacco, B. (1990). Oscillatory metastability in homogeneous and inhomogeneous flip-flops. *IEEE Journal of Solid-State Circuits, 25*(1), 254–264. doi:10.1109/4.50312

Rivera, D. E., Morari, M., & Skogestad, S. (1986). Internal model control 4. PID controller design. *Industrial & Engineering Chemistry Process Design and Development, 25*(1), 252–265. doi:10.1021/i200032a041

Rostami, M., Koushanfar, F., Rajendran, J., & Karri, R. (2013, November). Hardware security: Threat models and metrics. In *Proceedings of the International Conference on Computer-Aided Design,* (pp. 819-823). IEEE Press.

Roszkowska, E. (2011). Multi-criteria decision making models by applying the TOPSIS method to crisp and interval data. *Multiple Criteria Decision Making, 6*, 200-230.

Roy, P., Mahapatra, G. S., & Dey, K. N. (2017). An Efficient Particle Swarm Optimization-Based Neural Network Approach for Software Reliability Assessment. *International Journal of Reliability Quality and Safety Engineering, 24*(4), 1–24. doi:10.1142/S021853931750019X

Rührmair, U., Sölter, J., & Sehnke, F. (2009). On the Foundations of Physical Unclonable Functions. *IACR Cryptology ePrint Archive, 2009*, 277.

Rührmair, U., Devadas, S., & Koushanfar, F. (2012). Security based on physical unclonability and disorder. In *Introduction to Hardware Security and Trust* (pp. 65–102). New York, NY: Springer. doi:10.1007/978-1-4419-8080-9_4

Rupérez, M. J., Martín-Guerrero, J. D., Monserrat, C., & Alcañiz, M. (2012). Artificial neural networks for predicting dorsal pressures on the foot surface while walking. *Expert Systems with Applications, 39*(5), 5349–5357. doi:10.1016/j.eswa.2011.11.050

Saadat, H. (2002). *Power System Analysis*. New Delhi: Tata McGraw Hill Ltd.

Saber, A. Y., Yare, Y., Member, S., Venayagamoorthy, G. K., & Member, S. (2009). Economic Dispatch of a Differential Evolution Based Generator Maintenance Scheduling of a Power System. *Proceedings of IEE Power and Energy Society General Meeting Conference*, 1–8.

Sahib, M. A. (2015). A novel optimal PID plus second order derivative controller for AVR system. *J Eng Sc Technol., 18*(2), 194–206.

Sakata, K., Kobayashi, A., Fukaumi, T., Nishiyama, T., & Arai, S. (1995). *Solid electrolytic capacitor and method for manufacturing the same*. Google Patents.

Samanta, B. (2004). Gear fault detection using artificial neural networks and support vector machines with genetic algorithms. *Mechanical Systems and Signal Processing, 18*(3), 625–644. doi:10.1016/S0888-3270(03)00020-7

Samanta, B., & Al-Balushi, K. (2003). Artificial neural network based fault diagnostics of rolling element bearings using time-domain features. *Mechanical Systems and Signal Processing, 17*(2), 317–328. doi:10.1006/mssp.2001.1462

Saraiva, J. T., Pereira, M. L., Mendes, V. T., & Sousa, J. C. (2011). A Simulated Annealing based approach to solve the generator maintenance scheduling problem. *Electric Power Systems Research, 81*(7), 1283–1291. doi:10.1016/j.epsr.2011.01.013

Sarma, R., Bhargava, C., Dhariwal, S., & Jain, S. (2019). UCM: A Novel Approach for Delay Optimization. *International Journal of Performability Engineering, 15*(4).

Savelberg, H. H. C. M., & De Lange, A. L. H. (1999). Assessment of the horizontal, fore-aft component of the ground reaction force from insole pressure patterns by using artificial neural networks. *Clinical Biomechanics (Bristol, Avon), 14*(8), 585–592. doi:10.1016/S0268-0033(99)00036-4 PMID:10521642

Schaub, A., Danger, J. L., Guilley, S., & Rioul, O. (2018, August). An improved analysis of reliability and entropy for delay PUFs. In *2018 21st Euromicro Conference on Digital System Design (DSD)* (pp. 553-560). IEEE. 10.1109/DSD.2018.00096

Schmid, H., Björk, M. T., Knoch, J., Riel, H., Riess, W., Rice, P., & Topuria, T. (2008). Patterned epitaxial vapor-liquid-solid growth of silicon nanowires on Si (111) using silane. *Journal of Applied Physics, 103*(2), 024304. doi:10.1063/1.2832760

Schulze, A., Hantschel, T., Eyben, P., Verhulst, A. S., Rooyackers, R., Vandooren, A., ... Vandervorst, W. (2011). Observation of diameter dependent carrier distribution in nanowire-based transistors. *Nanotechnology*, *22*(18), 185701. doi:10.1088/0957-4484/22/18/185701 PMID:21415466

Seborg, D. E., Edgar, T. F., & Mellichamp, D. A. (1989). *Process dynamics and control*. New York: Willey.

Sengupta, A., & Roy, D. (2017). Protecting an intellectual property core during architectural synthesis using high-level transformation based obfuscation. *Electronics Letters*. doi:10.1049/el.2017.1329

Sengupta, A., Roy, D., Mohanty, S., & Corcoran, P. (2017). DSP Design Protection in CE through Algorithmic Transformation Based Structural Obfuscation. *IEEE Transactions on Consumer Electronics*, *63*(4), 467–476. doi:10.1109/TCE.2017.015072

Sepulveda, F., Wells, D. M., & Vaughan, C. L. (1993). A neural network representation of electromyography and joint dynamics in human gait. *Journal of Biomechanics*, *26*(2), 101–109. doi:10.1016/0021-9290(93)90041-C PMID:8429053

Seth, S. (2014). MExS A Fuzzy Rule Based Medical Expert System To Diagnose The Diseases. *IOSR Journal of Engineering*, *4*(7), 57–62. doi:10.9790/3021-04735762

Shamsuzzoha, M. (2015). A unified approach for proportional–integral– derivative controller design for time delay processes. *Korean Journal of Chemical Engineering*, *32*(4), 583–596. doi:10.100711814-014-0237-6

Shamsuzzoha, M., & Lee, M. (2007). IMC–PID controller design for improved disturbance rejection of time-delayed processes. *Industrial & Engineering Chemistry Research*, *46*(7), 2077–2091. doi:10.1021/ie0612360

Shanthi, S., Poovaragan, S., Arularasu, M., Nithya, S., Sundaram, R., Magdalane, C. M., ... Maaza, M. (2018). Optical, Magnetic and Photocatalytic Activity Studies of Li, Mg and Sr Doped and Undoped Zinc Oxide Nanoparticles. *Journal of Nanoscience and Nanotechnology*, *18*(8), 5441–5447. doi:10.1166/jnn.2018.15442 PMID:29458596

Siddiqui, M., & Çağlar, M. (1994). Residual lifetime distribution and its applications. *Microelectronics and Reliability*, *34*(2), 211–227. doi:10.1016/0026-2714(94)90104-X

Si, F., Romero, C. E., Yao, Z., Schuster, E., Xu, Z., Morey, R. L., & Liebowitz, B. N. (2009). Optimization of coal-fired boiler SCRs based on modified support vector machine models and genetic algorithms. *Fuel*, *88*(5), 806–816. doi:10.1016/j.fuel.2008.10.038

Simsek, M., Bennis, M., & Guvenc, I. (2015). Mobility management in HetNets: A learning-based perspective. *EURASIP Journal on Wireless Communications and Networking*, *2015*(1), 26. doi:10.118613638-015-0244-2

Singh, A., & Mourelatos, Z. P. (2010). On the Time-Dependent Reliability of Non-Monotonic, Non-Repairable Systems. *SAE International Journal of Materials and Manufacturing, 3*, 425-444.

Singh, A. K., De, B. P., & Maity, S. (2012). Design and Comparison of Multipliers Using Different Logic Styles. *International Journal of Soft Computing and Engineering*, *2*(2), 374–379.

Sinteza, K. I. M., & Galun-Lete, N. (2013). Synthesis, Characterization and Sensing Application of a Solid Alum/Fly Ash Composite Electrolyte. *Materiali in Tehnologije*, *47*(4), 467–471.

Skogestad, S. (2003). Simple analytic rules for model reduction and PID controller tuning. *Journal of Process Control*, *13*(4), 291–309. doi:10.1016/S0959-1524(02)00062-8

Solomon, R., Sandborn, P. A., & Pecht, M. G. (2000). Electronic part life cycle concepts and obsolescence forecasting. *IEEE Transactions on Components and Packaging Technologies*, *23*(4), 707–717. doi:10.1109/6144.888857

Sortrakul, N., Nachtmann, H. L., & Cassady, C. R. (2005). Genetic algorithms for integrated preventive maintenance planning and production scheduling for a single machine. *Computers in Industry*, *56*(2), 161–168. doi:10.1016/j.compind.2004.06.005

Sousa, L. A. (2003). Algorithm for modulo (2^n+ 1) multiplication. *Electronics Letters*, *39*(9), 752–754. doi:10.1049/el:20030467

Srinivasan, S., Gander, R. E., & Wood, H. C. (1992). A movement pattern generator model using artificial neural networks. *IEEE Transactions on Biomedical Engineering*, *39*(7), 716–722. doi:10.1109/10.142646 PMID:1516938

Srinivasan, V., & Weidner, J. W. (1999). Mathematical modeling of electrochemical capacitors. *Journal of the Electrochemical Society*, *146*(5), 1650–1658. doi:10.1149/1.1391821

Stringfellow, G. B. (1985). Organometallic vapor-phase epitaxial growth of III–V semiconductors. *Semiconductors and Semimetals*, *22*, 209–259. doi:10.1016/S0080-8784(08)62930-0

Stroud Charles, E. (2002). *A Designer's Guide to Built-in Self-Test*. Springer.

Subban, R. H. Y., & Arof, A. K. (2003). Experimental investigations on PVC-LiCF$_3$SO$_3$-SiO$_2$ composite polymer electrolytes. *Journal of New Materials for Electrochemical Systems*, *6*(3), 197–203.

Su, F. C., & Wu, W. L. (2000). Design and testing of a genetic algorithm neural network in the assessment of gait patterns. *Medical Engineering & Physics*, *22*(1), 67–74. doi:10.1016/S1350-4533(00)00011-4 PMID:10817950

Suh, G. E., & Devadas, S. (2007, June). Physical unclonable functions for device authentication and secret key generation. In *2007 44th ACM/IEEE Design Automation Conference* (pp. 9-14). IEEE.

Surawijaya, A., Anshori, I., Rohiman, A., & Idris, I. (2011, July). Silicon nanowire (SiNW) growth using Vapor Liquid Solid method with gold nanoparticle (Au-np) catalyst. In *Proceedings of the 2011 International Conference on Electrical Engineering and Informatics* (pp. 1-3). IEEE. 10.1109/ICEEI.2011.6021750

Suresh, K., & Kumarappan, N. (2012). Particle swarm optimization based generation maintenance scheduling using probabilistic approach. *Procedia Engineering*, *30*, 1146–1154. doi:10.1016/j.proeng.2012.01.974

Swidenbank, E., Brown, M. D., & Flynn, D. (1999). Self-tuning turbine generator control for power plant. *Mechatronics*, *9*(5), 513–537. doi:10.1016/S0957-4158(99)00009-4

Tabachnick, B. G., Fidell, L. S., & Osterlind, S. J. (2001). *Using multivariate statistics*. Allyn and Bacon.

Taylor, P., Ashayeri, J., Teelen, A., & Selenj, W. (1996). A production and maintenance planning model for the process industry. *International Journal of Production Research*, *34*(12), 37–41.

Tian, Z., Wong, L., & Safaei, N. (2010). A neural network approach for remaining useful life prediction utilizing both failure and suspension histories. *Mechanical Systems and Signal Processing*, *24*(5), 1542–1555. doi:10.1016/j.ymssp.2009.11.005

Toh, N., Naemura, Y., Makino, H., Nakase, Y., Yoshihara, T., & Horiba, Y. (2001). A 600-MHz 54-bit multiplier with rectangular-styled Wallace tree. *IEEE Journal of Solid-State Circuits*, *36*(2), 249–257. doi:10.1109/4.902765

Toso, M. A., & Gomes, H. M. (2014). Vertical force calibration of smart force platform using artificial neural networks. *Revista Brasileira de Engenharia Biomédica*, *30*(4), 406–411. doi:10.1590/1517-3151.0569

Tran, P. N., & Boukhatem, N. (2008, September). Comparison of MADM decision algorithms for interface selection in heterogeneous wireless networks. In *2008 16th International Conference on Software, Telecommunications and Computer Networks* (pp. 119-124). IEEE.

Trucco, G., & Liberali, V. (2011). Analog design issues for mixed-signal CMOS integrated circuits. In *Advances in Analog Circuits* (pp. 165–180). London, UK: InTech. doi:10.5772/15176

Umamaheshwari, G., Nivedha, M., & Prisci Dorritt, J. (2016). Design of tunable method for PID controller for higher order system. *Int. J. Engg. and Comp Sc, 5*(7), 17239–17242.

Valdés, M., Durán, M. D., & Rovira, A. (2003). Thermoeconomic optimization of combined cycle gas turbine power plants using genetic algorithms. *Applied Thermal Engineering, 23*(17), 2169–2182. doi:10.1016/S1359-4311(03)00203-5

Vanavil, B., Chaitanya, K. K., & Rao, A. S. (2015). Improved PID controller design for unstable time delay processes based on direct synthesis method and maximum sensitivity. *Int. J. Syst. Sci., 46*(8), 1349–1366.

Vara Prasada, R. R., Anjaneya, V. N., Sudhakar, B. G., & Murali, M. C. (2013). Power Optimization of Linear Feedback Shift Register (LFSR) for Low Power BIST implemented in HDL. *International Journal of Modern Engineering Research, 3*, 1523–1528.

Varde, P. (2010). Physics-of-failure based approach for predicting life and reliability of electronics components. *Barc Newsletter, 313*.

Venet, P., Perisse, F., El-Husseini, M., & Rojat, G. (2002). Realization of a smart electrolytic capacitor circuit. *IEEE Industry Applications Magazine, 8*(1), 16–20. doi:10.1109/2943.974353

Vichare, N. M., & Pecht, M. G. (2006). Prognostics and health management of electronics. *IEEE Transactions on Components and Packaging Technologies, 29*(1), 222–229. doi:10.1109/TCAPT.2006.870387

Vijayan, V., & Panda, R. C. (2012). Design of PID controllers in double feedback loops for SISO systems with set-point filters. *ISA Transactions, 51*(4), 514–521. doi:10.1016/j.isatra.2012.03.003 PMID:22494496

Virk, S. M., Muhammad, A., & Martinez-Enriquez, A. (2008). *Fault prediction using artificial neural network and fuzzy logic.* Paper presented at the IEEE Seventh Mexican International Conference on Artificial Intelligence (MICAI'08), Atizapan de Zaragoza, Mexico. 10.1109/MICAI.2008.38

Volkanovski, A., Mavko, B., Boševski, T., Čauševski, A., & Čepin, M. (2008). Genetic algorithm optimisation of the maintenance scheduling of generating units in a power system. *Reliability Engineering & System Safety, 93*(6), 779–789. doi:10.1016/j.ress.2007.03.027

Wagner, R. S. (1965). The vapor-liquid-solid mechanism of crystal growth and its application to silicon. *Transactions of the Metallurgical Society of AIME, 233*, 1053–1064.

Wallace, C. S. (1964). A suggestion for a fast multiplier. *IEEE Transactions on Electronic Computers, EC-13*(1), 14–17. doi:10.1109/PGEC.1964.263830

Wang, L., Barnes, T. J. D., & Cluett, W. R. (1995) New frequency domain design method for PID controllers. *IEE Proc. Control Theory Appl., 142*(4), 265–271.

Wang, X. (2009). *Intelligent modeling and predicting surface roughness in end milling.* Paper presented at the IEEE Fifth International Conference on Natural Computation (ICNC'09), Tianjin, China.

Wang, C. S., Wang, C. C., & Chang, T. R. (2013, July). Neural network evaluation for shoe insoles fitness. In *2013 Ninth International Conference on Natural Computation (ICNC)* (pp. 157-162). IEEE. 10.1109/ICNC.2013.6817962

Wang, K., Bannister, M. E., Meyer, F. W., & Parish, C. M. (2017). Effect of starting microstructure on helium plasma-materials interaction in tungsten. *Acta Materialia, 124*, 556–567. doi:10.1016/j.actamat.2016.11.042

Wang, L., Yang, Y., Dong, C., Morosuk, T., & Tsatsaronis, G. (2014). Multi-objective optimization of coal-fired power plants using differential evolution. *Applied Energy, 115*, 254–264. doi:10.1016/j.apenergy.2013.11.005

Wang, L., Yang, Y., Dong, C., Morosuk, T., & Tsatsaronis, G. (2014). Parametric optimization of supercritical coal-fired power plants by MINLP and differential evolution. *Energy Conversion and Management, 85*, 828–838. doi:10.1016/j.enconman.2014.01.006

Wang, Q. G., Hang, C. C., & Bi, Q. (1997). A frequency domain controller design method. *Transactions of the Institution of Chemical Engineers, 75*(1), 64–72. doi:10.1205/026387697523228

Wang, Q. G., Hang, C. C., & Yang, X. P. (2001). Single-loop controller design via IMC principles. *Automatica, 37*(12), 2041–2048. doi:10.1016/S0005-1098(01)00170-4

Wang, Q. G., Lee, T. H., Ho, W. H., Bi, Q., & Zhang, Y. (1999). PID tuning for improved performance. *IEEE Transactions on Control Systems Technology, 7*(4), 457–465. doi:10.1109/87.772161

Wang, Q. G., Zhang, Z., Astrom, K. J., & Chek, L. S. (2009). Guaranteed dominant pole placement with PID controllers. *Journal of Process Control, 19*(2), 349–352. doi:10.1016/j.jprocont.2008.04.012

Wang, Q., & Qu, G. (2018). A Silicon PUF Based Entropy Pump. *IEEE Transactions on Dependable and Secure Computing, 16*(3), 402–414. doi:10.1109/TDSC.2018.2881695

Wei, S., Meguerdichian, S., & Potkonjak, M. (2010). Gate-level characterization: foundations and hardware security applications. *Proceedings of the Design Automation Conference (DAC '10)*, 222–227. 10.1145/1837274.1837332

Wieczorek, W., Florjanczyk, Z., & Stevens, J. R. (1995). Composite polyether based solid electrolytes. *Electrochimica Acta, 40*(13-14), 2251–2258. doi:10.1016/0013-4686(95)00172-B

Wightwick, A., & Halak, B. (2016). Secure communication interface design for IoT applications using the GSM network. In *2016 IEEE 59th International Midwest Symposium on Circuits and Systems (MWSCAS)* (pp. 1-4). IEEE. 10.1109/MWSCAS.2016.7870010

Wind, S. J., Appenzeller, J., & Avouris, P. (2003). Lateral scaling in carbon-nanotube field-effect transistors. *Physical Review Letters, 91*(5), 058301. doi:10.1103/PhysRevLett.91.058301 PMID:12906636

Winter, D. A. (1990). Biomechanics and motor control of human movement Wiley. New York: Academic Press.

Wu, C.-H., Ho, J.-M., & Lee, D.-T. (2004). Travel-time prediction with support vector regression. *IEEE Transactions on Intelligent Transportation Systems, 5*(4), 276–281. doi:10.1109/TITS.2004.837813

Wysocki, P., Vashchenko, V., Celaya, J., Saha, S., & Goebel, K. (2009). *Effect of electrostatic discharge on electrical characteristics of discrete electronic components*. Paper presented at the Prognostics and Health Management Society Annual Conference of the Prognostics and Health, San Diego, CA.

Xie, M., & Lai, C. D. (1996). Reliability analysis using an additive Weibull model with bathtub-shaped failure rate function. *Reliability Engineering & System Safety, 52*(1), 87–93. doi:10.1016/0951-8320(95)00149-2

Xu, W., & Wang, W. (2012). *An adaptive gamma process based model for residual useful life prediction*. Paper presented at the IEEE Conference on Prognostics and System Health Management (PHM 2012), Beijing, China.

Xu, J., Saavedra, C. E., & Chen, G. (2011). An active inductor-based VCO with wide tuning range and high DC-to-RF power efficiency. *IEEE Transactions on Circuits and Wystems. II, Express Briefs, 58*(8), 462–466. doi:10.1109/TCSII.2011.2158713

Xu, T., & Potkonjak, M. (2016). Digital bimodal functions and digital physical unclonable functions: architecture and applications. In *Secure System Design and Trustable Computing* (pp. 83–113). Cham: Springer. doi:10.1007/978-3-319-14971-4_3

Yadav, O. P., Singh, N., Chinnam, R. B., & Goel, P. S. (2003). A fuzzy logic based approach to reliability improvement estimation during product development. *Reliability Engineering & System Safety*, *80*(1), 63–74. doi:10.1016/S0951-8320(02)00268-5

Yamazoe, N., & Shimizu, Y. (1986). Humidity sensors: Principles and applications. *Sensors and Actuators*, *10*(3-4), 379–398. doi:10.1016/0250-6874(86)80055-5

Yang, Y.-m., Ge, Z.-x., & Xu, Y.-c. (2008). *Fault diagnosis of complex systems based on multi-sensor and multi-domain knowledge information fusion.* Paper presented at the IEEE International Conference on Networking, Sensing and Control (ICNSC 2008), Sanya, China. 10.1109/ICNSC.2008.4525374

Yang, G.-B. (1994). Optimum constant-stress accelerated life-test plans. *IEEE Transactions on Reliability*, *43*(4), 575–581. doi:10.1109/24.370223

Yangping, Z., Bingquan, Z., & Dongxin, W. (2000). *Application of genetic algorithms to fault diagnosis in nuclear power plants.* Academic Press.

Yang, X., Xu, B., & Chiu, M. S. (2011). PID controller design directly from plant data. *Industrial & Engineering Chemistry Research*, *50*(3), 1352–1359. doi:10.1021/ie100784k

Yao, Y., Kim, M., Li, J., Markov, I. L., & Koushanfar, F. (2013, March). ClockPUF: Physical Unclonable Functions based on clock networks. In *Proceedings of the Conference on Design, Automation and Test in Europe* (pp. 422-427). EDA Consortium. 10.7873/DATE.2013.095

Ye, H., Lin, M., & Basaran, C. (2002). Failure modes and FEM analysis of power electronic packaging. *Finite Elements in Analysis and Design*, *38*(7), 601–612. doi:10.1016/S0168-874X(01)00094-4

Yi, Q., & Han, J. (2009, July). An improved design method for multi-bits reused booth multiplier. In *2009 4th International Conference on Computer Science & Education* (pp. 1914-1916). IEEE.

Yu, Q., Dofe, J., Zhang, Y., & Frey, J. (2017). Hardware Hardening Approaches Using Camouflaging, Encryption, and Obfuscation. *Hardware IP Security and Trust*, 135–163.

Yuan, X., & Yuan, Y. (2006). Application of cultural algorithm to generation scheduling of hydrothermal systems. *Energy Conversion and Management*, *47*(15-16), 2192–2201. doi:10.1016/j.enconman.2005.12.006

Zhang, J. (2013). FPGA IP protection by binding finite state machine to physical unclonable function. *Proc. 23rd Int. Conf. Field Program. Logic Appl. (FPL)*, 1-4. 10.1109/FPL.2013.6645555

Zhang, K., Sun, M., Lester, D. K., Pi-Sunyer, F. X., Boozer, C. N., & Longman, R. W. (2005). Assessment of human locomotion by using an insole measurement system and artificial neural networks. *Journal of Biomechanics*, *38*(11), 2276–2287. doi:10.1016/j.jbiomech.2004.07.036 PMID:16154415

Zhang, W., Xi, Y., Yang, G., & Xu, X. (2002). Design PID controllers for desired time-domain or frequency-domain response. *ISA Transactions*, *41*(4), 511–520. doi:10.1016/S0019-0578(07)60106-2 PMID:12398281

Zhao, F., Chen, J., & Xu, W. (2009). Condition prediction based on wavelet packet transform and least squares support vector machine methods. *Journal of Process Mechanical Engineering*, *223*(2), 71–79. doi:10.1243/09544089JPME220

Zhao, H., Ru, Z., Chang, X., & Li, S. (2015). Reliability Analysis Using Chaotic Particle Swarm Optimization. *Quality and Reliability Engineering International*, *31*(8), 1537–1552. doi:10.1002/qre.1689

Zheng, Y., Arasu, M. A., Wong, K. W., The, Y. J., Suan, A. P. H., Tran, D. D., . . . Kwong, D. L. (2008, February). A 0.18 m CMOS 802.15. 4a UWB transceiver for communication and localization. In 2008 IEEE International Solid-State Circuits Conference-Digest of Technical Papers (pp. 118-600). IEEE.

Zhou, H. (2017). Structural Transformation-Based Obfuscation. Hardware Protection through Obfuscation, 221–239.

Zhuang, M., & Atherton, D. P. (1993). Automatic tuning of optimum PID controllers. *IEE Proc. D-Control Theory Appl.*, *140*(3), 216–224.

Zhu, H., Li, L., Zhao, Y., Guo, Y., & Yang, Y. (2009). CAS algorithm-based optimum design of PID controller in AVR system. *Chaos Solutions Fract.*, *42*(2), 792–800. doi:10.1016/j.chaos.2009.02.006

Ziegler, J. G., & Nichols, N. B. (1942). Optimum Settings for Automatic Controllers. *Trans. of ASME*, *64*, 759–768.

Zimmermann, H. J. (1991). *Fuzzy Set Theory and its Applications* (2nd ed.). Kluwer Academic Publishers.

About the Contributors

Cherry Bhargava is working as an associate professor and head, VLSI domain, School of Electrical and Electronics Engineering at Lovely Professional University, Punjab, India. She has more than 14 years of teaching and research experience. She is PhD (ECE), IKGPTU, M.Tech (VLSI Design & CAD) Thapar University and B.Tech (Electronics and Instrumentation) from Kurukshetra University. She is GATE qualified with All India Rank 428. She has authored about 50 technical research papers in SCI, Scopus indexed quality journals and national/international conferences. She has seven books related to reliability, artificial intelligence and digital electronics to her credit. She has registered three copyrights and filed two patents. She is recipient of various national and international awards for being outstanding faculty in engineering and excellent researcher. She is an active reviewer and editorial member of various prominent SCI and Scopus indexed journals. She is a lifetime member of IET, IAENG, NSPE, IAOP, WASET and reliability research group. Her area of expertise includes reliability of electronic systems, digital electronics, VLSI design, artificial intelligence and related technologies.

* * *

Mohammad Hossein Ahmadi received his PhD degree from University of Tehran in 2016. Presently, he works as an Assistant Professor at Shahrood University of Technology in the department of Mechanical Engineering. He published more than 170 articles in international journals. Also, he was considered as one of the top 1 percent of most cited researchers in 2017 and 2018. His interest's fields include Artificial intelligence methods, Optimization, Renewable Energy and Energy systems.

Shamik Chatterjee was born in 1988 at Dhanbad, Jharkhand, India. He received the B. Tech. degree in electrical engineering from West Bengal University of Technology, Kolkata, West Bengal, India in 2011 and M. Tech degree in power system engineering from West Bengal University of Technology, Kolkata, West Bengal, India in 2014 and he has received his PhD. Degree from Indian Institute of Technology (Indian School of Mines), Dhanbad. His area of research includes power system operation and control, controller design, evolutionary programming.

Hanumant P. Jagtap is research Scholar of Savitribai Phule Pune University and working Assistant Professor in the department of mechanical engineering. His area of interested are reliability engineering, availability analysis, maintenance optimization and condition monitoring.

Bhavana Jangid was born in 1994 at Rajasthan, India. She received the B.Tech degree in Electrical and Electronics Engineering from RTU, Kota, Rajasthan, India in 2014 and M.Tech degree in Power Systems from Thapar University, Patiala, Punjab, India in 2017. Her research area includes power system optimization, power electronics converter.

Vikram Kumar Kamboj presently working as Associate Professor and Head of Department (Power Systems) in School of Electronics and Electrical Engineering at Lovely Professional University, Phagwara, Punjab, India. He received his Bachelor of Engineering (Instrumentation and Control Engineering) and Master of Technology (Power Systems Engineering) degreewith honours and awarded doctorate degree in 2017. His current research work focuses on Power System Planning and Optimization, Optimal Scheduling and Dispatch of power generating units, Renewable Energy and Smart Grids System, Meta-heuristics and memetic algorithms. His long-term research focus is on Multi-disciplinary design and Optimization, Optimal utilization of Renewable Energy Sources for power generation, evolutionary programming, Artificial Intelligence, Multi Objective Optimization, Wireless Body Area Network, Brain-Machine interfacing and control through meta-heuristics search algorithms and Prosthesis design & control using artificial Intelligence.

Abhishek Kumar completed graduation in Electronics and Communication Engineering from Institution of Engineers(India) and Master in Electronics Engineering from University of Mumbai. He is pursuing Doctorate from Lovely professional university, His research interest includes CMOS VLSI Design, Hardware Security, Crypt analysis.

Ravinder Kumar is presently working as an Associate Professor in Mechanical Engineering Department, Lovely Professional University, Punjab India. His area of research is related to energy conservation, thermal engineering, thermodynamic modeling, reliability engineering and optimization of thermal system.

V. Mukherjee was born in 1970 at Raina, Burdwan, West Bengal, India. He received his graduation in electrical engineering and post graduation in power system from B.E. College, Shibpur, Howrah, India and B.E. College (Deemed University), Shibpur, Howrah, India, respectively. He received his Ph.D. degree from NIT, Durgapur, India. Presently, he is an associate professor in the department of electrical engineering, IIT (ISM), Dhanbad, Jharkhand, India. His research interest is application of soft computing intelligence to various fields of power systems.

Jyotirmoy Pathak is pursuing PhD from lovely professional university. His research area includes VLSI design for circuit and system.

Dipen Rajak has a doctorate in Material and Design Engineering from Indian Institute of Technology (ISM) Dhanbad-India at the age of 27 Year. The academic achievements are manifested in his original research contributions through over 30 papers published in refereed international journals (SCI & Scopus), 05-Patent,01-Design, 01-book chapter, and 01-book.These contributions are well reckoned as shown through numerous citations to his papers. Dr. Rajak has worked as a Conference Chair in "ICMMSE: IR-2018". He has received numerous awards "Young Scientist Awards" 2019, Kasetsart University, Bangkok-Thailand, Young Scientist on Research Excellence and Academic Awards-2019, Mumbai and "Excellence in Research Awards" 2018 Delhi-India, "Outstanding Reviewer Award" from

Elsevier journals in 2016 and "International Travel Grant", "Best Session Paper Award" in International Conferences-2015 in Thailand". Dr. Rajak is associated with foreign Universities and National & International societies, IEI, IIM, ISTE, IETI, DAAAM, IAAM, ISAET, NASS along with that he is an editorial board member and reviewer of refereed international journals. Dr. Rajak is currently working as Assistant Professor of Department of Mechanical Engineering at Sandip Institute of Technology & Research Centre, Nashik. Also, serving as a Nashik-Branch President for International Engineering and Technology Institute, Hong Kong, and the recent work consists of development of aluminium foams for different applications where low density and high specific energy absorption is required in automobile, aerospace and thermal management sectors.

Kamalpreet Sandhu works as Assistant Professor at Lovely Professional University, Phagwara, Punjab. He Did Master's in CAD/CAM from Thapar institute of engineering and technology, Patiala and Bachelor in Mechanical Engineering from Punjabi University, Patiala. . He is an active researcher in the area of Additive Manufacturing (AM) and Bio-mechanics area. His primary focus is on the development of AM products and their applications in the healthcare sector.

Rajkumar Sarma received his B.E. in Electronics and Communications Engineering from Vinayaka Mission's University, Salem, India & M.Tech degree from Lovely Professional University, Phagwara, Punjab and currently pursuing PhD from Lovely Professional University, Phagwara, Punjab. He is an Assistant Professor in the School of Electronics and Electrical Engineering, Lovely Professional University, Punjab since July 2012. His research interests include Analog and Digital VLSI design, Prototype development using FPGA etc. The author has around 15 research publication.

Pardeep Kumar Sharma is working as an assistant professor at Lovely Professional University, Punjab, India. He has more than 13 years of teaching experience in the field of applied chemistry, experimental analysis, design of experiments and reliability prediction. He is currently submitted PhD thesis at Lovely Professional University. He has authored about twenty research papers in SCI, Scopus indexed quality journals and national/international conferences. He has two books to his credit, in the field of reliability. He has filed two patents and two copyrights. He is recipient of various national and international awards. He is an active reviewer of various indexed journals.

Suman Lata Tripathi has completed her Ph.D in the area of microelectronics and VLSI from MNNIT, Allahabad. She did her M.Tech in Electronics Engineering from UP Technical University, Lucknow and B.Tech in Electrical Engineering from Purvanchal University, Jaunpur. She is associated with Lovely Professional University as Professor with more than seventeen years of experience in academics. She has published more than 35 research papers in referred journals and conference. She has organized a number of workshops, summer internships and expert lectures for students. She has worked as session chair, conference steering committee member, editorial board member and reviewer in international/ national IEEE Journal and conferences.

Index

A

accelerated life testing 38, 81, 85
accuracy 36, 39, 58, 81, 83, 98, 105, 120, 129, 140, 151, 161, 228, 265
active inductor 291-293, 296
amorphous 45, 50-52, 113
Arrhenius equation 38
artificial bee colony 264
artificial intelligence 14, 35, 38, 81, 85-86, 92, 97-98, 107, 142, 224-226, 228, 230, 263
artificial neural networks 14, 38, 81, 100, 128, 132, 150
automatic voltage regulator 262-263

B

BIST 175, 180-182, 184-185, 189

C

cadence spectre tool 239
carbon nanotube 56
composites 45, 103, 105, 110
crystalline 45, 50-52, 60, 100, 103, 105, 113

D

DC motor 217
DFT 175, 184
DHT11 12, 14, 81, 84, 90-91, 93-94
DSP architecture 168-169

E

estimation 14, 40, 81-83, 119, 231
experimental 14, 35, 38-40, 83-84, 91, 97, 101

F

fabrication 28, 55-56, 58-59, 100, 102, 120, 165, 177-178, 245
failure modes 8, 17
field effect transistor 56, 59, 61
Field Programmable Gate Array (FPGA) 176

G

gait parameters and patterns 124, 128, 133

H

hardware security 252, 258
hardware security module 252, 258
high speed 101, 176
humidity sensor 12-14, 81, 84, 97, 100, 106, 118-120

I

integrating processes 192
interface management 224, 229, 233-235

L

Linear Quadratic Regulator 262

M

modelling 26, 29, 31, 36-38, 85-86, 88, 92, 94, 98, 100-101, 107, 109, 119, 136, 266
MOSFET 64
MTBF 5-6, 15, 30, 83-84, 99

Recommended Reference Books

Premier Reference Source

Cases on Immersive Virtual Reality Techniques

ISBN: 978-1-5225-5912-2
© 2019; 349 pp.
List Price: $215

Critical Explorations

Cloud Security

Concepts, Methodologies, Tools, and Applications

Information Resources Management Association

ISBN: 978-1-5225-8176-5
© 2019; 2,218 pp.
List Price: $2,950

Premier Reference Source

Smart Devices, Applications, and Protocols for the IoT

ISBN: 978-1-5225-7811-6
© 2019; 317 pp.
List Price: $225

Premier Reference Source

Innovative Solutions and Applications of Web Services Technology

ISBN: 978-1-5225-7268-8
© 2019; 316 pp.
List Price: $215

Premier Reference Source

Code Generation, Analysis Tools, and Testing for Quality

ISBN: 978-1-5225-7455-2
© 2019; 288 pp.
List Price: $205

Research Insights

Ambient Intelligence Services in IoT Environments

Emerging Research and Opportunities

ISBN: 978-1-5225-8973-0
© 2019; 200 pp.
List Price: $195

Do you want to stay current on the latest research trends, product announcements, news and special offers?
Join IGI Global's mailing list today and start enjoying exclusive perks sent only to IGI Global members.
Add your name to the list at **www.igi-global.com/newsletters.**

Ensure Quality Research is Introduced to the Academic Community

Become an IGI Global Reviewer for Authored Book Projects

Premier Reference Source
Emerging GIS Applications for Emergency and Disaster Management

Premier Reference Source
Managerial Strategies and Green Solutions for Project Sustainability

Premier Reference Source
Comparative Approaches to Using R and Python for Statistical Data Analysis

Premier Reference Source
Solutions for High-Touch Communications in a High-Tech World

The overall success of an authored book project is dependent on quality and timely reviews.

In this competitive age of scholarly publishing, constructive and timely feedback significantly expedites the turnaround time of manuscripts from submission to acceptance, allowing the publication and discovery of forward-thinking research at a much more expeditious rate. Several IGI Global authored book projects are currently seeking highly-qualified experts in the field to fill vacancies on their respective editorial review boards:

Applications and Inquiries may be sent to:
development@igi-global.com

Applicants must have a doctorate (or an equivalent degree) as well as publishing and reviewing experience. Reviewers are asked to complete the open-ended evaluation questions with as much detail as possible in a timely, collegial, and constructive manner. All reviewers' tenures run for one-year terms on the editorial review boards and are expected to complete at least three reviews per term. Upon successful completion of this term, reviewers can be considered for an additional term.

If you have a colleague that may be interested in this opportunity,
we encourage you to share this information with them.

IGI Global Proudly Partners With eContent Pro International

Receive a 25% Discount on all Editorial Services

Editorial Services

IGI Global expects all final manuscripts submitted for publication to be in their final form. This means they must be reviewed, revised, and professionally copy edited prior to their final submission. Not only does this support with accelerating the publication process, but it also ensures that the highest quality scholarly work can be disseminated.

English Language Copy Editing

Let eContent Pro International's expert copy editors perform edits on your manuscript to resolve spelling, punctuaion, grammar, syntax, flow, formatting issues and more.

Scientific and Scholarly Editing

Allow colleagues in your research area to examine the content of your manuscript and provide you with valuable feedback and suggestions before submission.

Figure, Table, Chart & Equation Conversions

Do you have poor quality figures? Do you need visual elements in your manuscript created or converted? A design expert can help!

Translation

Need your documjent translated into English? eContent Pro International's expert translators are fluent in English and more than 40 different languages.

Email: customerservice@econtentpro.com   www.igi-global.com/editorial-service-partners

Printed in the United States
By Bookmasters